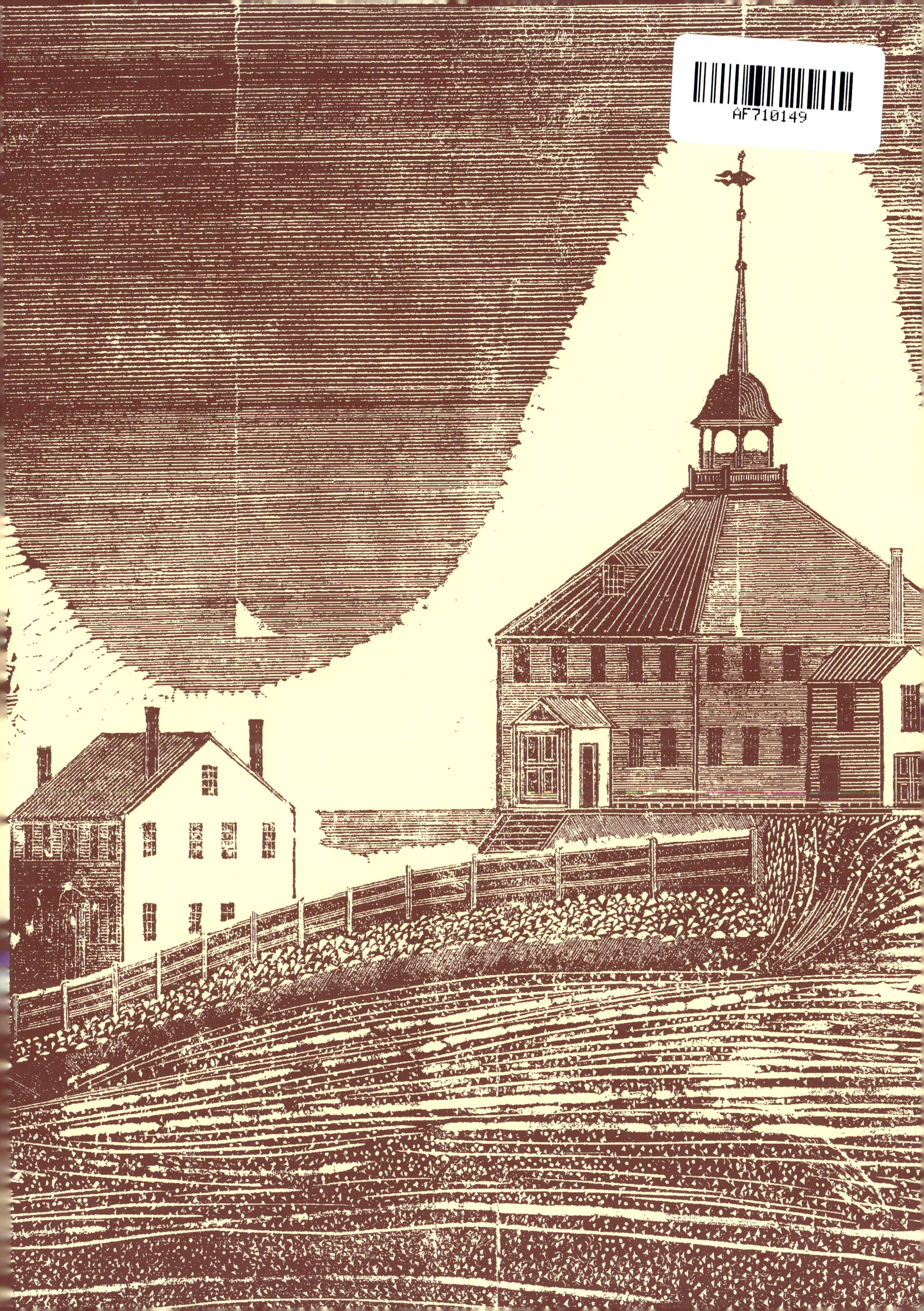

AF710149

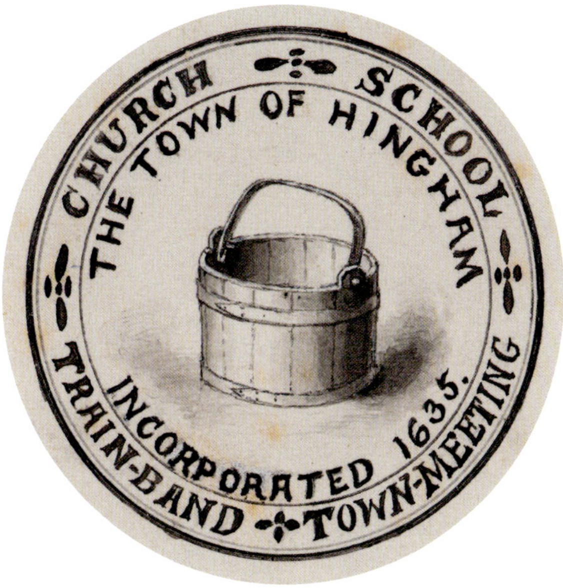
CHURCH
SCHOOL
THE TOWN OF HINGHAM
INCORPORATED 1635.
TRAIN-BAND
TOWN-MEETING

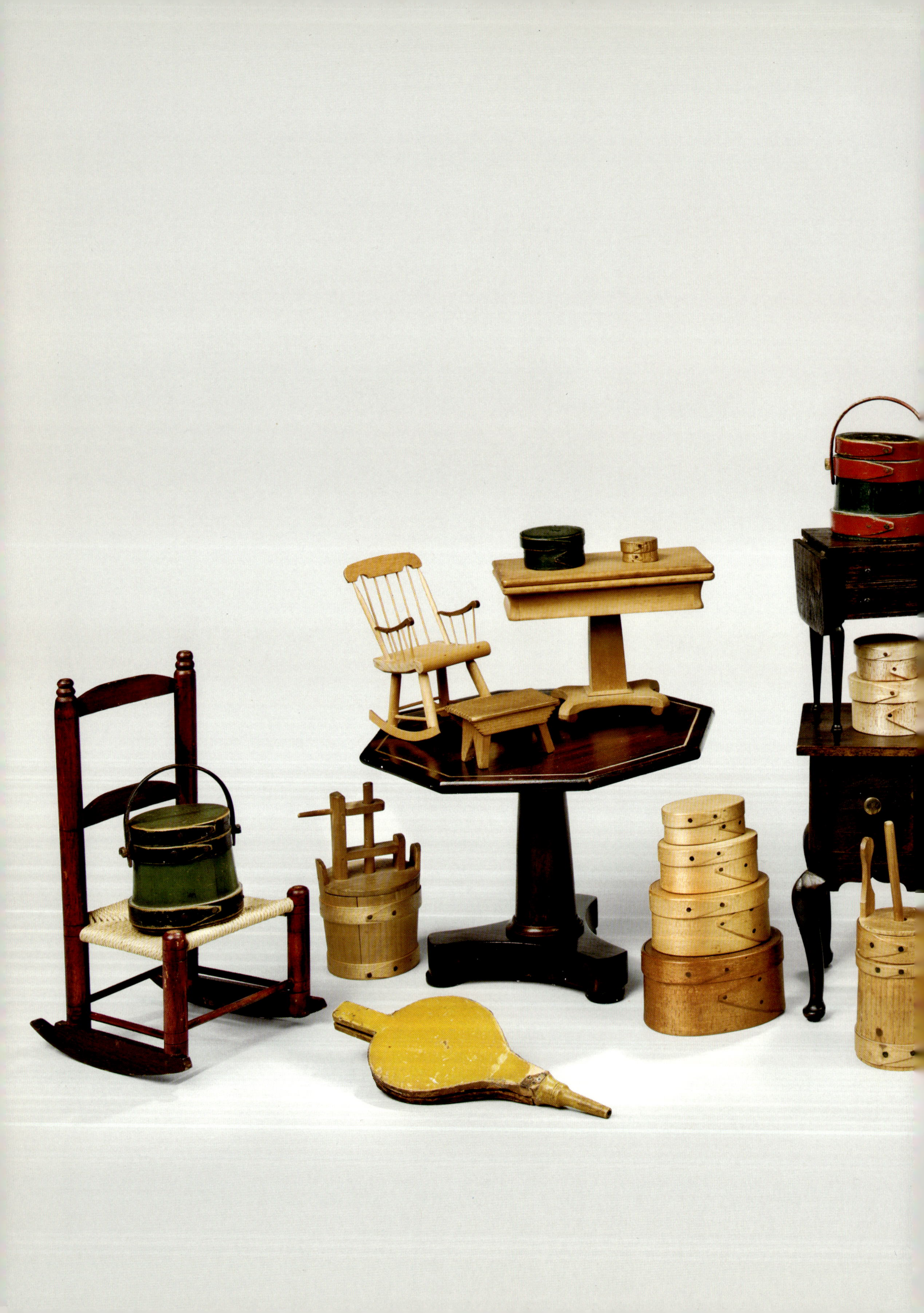

Bucket Town

Woodenware and Wooden Toys of Hingham, Massachusetts, 1635–1945

Derin T. Bray

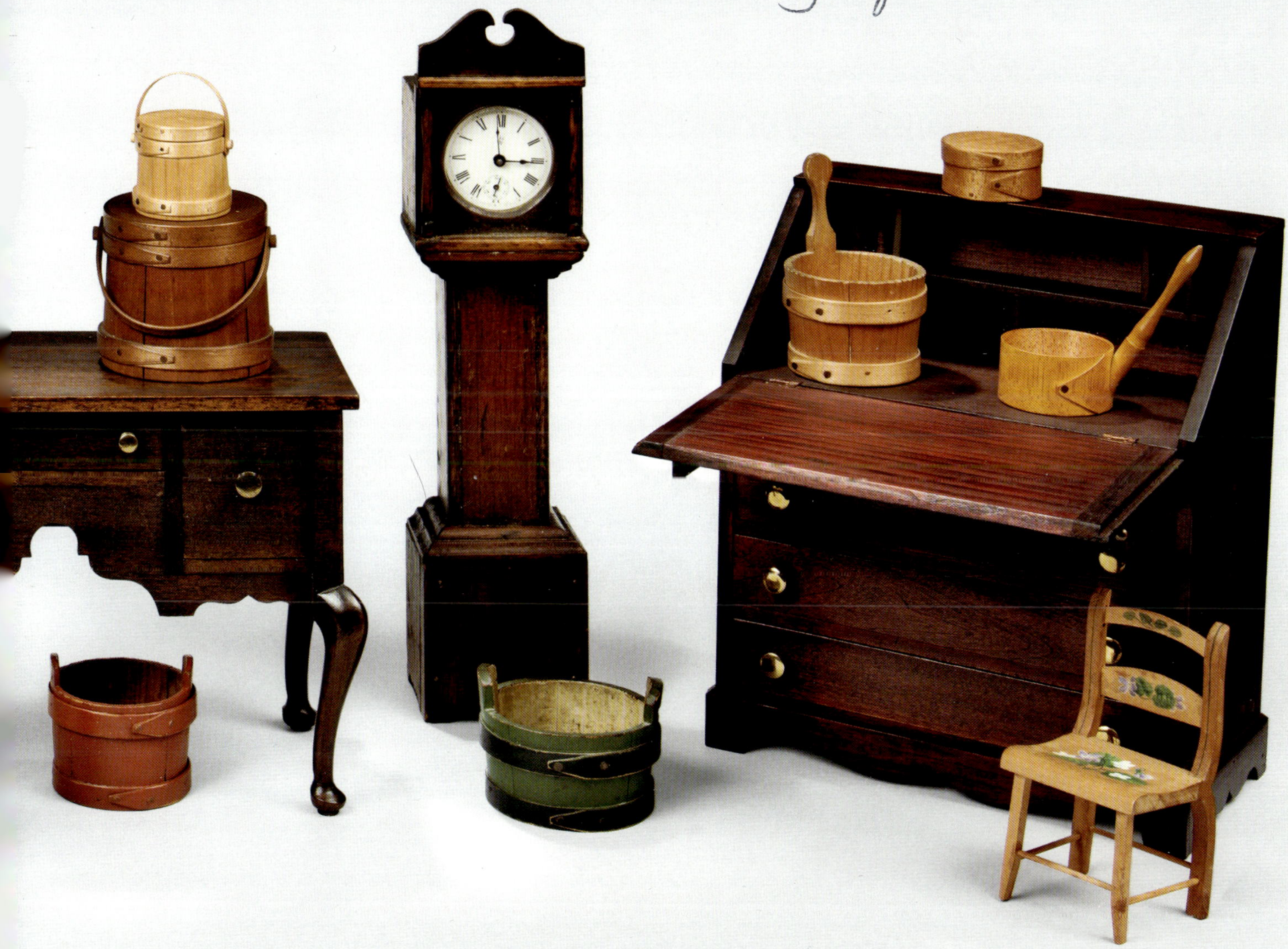

Photography by Gavin Ashworth

Hingham Historical Commission
Hingham, Massachusetts

Published by Hingham Historical Commission
210 Central Street
Hingham, Massachusetts 02043-2759
www.Hingham-ma.gov

Copyright © 2014 by Peter W. Hersey
www.buckettown.com

ISBN: 978-0-615-97537-5

Library of Congress Cataloging-in-Publication Data

Bray, Derin.
Bucket Town : woodenware & wooden toys of Hingham, Massachusetts, 1635-1945 / Derin T. Bray.
pages cm
"Issued in connection with an exhibition held in 2014 at Old Sturbridge Village, Sturbridge, Massachusetts"--Provided by publisher.
Includes bibliographical references and index.
ISBN 978-0-615-97537-5 (hardcover : alkaline paper) -- ISBN 0-615-97537-2 (hardcover : alkaline paper)
1. Treenware--Massachusetts--Hingham--History--Exhibitions. 2. Coopers and cooperage--Massachusetts--Hingham--History--Exhibitions. 3. Wooden toys--Massachusetts--Hingham--History--Exhibitions. 4. Pails--Massachusetts--Hingham--History--Exhibitions. 5. Americana--Massachusetts--Hingham--History--Exhibitions. 6. Folk art--Massachusetts--Hingham--History--Exhibitions. 7. Hingham (Mass.)--Social life and customs--Exhibitions. I. Title.
TT200.B73 2014
745.592074'74482--dc23
2014009592

Produced by Vern Associates, Inc. Amesbury, Massachusetts
www.vernassoc.com
Designed by Peter M. Blaiwas
Edited by Ann Twombly
Layout by Glenna Collett
Production by Susan McNally
Printed and bound by Puritan Capital, Hollis, New Hampshire, U.S.A.

Jacket front: Reuben Hersey toy & boxmaking shop, Hersey Farm Collection

Jacket back: Interior of the Reuben Hersey toy & boxmaking shop, Hersey Farm Collection

Half-title page:
Design sketch for Hingham town seal by Wallace Corthell, ca. 1895, courtesy of the Hersey family

Title page:
Dumb betty (pl. 13.2), bellows (fig. 3.16), tilt-top table (pl. 18.4), rocking chair (pl. 12.4), card table (pl. 12.4), box (fig. 3.21), tiny oval box, stack of oval boxes by Samuel Hersey, work table (pl. 19.3), two round boxes by Nelson Corthell, churn (pl. 13.2), buckets (pl. 12.2), low boy by Loring H. Cushing, red tub by Samuel Hersey, clock (pl. 16.5), piggin (pl. 13.2), desk (pl. 17.6), side chair (fig. 2.9), courtesy of the Hersey family
Rocking chair (pl. 17.6), footstool (pl. 12.4), courtesy of Hingham Historical Society
Buckets (pl. 11.7), dipper (pl. 13.3), box (pl.12.2), courtesy of David A. Schorsch
Tub (fig. 2.6), courtesy of Peter and Diana Bennett

CONTENTS

FOREWORD

The first official history of Hingham was published in 1827, a robust, scrupulous work by local lawyer and antiquarian Solomon Lincoln. He and many of his contemporaries—Beals, Cushings, Herseys, Spragues, Towers, and others—had good reason to count the town's already lengthy history a part of their own. Their ancestors, English immigrants who threw down tenacious roots in the 1620s and 1630s, were among Hingham's earliest settlers. While not the young nation's first town history, Solomon Lincoln's was certainly in the vanguard, and it set a precedent; Hingham has since expressed a notable mindfulness of its past in five more careful and accomplished historical town studies, all but one of which were published by the Town of Hingham itself.

This volume adds a seventh account of Hingham's history and a fresh perspective. *Bucket Town* is like its predecessors in significant ways, and has benefited greatly from those parallels. Like Solomon Lincoln's work, *Bucket Town* owes its conception and realization to a descendant of one of Hingham's earliest families, and its publisher, the Hingham Historical Commission, is the same municipal agency that in 1993 printed Hingham's previous town-issued history, Hart and Hart's elegant *Not All Is Changed: A Life History of Hingham*. But there the similarities end. *Bucket Town*, the first study of Hingham in twenty years, plants its flag in new, fertile ground: It takes as its subject the town's long, deep traditions of coopering and toymaking.

For three centuries, generations of skilled, entrepreneurial Hingham craftsmen forged a local economy lucrative and enduring enough to make their town synonymous with their best-known product—buckets. Boys learned the craft from their fathers, uncles, and grandfathers, and an extraordinary cooperative network of industry formed based largely on bonds of family. It profoundly influenced Hingham's historical trajectory and its culture.

These craftsmen first worked in tiny backyard shops, cramped and efficient as ship's cabins, and sold their goods to the Boston market. In the nineteenth century they built large factories, installed steam-powered machinery, hired dozens of workers, and sold their buckets and toys worldwide. What remained constant, even in the face of mechanization, was the aesthetic appeal of the objects designed and crafted by Hingham's master coopers. From utilitarian buckets, covered pails, canteens, and pantry boxes to charming and whimsical doll furniture and miniature woodenware, the best examples hew to standards of proportion, color, and finish that delight even the most discerning eye.

For more than three years Derin Bray has combed the collections of museums, historical societies, and private individuals all over the country in search of the finest surviving examples of Hingham woodenware and toys, the tools that made them, and the account books, letters, newspapers, handbills, probates, and photographs that document their history. Under the auspices of Peter W. Hersey, a tenth-generation Hingham resident, Derin has pursued endless leads, interviewed descendants, consulted experts, and read widely. The result is a seminal history of Hingham's most iconic industries and the first effort to document the exuberant range of forms, colors, and decoration produced by Hingham's master coopers and toymakers.

A final important parallel between Solomon Lincoln's 1827 history and this volume exists. In his work, Lincoln strove to "preserve . . . whatever is worthy of preservation in the history of the town, or in the character of individuals." This is *Bucket Town*'s objective—and achievement—as well. *Bucket Town,* however, redefines old notions of historical value. By illuminating the beauty in these utilitarian objects and the skill and industry of their yeoman makers, it proves Hingham's craft culture supremely worthy of preservation, too.

Alexander Macmillan
Hingham Town Historian

ACKNOWLEDGMENTS

Bucket Town was made possible through the assistance and encouragement of numerous individuals and institutions. At the top of the list is Peter W. Hersey. He first conceived of the book in 2007, following the discovery of a nineteenth-century toy and boxmaking shop on Hersey Farm in Hingham. His passion for local and family history motivated this study. The Hersey family's sole support of the research and writing made it a reality.

The Hingham historian John P. Richardson (1935–2011) was instrumental throughout the course of the project. With the assistance of Frank Tipping, he meticulously cleaned, catalogued, and organized thousands of items from Hersey Farm. This extraordinary body of material provided the basis for Chapter 3. John's vast collection of local objects and manuscripts—dubbed the Old Fort House Archives & Historical Repository—was an excellent source for information on coopering and toymaking. Many pieces bearing his handwritten tags are referenced or illustrated throughout the book. They are a testament to his commitment to studying and preserving Hingham history. John's untimely death in 2011 was a tremendous loss to everyone.

A debt of gratitude is owed to the Hingham Historical Commission. In 2013 the Commission voted unanimously to publish *Bucket Town*, recognizing its value as an important educational tool and work of cultural preservation. Thanks are due to Chair Tom Willson, Vice Chair Bob Curley, Administrator Andrea Young, Arthur Choo Jr., Michael Dyer, Huck Handy, Steve Swett, and Beth Rockoff. Several individuals also lent their expertise to the fundraising committee. They include committee chair Virginia Tay, Sarah Dewey, Larry Lindner, Jim Macedo, Mark Murphy, and Mary E. Tondorf-Dick. Former chair of the Commission and current Town Historian Alexander Macmillan was an early champion of the project.

Many talented individuals participated in the production stage of the book. I am especially indebted to Peggy Hodges for the countless hours she spent helping to write and rewrite the chapter essays and catalog entries, bringing much of the text to life. I thank Bob Trent and Peter W. Hersey for their insightful comments on the draft of the manuscript. The stunning images are the work of photographer Gavin Ashworth, who shot around the clock to meet a tight deadline. Margot Cheel and Andrew Davis also provided beautiful photographs. Vern Associates went above and beyond the call of duty to pull it all together; executive director Brian Hotchkiss skillfully guided the production and editing of the manuscript from beginning to end; creative director Peter Blaiwas transformed the words and images into a beautifully designed book; Ann Twombly edited the manuscript with painstaking attention to detail; Susan McNally oversaw the printing at Puritan Capital in Hollis, New Hampshire.

Special recognition is due to Old Sturbridge Village for mounting the exhibition accompanying the publication. President and CEO Jim Donahue and vice president Ed Hood were vital in moving the exhibition forward. Equal praise must be given to Christie Jackson, senior curator of American decorative arts. Christie gave her steadfast support throughout the production of the book and, as co-curator of the exhibition, played a critical role in shaping the look and concept of the show. It would not have been possible without her. Tom Kelleher, curator of mechanical arts, provided wonderful insights into the trade of coopering and helped identify early agricultural and coopering tools at Hersey Farm. Tom also masterfully recreated a handle for the Jared Lane pail (pl. 2.1). The following individuals also made important contributions to the project: Jeannette Robichaud, library assistant; Rebecca Beall, collections manager and curator of textiles; Bruce Craven; exhibit designer; Doreen St. John, art director; Deb Friedman, senior vice president; Greg Dixon, vice president, marketing and communications; and Kate Swisher, curatorial fellow.

Many people were invaluable to the development of *Bucket Town*. I thank the following individuals for sharing information and research leads, opening their homes and personal collections, or contributing to the project in other meaningful ways: Robert Adam, Andy Addoms, Gerry and Dick Ainslie, Tom and Sandra Ammann, Tim Andreadis, Janet Hersey Andrews and Rodney E. Andrews, Jeanmarie Andrews, George Badgio, Stephen and Harriet Baker, Laura Beach, Tom Belyea, Gerry and Rindy Bennett, Peter and Diana Bennett, Bickford Woodworking (Peter B. Bickford), Dave Blowers, Cheryl Bonin, Joe Borsari, John Brandt, Ernie Britt, Jackie Britton, Norman Brosterman, Gus and Trisha Brown, Joyce K. Brown, Joan Burrell, Michael Burrey, Brad Cass, Richard Chaisson, Concrete Construction (Edward Duffy), Robert M. Conrad, Debbie Cook, Nancy Fee Cook, Larry Corthell, Dennis Cuddy, John L. Curtis, Michael Cushing, Jerry Dalton, Carl and Ellen Davis, Betsy Dean, Mary Deisenroth, Bob Delorey, Frank Derubis, Marcello DeSouza, Maryellen Dever, Michelle DiVincenzo, Marion Douglas, Eric Dresser, Chief Mark Duff (Hingham Fire Department), Christopher Duvall, E. F. Anthony, Inc. (Eric Anthony), Diane Elliott, Adam Erby, John Fawcett, Kathy Fearing, Ann Smith Finn, Jack Flanagan, Paul Foley, Peter Follansbee, Chad and Diane Forman, Fred Gaede, Joseph Garver, Ebenezer and Diana Gay, Ebenezer R. Gay, Wayne and Heather Graf, Daniel Haff, Bonnie Benson Hanson, Tara Hanson, Lorena L. Hart, Teddy Helfrich, Bill Henderson, Galen H. Hersey, John R. Hersey Jr., Lucille Hersey, Patricia Hersey, Stanley L. Hersey, David Hewitt, Kathy Hind, *Hingham Journal* (Mary Ford), Hingham Light Plant (Paul Heanue), Hingham Then and Now (Stephen Dempsey), Diane Hoffman, Larry Holbrook, Nathan Hood, Carl Howk, Carolyn and Dick Housman, Jay Ippolito, Matthew Isenburg, Amanda Jacobs, David Jameson, Craig and Alison Jewett, Tom Jones, Joseph T. Cazeault & Sons, Inc. (Christopher Cazeault), Matthew Keagle, Joy Kelleher, Dale Kelley, Bud and Patricia C. Kenny and Della (the mule), Jonathan Kerner, Frank Klay, Arthur and Eleanor Kress, Kathleen Krucker, Sondra Krueger, Selectman Irma Lauter (Town of Hingham), Peter J. Lemonias, Charles E. Leverone, Gail F. Lincoln, Robin Litchfield, Marilyn Hersey MacDonald, Paul and Lindsay Malone, Pamela Marcella, David Margetts, Judy Margetts, David Mason, William Massa, Bruce McClintock, Michael McClintock, Dan McKinnon, Kimi McLean, Dennis R. Means, Ann Meehan, Marilee Meyer, M. Stephen Miller, Brian Mills, George Moller, Lynn Morin, Bob and Peggy Mosher, Richard Mueller, Edgar Murphy, Robert Mussey, Mohsen Nasr, Gary D. Naylor, Paul Neuman, Rob Niederman, Paul O'Donnell, Judy O'Donovan, Marc Olimpio, Jim O'Toole, Maryellen and Brendan O'Toole, Sal Palumbo, Parnell Painting (Alan Parnell Jr.), Robert A. Pattison Sr., Michael Payne, Clark Pearce, Elizabeth Putnam, Glenn Ralston, John Rees, Helena Richardson, Madelia Ring, Michael Rosenbloom, Bob B. Rosenthal, Sandra Russell, Pat Saef, Scott Sanborn, Charles Santore, Ken and Janice Sargent, Marty Saunders, Norm Schaut, Joel Schiff, Edward and Joyce Schreier, Jennifer Schwartz, Phil and Jane Shute, Valerie Slavin, Faith Smith, Jim Sneed, Elizabeth Stillinger, John and Barbara Stoddard, Tom and Myrna Stone, Matthew Stowell, Bob Studley, Michael Studley, Tom and Avis Studley, Charles Templeton, Frank A. Tipping III, Don Troiani, Edward "Tuck"

Wadleigh Sr., John Walker, Perch Walker, Leslie and Peter Warwick, Steven Weiss, George Wetzel, Tom Whalen, Wendy Whitman, Dr. Christopher Whittle, Bob Wilkins, Marjorie Winnie, Mickey Wnek, Gwen Woolley, Nan Wolverton, Charlie and Rosie Woodward, and Rick Young.

Museums, historical societies, and libraries were essential to the success of this project. I am grateful to the following institutions and staff who offered their assistance: American Antiquarian Society, Thomas Knoles, Jackie Penny; American Folk Art Museum, Stacy Hollander; American Society of Arms Collectors, John Spangler; Ancient and Honorable Artillery Company of Massachusetts, Charles Fazio; Athol Historical Society, Debra Ellis; Bare Cove Fire Museum, Dave Clark, Geri Duff; Boston Athenaeum, Patricia Boulos, Stanley Cushing, Carrole Morini, Mary Warnement; Boston Children's Museum, Rachel Farkas, Jennifer Jensen; Boston Herald, James Mahoney, Martha Reagan; Boston Public Library, Karen Shafts, Jane Winton; Bostonian Society, Laura Hoff, Elizabeth Roscio; Canadian Centre for Architecture, Renata Guttman; Cohasset Historical Society, Lynne DeGiacomo; Concord Museum, David Wood; Connecticut Valley Historical Museum, Guy McLain; Delaware Historical Society; Jennifer Potts; Dennys River Historical Society, Rebecca Weston Hobart; Denver Museum of Miniatures, Dolls, and Toys, Jillian Anderson; Duxbury Rural and Historical Society, Patrick Browne; Dyer Memorial Library, Joice Himawan; First Parish, Hingham, Tom Willson; Fort Ticonderoga, Christopher Fox; Fruitlands Museum, Roben Campbell, Michael Volmar; Hanover Historical Society, Barbara Barker, Dick Bancroft, Judy Grecco, Rachel Lovett; Heritage Museums & Gardens, Jennifer Madden; Hingham Historical Society, Michael Achille, Suzanne Buchanan, Mary Fitzmaurice, Virginia Giusti-McLaughlin, Bruce Madio, Faith Smith, Michael Shilhan; Hingham Public Library, Dennis Corcoran, Anne Dalton; Historic Deerfield, Josh Lane, Ned Lazaro, Phil Zea; Historic New England, Nancy Carlisle, Lorna Condon, Jeanne Gamble, Kristen Weiss; Historical Society of Old Abington, Pam Whiting; Jones River Village Historical Society, Norman Tucker; Marshfield Historical Society; Mass Audubon, Amy Montague; Massachusetts Archives, Jennifer Fauxsmith; Massachusetts Historical Society, Anne Bentley, Anna Cooke, Peter Drummey, Dennis Fiori; Massachusetts National Guard Military Museum and Archive, Col. Leonid Kondratiuk; Milton Historical Society, Steve Kluskens; Museum of the American Revolution, R. Scott Stephenson; Museum of Fine Arts, Boston, Dennis Carr, Marta Fodor; Museum of the City of New York, Donald Albrecht, Jim Tottis; Mystic Seaport, Maribeth Bielinski; National Building Museum, Nancy Bateman, Chrysanthe Broikos, Martin Moeller; New Bedford Whaling Museum, Stuart Frank; New England Historic Genealogical Society, Marie Daly, Rhonda R. McClure, Timothy Salls; New York Historical Society, Margi Hofer, Eleanor Gillers; Newton Historical Society, Thomas McGinley; Norwood Historical Society, Barbara Rand; Old Colony Historical Society, Jane Hennedy; Peabody Essex Museum, Dean Lahikainen, Dave O'Ryan; Philadelphia Museum of Art, David Barquist, Lucy Peterson; Pilgrim Hall Museum, Stephen O'Neill; Plympton Historical Society, Chris Maiorano; Plymouth Antiquarian Society, Donna Curtin; Quincy Historical Society, Ed Fitzgerald, Corinne Waite; Randolph Historical Society, Henry Cooke; Rehoboth Antiquarian Society, Laura Napolitano; Rochester Historical Society, Sue LaFleur; Rosalie Whyel Museum of Doll Art, Jill Gorman, Rosalie Whyel; Scituate Historical Society, David Ball, Betty Miessner, Carol Miles, Ron Swan; Shaker Museum and Library, Jerry Grant; Shelburne Museum, Jean Burks, Jacquelyn Oak, Barbara Rathburn; Strawbery Banke, Ron Raiselis; Smithsonian, Nan Card, Richard Stamm, Bill Yeingst; The Colonial Williamsburg Foundation, Tara Chicirda; The Henry Ford, Donna Braden; The Hyde Collection, Jayne Stokes; The Jonathan Belcher House, Ellen D'Isidoro; The Society of the Cincinnati, Emily Schulz; The Strong National Museum of Play, Patricia Hogan, Carol Sandler; The Tower Genealogical Society, Patricia Brown; The Toy Museum of New York, Marlene Hochman; Town of Hingham, Eileen McCracken; Tides

Institute, Hugh French; Toy & Miniature Museum of Kansas City, Kristie Dobbins; U.S. Naval Observatory and Library, Sally Bosken; Van Cortlandt House Museum, Laura Carpenter; Victoria and Albert Museum, Glenn Adamson, Nick Humphrey; Vineland Historical Society, Patt Martinelli; Wellesley Historical Society, Kathleen Fahey; Weymouth Historical Society, Jim Palmieri, Debbie Sullivan; Winslow House Museum, Kathleen O'Connor; Winterthur Museum, Garden and Library, Ritchie Garrison, Emily Guthrie, Brock Jobe, Rich McKinstry, Lisa Minardi, Jeanne Solensky, Matt Thurlow, Ann Wagner; Yale University, Ned Cooke, John Demos.

Bucket Town benefited greatly from dealers and auction houses who helped locate examples of toys and woodenware, sometimes in their own collections. I thank the following businesses and individuals for their assistance: Charles Adams; Antique Associates at West Townsend, David Hillier; Judith Armitstead; Axtell Antiques, Richard Axtell; Bill Bakeman; Philip Balcom; Lori Baxter and Scott Zubee; Gerald Bell; Bob Jessen & Jim Hohnwald; Ruth Briggs; Broadcove Auctions, Norman McCulloch, Judy McCulloch, Scott M. McCulloch; Arthur Burrows; Henry T. Callan; Cathy Cosentino; Barbara R. Coulter; Cotie Cowgill; R.E. Crawford; Christie's, Andrew Holter, Hillary Mazanec; Tom Clark; Conestoga Auctions; Jim Conrad; Cowan's Auctions; Brian Cullity; Devin Moisan Auctioneers, Devin Moisan; Doris Marks Antiques; Arthur Garrity; Garth's Auctions, Inc., Andrew Richmond; Gary Sullivan Antiques, Matt Buckley, Gary Sullivan; Michelle Genereux; Ellanna Gish; Gould Auctions, Timothy J. Gould; Griffiths Antiques; Mark Hall; John Hammer; Hap Moore Antiques Auctions, Hap Moore; Pat Hatch; Heritage Auctions; Hope Antiques, Terry and Renee Cormier; David Hutton; Jewett-Berdan Antiques, Tom Jewett, Butch Berdan; Thomas M. Kelleher; Ken and Susan Scott Antiques; Matt King; Barbara M. Lemme; Kurt Mansbach; Paul D. Marinucci; Martha Boynton Antiques, Martha Boynton; Newsom & Berdan, Michael Newsom, Betty Berdan; Hilary and Paulette Nolan; North River Antiques Center, Leslie Molyneux; Northeast Auctions, Ron Bourgeault; Noel Barrett Auctions, Noel Barrett, Jack Borelli; Olde Hope Antiques, Pat Bell; Pam Boynton Antiques, Pam Boynton; Shirley Parkhurst; Raynor's Historical Collectible Auctions; Stephen B. Rogers; Kathy Schoemer; David Schorsch; Mike and Lucinda Seward; Skinner Auctioneers and Appraisers, Inc., LaGina Austin, Chris Barber, Robert Cheney, Paul Dumanoski, Stephen Fletcher, Kathryn Gargolinski, Karen Keane, Kelli Lucas Stewart; Starr Antiques, Sally Starr; Bill Steere; Margaret Sutton; Anne Timpson; Stonehouse Antiques, Marie Anderson; Times Past Antiques, Carol Franzosa; Danny Wahl; Susan Walter; Willis Henry Auctions, Willis and Karel Henry.

The role of the Hersey family in making a success of *Bucket Town*—both the book and the exhibition—cannot be overstated. I am deeply indebted to Peter and Judi Hersey for their extraordinary support, encouragement, and enthusiasm through all phases of the project. I would also like to extend a special thank-you to Kristin Hersey, Keith Roche, and Evan Hersey Roche; Khara (Hersey), Joseph, Jacob, Anna, and Jonah Houston; Paul Hersey; Beth Hersey; and Heather, Keith, August, and Ellie Johnson.

Writing a book is like running a marathon. My heartfelt thanks go to my friends and family for their love and support as they cheered me to the finish line. I am especially grateful to Doug and Alesia Bray; Danielle, Josh, and Carter Tyson; Brianne, Pat, Grace, and Brayden Wagner; Seth and Kelly Bray; John and Paula Sivulka; Emily, Alan, Malena, and Luca Pereira; Mary, Craig, and Quinnlyn Atkinson; and Kim Sivulka.

Finally, I would like to thank my wife Bethany, a constant source of inspiration and the most remarkable person I know. She joined me deep in the trenches during many long days and late nights, helping to research, write, edit, and review photography and design work. This book would not have been possible without her.

Derin T. Bray

CONTRIBUTORS LIST

Reuben Hersey Society

Rodney E. and Janet Hersey Andrews
In memory of Reuben Woodrow and Eleanor Lula Hersey
Peter Woodrow and Judith Ann Hersey
Paul Edmund Hersey and Elizabeth Hersey
Hingham Historical Commission

Hingham Historical Commission, In memory of John Peabody Richardson,
the spirit of historic preservation in Hingham

Toymaker Circle

Jane Carr, in honor of Suzanne Buchanan, Director, Hingham Historical Society
Talbert E. and Irma H. Lauter
Alexander Macmillan

Bucket Guild

In honor of Kristin A. Hersey and Keith M. Roche, Evan Hersey Roche;
Khara M. and Joseph A. Houston,
Jacob Joseph, Anna Marie, and Jonah Peter Houston

In honor of Heather L. and Keith H. Johnson,
August Reuben and Ellie Monique Johnson

Tom and Diane Willson

Benefactors

Thomas and Kathleen Bright
Bill and Edna English
Kevin and Rita Gill
In memory of Mr. and Mrs. Gordon T. Hersey
Virginia E. Tay

Patrons

In memory of George F. Bennett
Henry and Lucinda Day
Deborah and Philip Edmundson
Galen H. and Myrla E. Hersey
Roger L. Hersey Jr.
James and Bernadette Ippolito
Charles and Jane Malme
William S. and Katharine W. Reardon
David A. Schorsch
Dr. Edward and Joyce Schreier
Gary Sullivan
Tower Genealogical Society
Andrea and Randy Young

Sponsors

Carolyn and Bill Aliski
Deirdre and Dave Anderson
In memory of George and Helen Bennett
Thomas and Christine Berk
Kathi and Bryce Blair
Derin and Bethany Bray
Broad Cove Auctions, Hingham
Christopher Cazeault
Charles Clapp III and Martha Clapp
In memory of Ebenezer Gay
Phyllis P. Godwin
Warland R. Hersey and Janice Hersey Baker
Hingham Public Library
Bonnie Hobbs and John Riley
Roger and Nancy Hoit
In memory of Bruce Kelly
The von der Luft Family
Jim and Eileen Macedo
Peggy and Bob Mosher
Thomas L. P. O'Donnell
Per and Kathryn Ostman
The Stowell Family
Talbots
Jeff and Gail Tocchio
Amy and Matthew Torrey
Deborah Zildjian

Subscribers

Carl and Susan Achille
Richard and Nancy Adams
Deborah Allinson
Michael Argiro
Isabel and Jonathan Asher
Erika Barnet
Ellen Silva Barry
Dana S. and Virginia Baxter
Susan and Charles Berry
Peter and Johanna Bickford
Audra and Tripp Boyle
Suzanne Buchanan
Laura Hersey Bustamante
Jane C. Carr and Andrew Hertig
Cynthia Chace
David and Mary Ann Chase
Arthur and Theresa Choo
Terry and Judy Clarke
Holly Ann Constant
Joan M. Cook
Nancy Fee Cook
In honor of John Patrick Crowley
Margaret Curley
Robert and Kathleen Curley
Brian and Margaret Curtis
Anne M. and John T. Dalton
George and Elizabeth Danis
William R. Davis
Maryellen Dever
Sarah Dewey and Rheinhold Hesse
David Donahue and Julie Cherau
Peg Doyle
Mark and Geri Duff
Gordon Dunley
Lightning Dunley
Joseph and Rose Durkin
Michael Dwyer and Kathie Mignacca
The Dyer Family
Mary Ann Eckhardt
The Reverend Robert Edson
Catherine J. Eisenmann
Elizabeth A. Eldredge
John and Sandy Emler
Doris and John Falvey Family
Gail Kingston Faring
Peter and Jacqueline Fee
Betty and Tino Fernandes
Mary D. Fitzmaurice
James and Dorothy Flett
In honor of Scott C. Ford
Bill and Jean Garrity
Roy and Kristen Giarrusso

Denise Gilardone
Kevin and Rita Gill
Joel K. Goloskie, Esq.
Jack and Elaine Gomez
Mary Griffin
Alexis Guise and Paul Scanlon
Mrs. and Mr. Jay Huck Handy
Marie D. and Harry J. Hanlon
Susan and Carl Harris
Richard W. Hatch and Martha Engstrom
Charlie and Deb Hayes
Steve Swett, In honor of Dorothy W. Hersey
Elizabeth Hersey
Lynne Hersey
Paul E. Hersey
In honor of Peter W. Hersey
Stanley and Betty Hersey
Stanley "Link" and Alison Hersey Jr.
Tom and Regina Hickey
Cynthia and Henry Hidell
Hingham Historical Society
Herb and Jan Hirsch
Dick and Carolyn Housman
Annabel A. Huntress
August Reuben Johnson
Ellie Monique Johnson
Louise D. Johnson
Dr. Sherry E. Johnson
Gerald and Linda Jones
Reverend William P. Joy
Karl M. Julian and Lucille Julian Hersey
Rev. Dr. Katherine T. Kallis
Daniel and Daphne Keeler
Mr. and Mrs. Joseph Kringdon
Irma and Talbert Lauter, In honor of Hingham Public Library
Monique Lehner
The Lehners
Peter J. Lemonias
Donald Lincoln
Richard W. Lincoln
Constance and Lawrence Lindner
Gar Linscott
Edwin R. and Angela M. Lofgren
In memory of Bill and Bobbie Lord
Louise Lincoln MacKinnon
Frank and Kate Mahony
Lindsay and Paul Malone
Dorothy Mareb
Fiona and Rick Marquis
In honor of Polly Hersey Martin and Kenneth Harold Martin
Eva and Tom Marx
Rick Mattila and Sharon Prehn
Bill Meahl
Wilfried and Judith Mehring
Jo Ann and Jim Mitchell
Lee Mount and Diana Charles
William Perkinson Mullin
Mark and Susan Murphy
Anne and Nestor Nicholas
Mary O'Connell and Terry Troyer
Necia and Ray O'Neill
John and Joan Ostheimer
Maryellen and Brendan O'Toole
The Paget Family
Dorothy Palmer
Christine and Michael Puzo
Carol and Tom Pyles
George D. Reardon Jr.
Mary Chris Richmond
Meg and Brooks Robbins
Chrissy and Jason Roberts
In memory of Betsy C. Robinson
Mark and Beth Rockoff
Louisa W. and David C. D. Rogers
In memory of Eugene Rust
Mr. and Mrs. Keith Rutherford
Mr. and Mrs. William D. Sarni
Marty Saunders
Carolyn and Bill Schrader
Philip and Jane Shute
In memory of Deborah Jenkins Simonson
Zoe and Lena Skiadas
Michael and Laura Slygh
Nancy and Marc Snyder
Charlie, Paula, and Jacquie Souther
Matthew H. Stowell
Bob and Dorothy Stuart
Michael and Christine Studley
Claire and Eleanor Summers
Nancy Sundheimer and Jonathan Boughton
The Allen Swartz family and Ann Levin and family
In honor of the publication of *Bucket Town*, Chaired by
Virginia E. Tay, of the Hingham Historical Commission
Bucket Town Advisory Committee
Maureen E. Taylor
Thomas and Ellen Thurston
Frank and Patricia Tierney
Hon. and Mrs. Paul E. Troy
Neal and Melissa Tully
Cap Vinal
Stefan and Arlyn Vogelmann
Richard Walker
Denny and Sally Weston
In memory of Bernays Wilder
In honor of James Wilder
In honor of Michael Wilder
In honor of William Wilder
Charles and Rose Woodard
Christopher Yule
Anonymous (4)

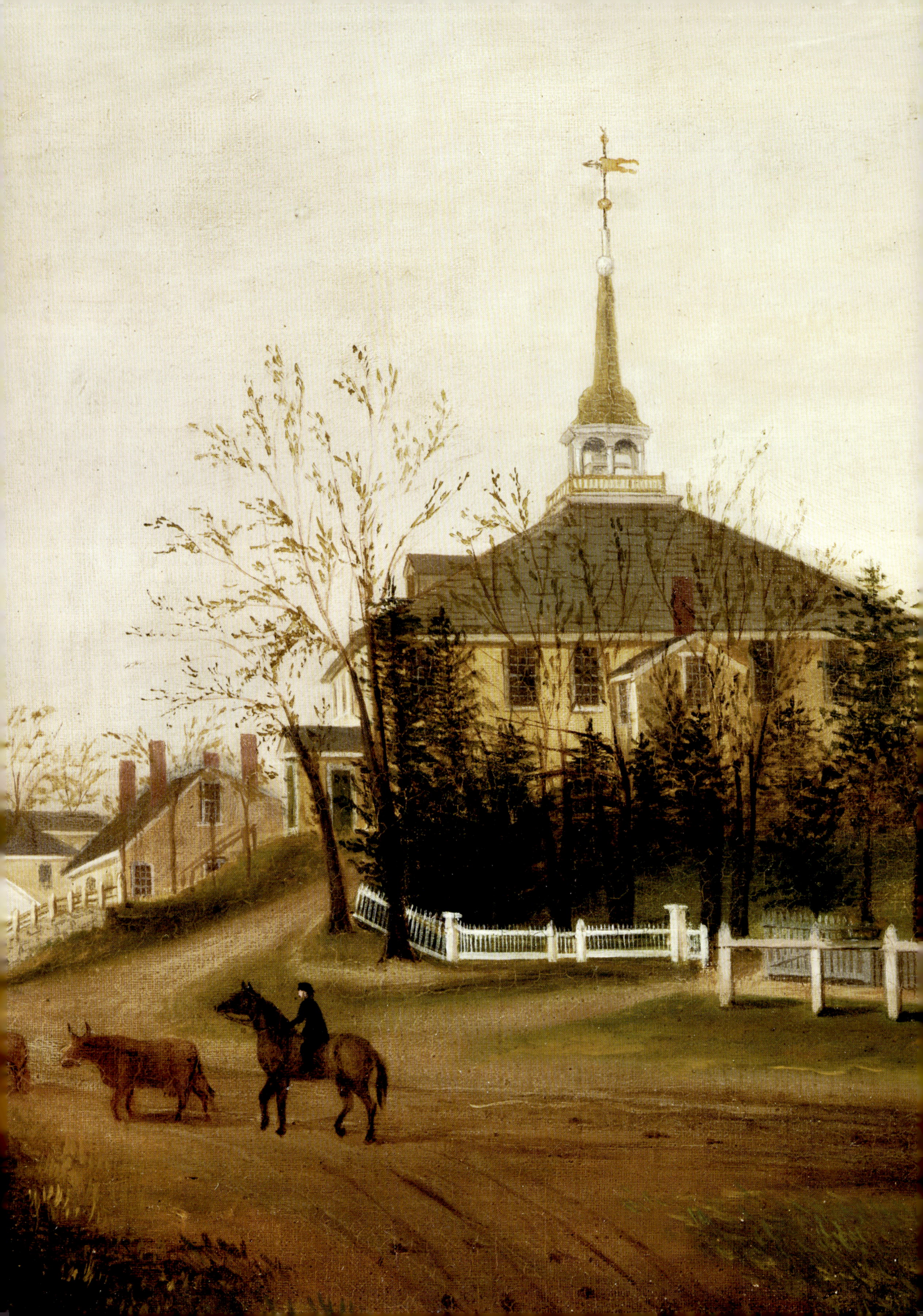

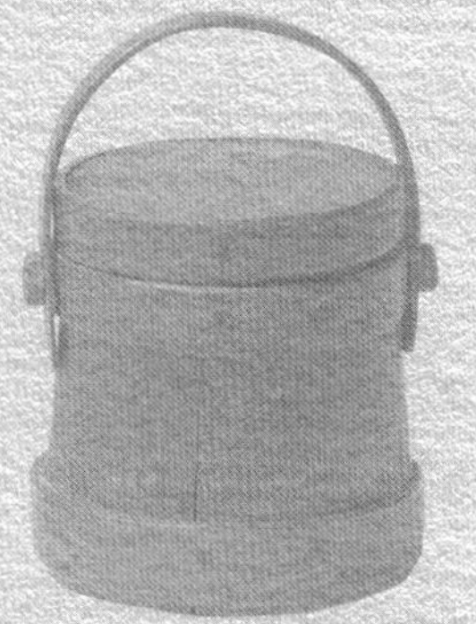

INTRODUCTION

FROM THE WINDING CARRIAGE PATHS OF WORLD'S END, THE CASUAL STROLLER in Hingham can view salt marshes, rocky beaches, meadows, woodlands, and gracefully landscaped glacial drumlins—all against the sweeping backdrop of Boston's dramatic skyline (fig. 1). Proximate and yet worlds apart, bucolic Hingham and urban Boston have drawn from each other's resources for centuries. Over time the nature of that exchange has inevitably shifted; today it turns on Hingham's much-prized reputation as a pastoral retreat from the pace and grind of city life. World's End itself is a product of this later relationship: the abandoned dream of a wealthy nineteenth-century Boston businessman, designed by the noted landscape architect Frederick Law Olmsted (1822–1903) and intended for a residential subdivision.[1] Yet from the earliest years of settlement until well after the end of the Civil War, Hingham's chief trade with nearby Boston was not in lifestyle aspirations or recreational opportunities but in woodenware—boatloads, cartloads, wagonloads of Hingham-made products from lowly dippers and pails to finely crafted boxes, charming miniatures, and a remarkable variety of toys. From Boston

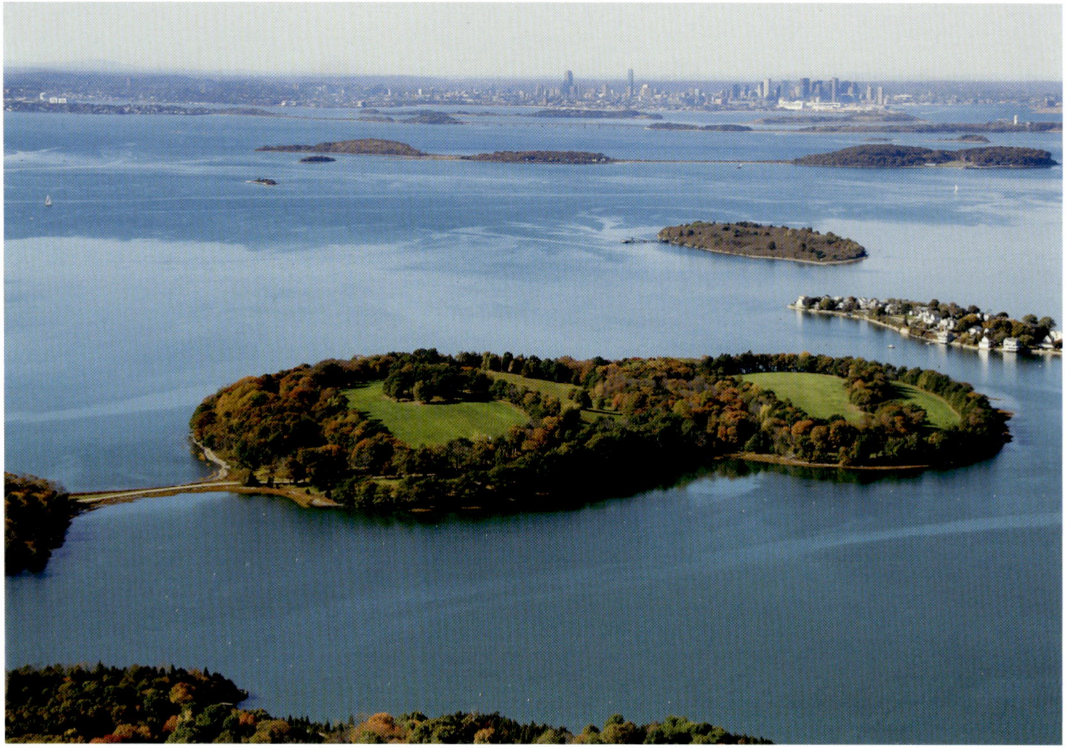

Figure 1
World's End looking northwest toward Boston. Margot Cheel Photography

and other New England ports Hingham woodenware and toys were traded throughout the country, to the West Indies, Canada, England, and France.

Little evidence of this long-lived coopering and toymaking activity is visible in Hingham today; its small shops and factories are gone—burned, demolished, repurposed, blown over, or lost to benign neglect. Local memory, too, lacking tangible reminders, has largely forgotten its industrial past. Nevertheless, Hingham's craft traditions stand equally with the community's earliest and most valued institutions—social order, faith, and learning—whose more enduring symbols of church and school are now the town's defining landmarks. Each of these forces, in tandem and in turn, powerfully shaped the town's character and helped determine its fortunes over the centuries.

First called Bare Cove, the town of Hingham was gradually settled between 1630 and 1640, as groups of families trickled in—initially from the West Country in England, then from the East Anglian counties of Norfolk (home to Hingham's namesake), Suffolk, and Essex. These were disparate populations, and they hewed to their own: the members of each small, successive wave of settlers knew each other by bonds of blood, marriage, class, local dialect and custom, and religion. In the new Hingham they formed a community by no means homogeneous. The West Country folk were primarily seafarers and farmers—possibly stragglers from earlier failed fishing colonies to the north—and were largely indifferent to the Church of England and the Protestant reform movement that brought the later East Anglian families to Hingham. These reform-minded Old Hinghamites arrived in a series of organized, well-funded, and competently led migrations; they were also well-to-do and educated, a further deep point of division between the two groups. Nevertheless, a succession of remarkable early Puritan leaders effected a policy of political authority and religious inclusiveness that made Hingham one of the more cohesive communities of the Massachusetts Bay Colony.[2]

The Massachusetts General Court's renaming of Bare Cove as Hingham in 1635 coincided with a rush of Puritan transplants from County Norfolk; by 1640 between 30 and 40 percent of Hingham's seven hundred inhabitants came from that area.[3] The town's political and religious power inevitably shifted into the hands of the newcomers—though not without conflict. Resentment flared as West Country town officers lost their seats to the new majority. The first pastor of Hingham Parish, the Reverend Peter Hobart, exercised a potent blend of diplomacy and practical evangelism, however, that brought a full 90 percent of all Hingham families into church membership in 1640—this compared to 50 percent participation in Boston Congregational churches, and a wretched 25 percent membership average throughout the colony.[4] With nearly everyone in the pews, Hobart's scripture-based sermons emphasizing Godly community, neighborly love, and obedience to God's laws reached most ears—and solidified Hobart's larger political clout.

Gifted with bountiful natural resources and a fine harbor, Hingham prospered. Lumbering brought an early abundance of rare cash to the town; petitions to the General Court to buy up portions of adjacent towns to acquire even more land, however, were denied. Conflicts with the colonial government, rooted in Hingham's wealth and religio-political differences between the Reverend Hobart and Governor John Winthrop, were chronic and not beneficial. Prominent townsmen served out jail sentences; some—including Hobart himself—were slapped with heavy fines.[5] Nevertheless, Hobart's fierce insistence on local religious and political authority strengthened and stabilized Hingham's community, if only because the townsmen found a meddling government a greater threat than their neighbors' nonconformity.

Hingham's remarkable insularity also protected it from internal dissent; between 1640 and 1660 as few as eight new families were admitted to the town.[6] Given Hingham's great care in taking in only those whose values and social standing were compatible with its residents', it is perhaps no surprise that nearly all the leading families associated with Hingham's long coopering tradition were established in Hingham by 1660. These included the Lincolns, Cushings, Fearings, Burrs, Gardners, Whitons, Towers, Lanes, Lorings, Wilders, and Herseys. During this period of minimal growth in the town rolls, some alienated landowners moved away; a few of these first families—notably the Cushings, Herseys, Lincolns, and Whitons—bought and consolidated these available tracts of land, extracting their wealth through farming and lumbering and settling their heirs on them, creating a class of prosperous, influential, landed families in Hingham.[7] As lumbering gave way to large-scale farming, expanded trading and early industry, wealth concentrated further in these families.

From the time of its incorporation in 1635, the political and religious center of Hingham was its meetinghouse. Today the Old Ship Church (fig. 2) still embodies the traits of faith and determination the town founders understood had allowed them to prevail—over a perceived wilderness and very real hardship. The oldest continuously active Anglo-American place of worship in the country, the Old Ship, built in 1681 to replace an original, fortlike structure, strikes the viewer as a formidable bastion; sited on a hilltop in the center of town, it is a stunning achievement of scale and construction for its time.[8] The church's massive, high-arching oak-beamed ceiling and carved struts and braces, reminiscent of the upturned hull of a great ship, suggest both grandeur and providential protection.

The human link between Hingham's most notable landmarks, the Old Ship and the Derby School, was the Reverend Ebenezer Gay. Minister of Hingham's church from 1718 until his death at age ninety-one in 1787, Dr. Gay preached and lived a religious liberalism that eased the town's strained bonds of community during the Great Awakening and the Revolution—periods of acute religious and political schism.[9] His hardened appearance—if his portrait (fig. 3) is to be believed—belied a spirit of great humaneness and discretion.

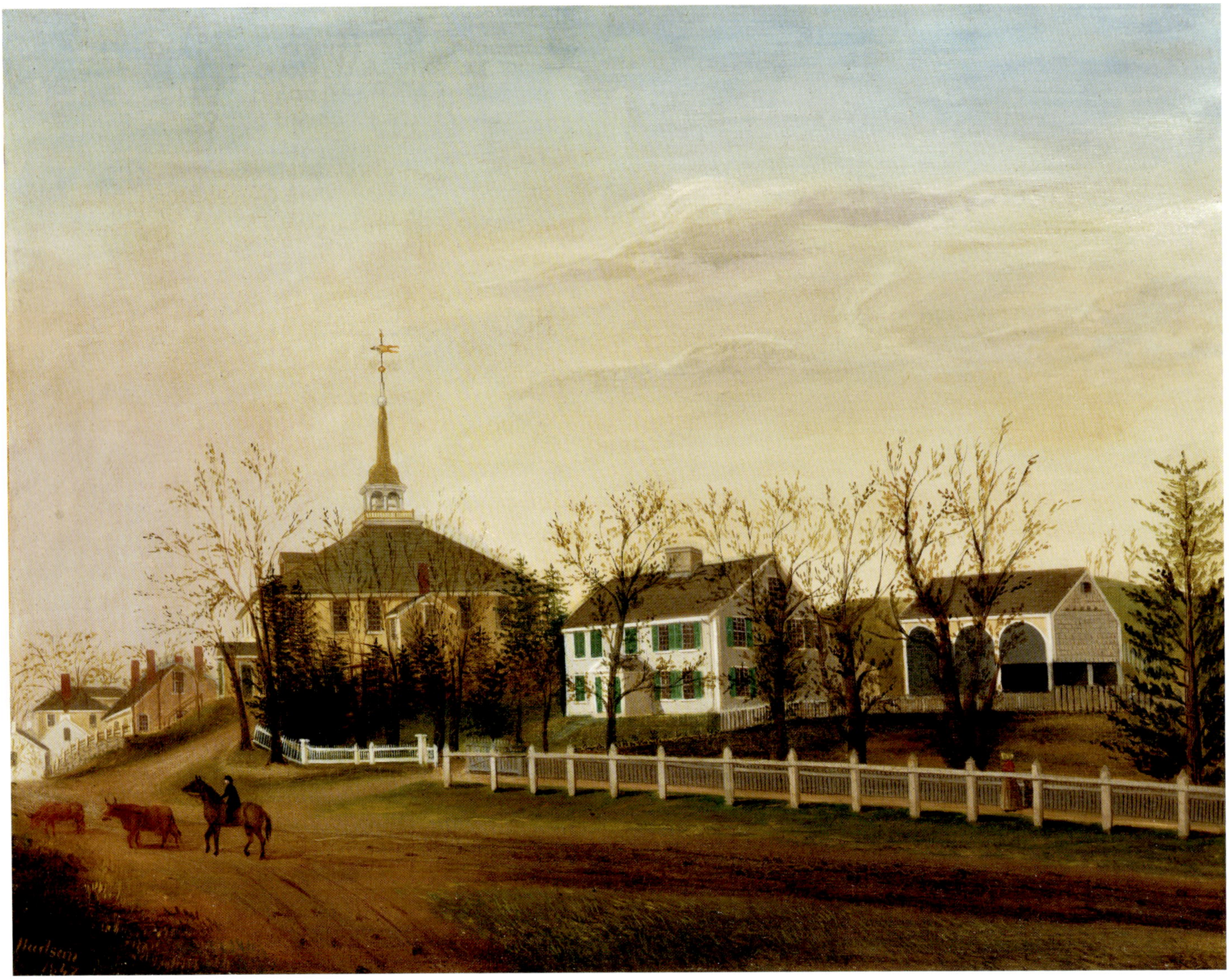

Figure 2

The Old Meetinghouse, by William Hudson, Hingham, 1847. Oil on canvas; H 23, W 30. Courtesy of First Parish, Hingham

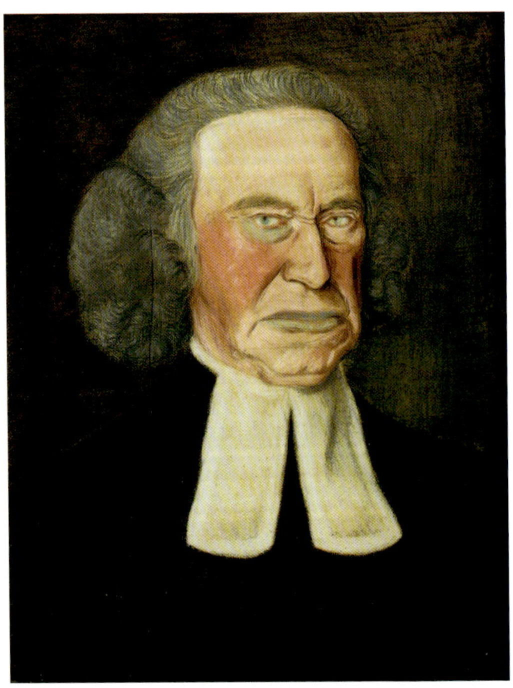

Figure 3

Portrait of the Reverend Ebenezer Gay, by John Hazlitt, Hingham, ca. 1785. Pastel on canvas; H 20, W 16. Collection of Diana R. Gay. Andrew Davis Photography

In 1783 the wealthy widow Sarah Hersey Derby sought the counsel of Dr. Gay in settling the estate of her first husband, the Hingham physician Ezekiel Hersey. Hersey's will gave one thousand pounds to Harvard College to endow a chair of anatomy and physic—the first funds given toward the establishment of Harvard Medical School. Gay's further influence led Sarah Derby to leave the bulk of her remaining estate from her first husband to found a private academy in Hingham, replacing a tax-supported college preparatory school that had closed a few years earlier.[10]

The Derby School opened its doors in 1791, the year after Sarah's death. Boys and girls attended classes on separate floors, where they undertook studies appropriate to their gender

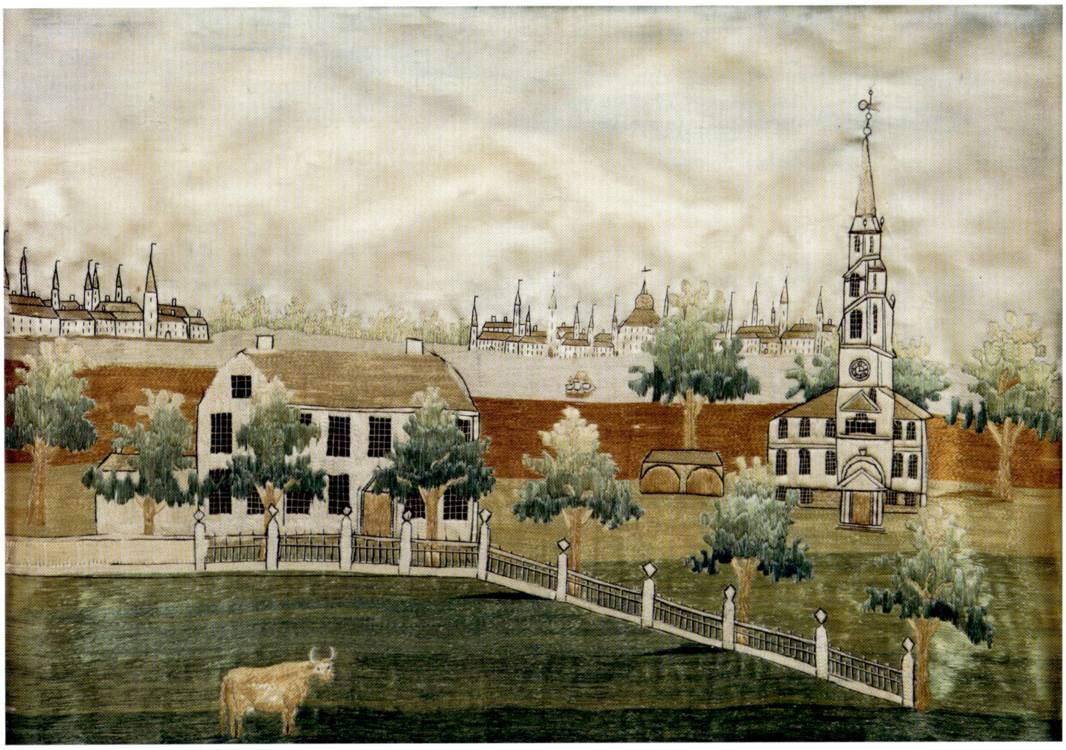

Figure 4
Silkwork picture by Lydia Cushing, Derby School, Hingham, ca. 1798. Silk and watercolor on silk; H 16, W 22. According to family tradition, this embroidery depicts the Boston Common, John Hancock's house, and the Park Street Church in Boston. Courtesy of Hingham Historical Society

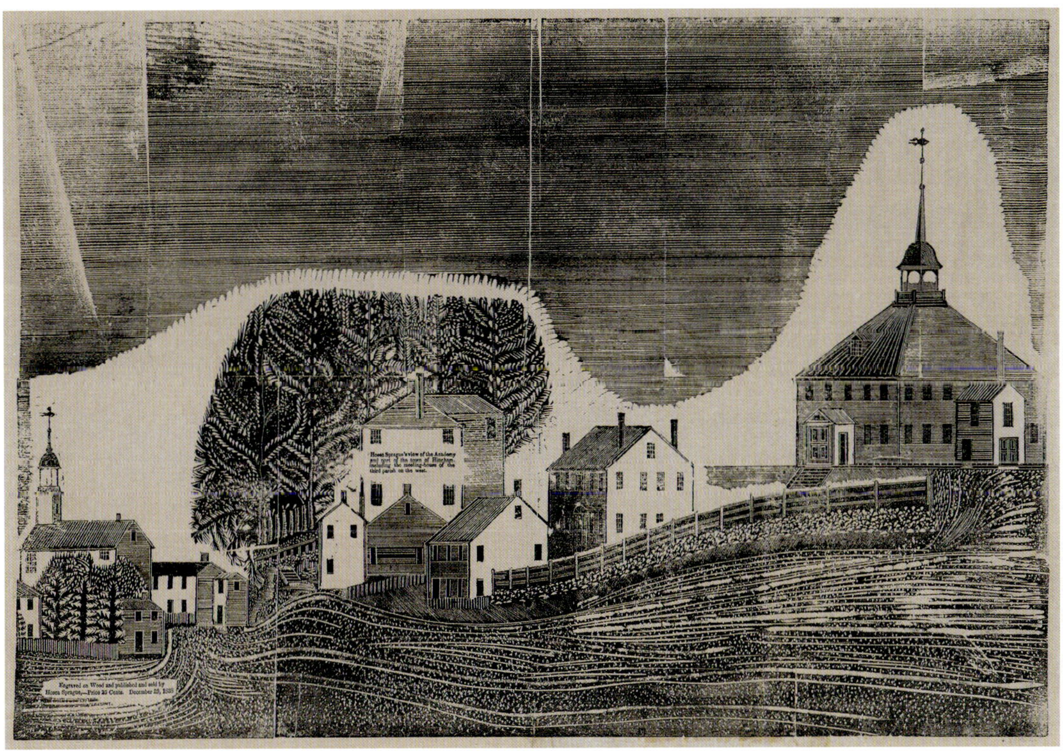

Figure 5
Hosea Sprague's View of the Academy and Parts of the Town of Hingham, Including the Meeting-House of the Third Parish on the West, drawn, engraved, and printed by Hosea Sprague, Hingham, 1839. Woodcut engraving on paper; H 21 ½, W 14 ⅜. Courtesy of Old Sturbridge Village

and status. Young men—many of whom attended the Derby School expressly to prepare for Harvard College—trained in Greek, Latin, geography, mathematics, and the practical sciences; young ladies improved their minds and their marriageability with English, French, and the decorative arts—including elaborate needlework pictures the girls' families clearly prized, as evidenced by the large number that survive (fig. 4).[11] In 1818, to accommodate a growing student body, a grand clapboarded hall was built for the considerable sum of $3,930.10, its

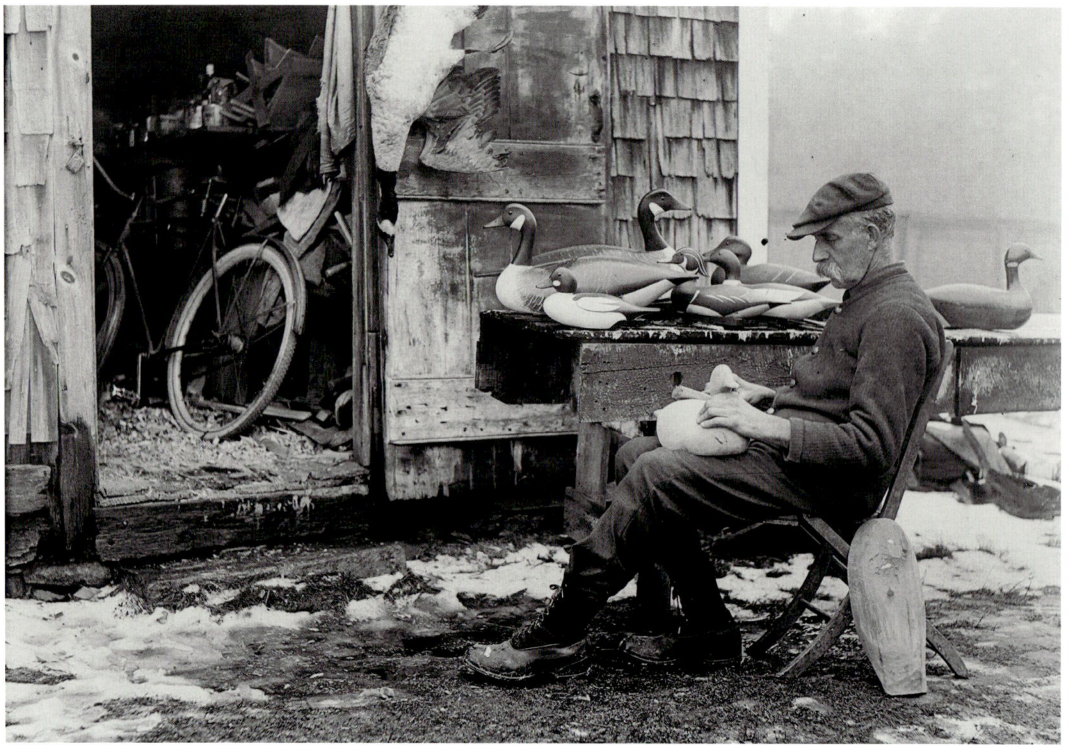

Figure 6
Joseph Lincoln sanding a decoy in front of his workshop, photograph by Leslie Jones, Hingham, 1926. Courtesy of Boston Public Library, Leslie Jones Collection

style reflective of the Republican ideals of order, rationality, and learnedness.[12] Set on the same hilltop as the Old Ship Church, it was the largest secular building in Hingham at the time; today it houses the Hingham Historical Society, home to an important collection of local art and artifacts. The Derby School, now known as Derby Academy, continues on in modern facilities—the oldest coeducational institution in the nation (fig. 5).

Hingham's quiet monument to its deep craft culture lies less than a mile from its celebrated counterparts. Hersey Farm, established around the time of the Revolutionary War, survives on its original eighteen-acre site with many of its outbuildings still intact—some with their original contents: plows, wagons, stoneboats, and other workaday objects essential to eighteenth- and nineteenth-century agricultural life. On a hillside at the back of the property stands a small shingled workshop (see fig. 3.13) once occupied by multiple generations of Hersey coopers and toymakers—pioneers in Hingham's most famous early industries. From the 1830s to about 1900, Hersey men crafted thousands of buckets, tubs, boxes, miniature woodenware, and toys in this modest space, all destined for useful or treasured lives in nearby and far-flung households. Incredibly, defying all the vagaries of time, the Hersey shop was discovered in 2007 with a wealth of early coopering and toymaking tools and materials still in situ—a remarkably preserved remnant of the many shops once active in Hingham.

A second small structure linked to Hingham's coopering history was occupied by Joseph W. Lincoln (1859–1938), a craftsman of protean abilities who worked there through the late nineteenth and early twentieth centuries. A boatbuilder, carpenter, and repairer of bicycles, antique clocks, and furniture—not to mention a cultivator, with his brother, of prizewinning dahlias—Joe Lincoln is recognized today as one of America's great master bird carvers. According to family tradition, his workshop was first used in nearby Norwell as a cowshed in the late 1700s; it was moved and then moved again to the side of a shingled Cape on Accord Pond, where Lincoln's grandfather Joseph Whiting (1777–1857) used it for coopering. Two generations later Whiting's grandson Joe Lincoln claimed the

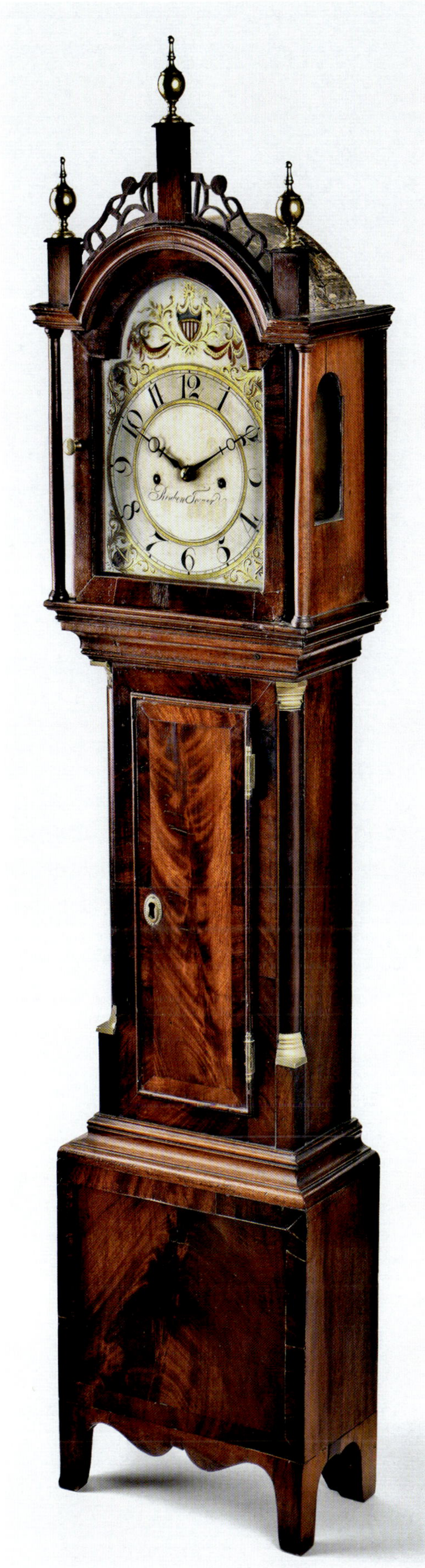

Figure 7
Dwarf clock, works by Reuben Tower, Hingham, case attributed to Abiel White, Weymouth, 1815–25. Mahogany, mahogany veneer, and white pine; H 50, W 11 ½, D 7 ¾. Courtesy of Winterthur Museum

shop—carving decoys out of cedar cut, perhaps, from the same stands his grandfather had used, and working at the same bench (fig. 6).[13] When the shop was cleared after Joe's death in 1938, the family discovered a trove of old woodenware, coopering tools, and materials.[14] Today Joe Lincoln's ducks, geese, and other decoys are considered masterpieces of American folk art, much sought after by collectors and museums.

Hingham's pervasive coopering and toymaking traditions existed alongside the practice of other skilled crafts as well. Reuben Tower (1795–1881) descended from three generations of woodenware makers, but following the death of his father in 1807, he broke with tradition and learned the trade of clockmaking; he became known throughout the South Shore for his handsome tall clocks, shelf clocks, banjo clocks, and other forms.[15] A particularly elegant example of a dwarf clock—made with the same proportions as its taller counterparts—boldly depicts his name at the center of the dial (fig. 7). Tower's sons were also exceptional craftsmen and mechanics; William S. Tower became a noted cooper and wood turner at Wilder's bucket factory, later opening Hingham's first toy factory with the assistance of two of his brothers.[16]

Even the fine arts intersected with Hingham's coopering industry. The self-taught painter Isaac Sprague (1811–95), the son of a boxmaker, achieved considerable fame in the nineteenth century for his precise and aesthetically pleasing landscape, botanical, and ornithological watercolors. His lively, delicate illustrations were published in some of the era's definitive texts on flora and fauna—including exploratory expeditions—and U.S. War Department reports.[17] In 1843 Sprague joined John James Audubon going up the Missouri River to sketch birds, mammals, flowers, and some scenes; his small traveling workbox (figs. 8, 9), painted red and with a domed top, may have accompanied him on the journey west. Discovered in 1969 among

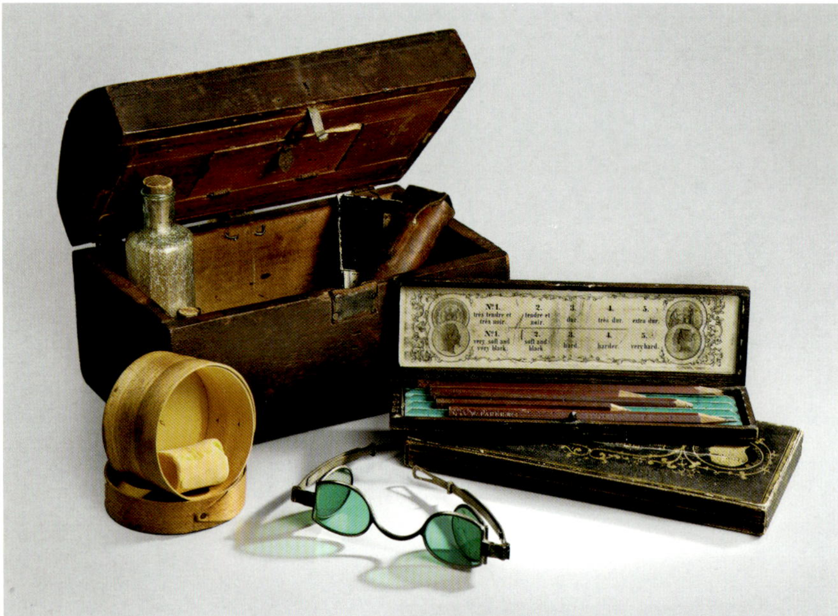

Figure 8

Traveling workbox with art supplies used by Isaac Sprague, Hingham, 1830–80. Tulip-poplar; H 4 5/8, W 8, D 4 5/8. Courtesy of the Hersey family

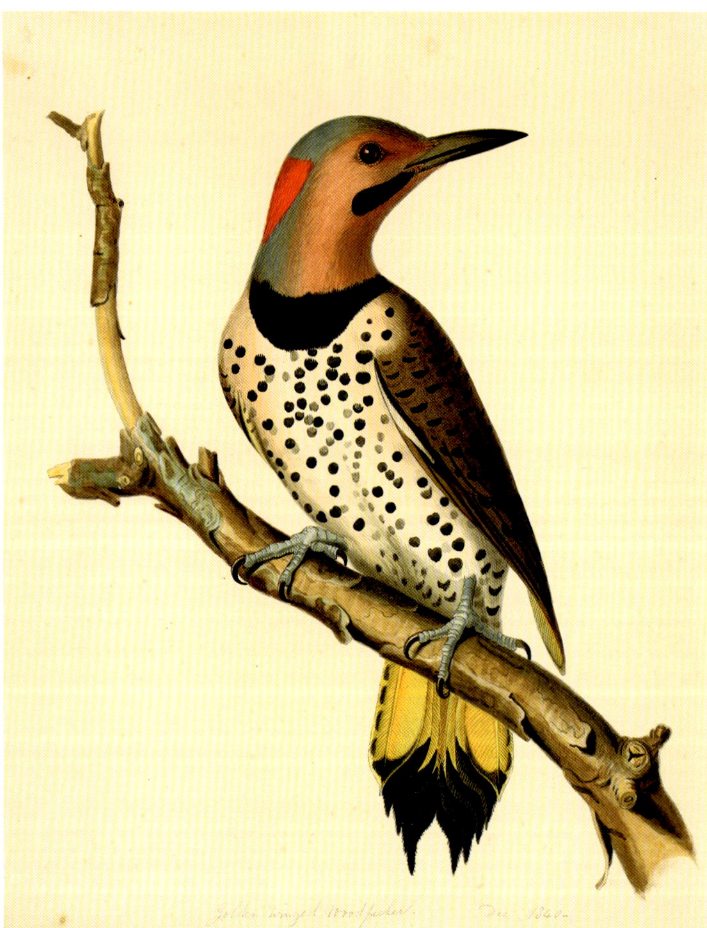

Figure 9

Golden winged woodpecker, by Isaac Sprague, Hingham, 1840. Watercolor on paper; H 11 13/16, W 9 1/2. Courtesy of Boston Athenaeum

the possessions of the Sprague family, the box still contains Isaac's green-tinted sunglasses—to aid in painting or shooting specimens—and other artist's supplies. A cake of fine yellow ochre pigment was found stored in a small round box made by a local cooper, Cotton Hersey (1792–1881).

In the 1893 *History of Hingham*, the town historian George Lincoln wrote of coopering, "To give a full account of this industry, of those who were engaged in it, and the many little workshops once so plentifully scattered throughout the town, would fill a volume."[18] This is an initial contribution to Lincoln's proposed volume; much remains for scholars to debate and explore, and many more discoveries are certain to come to light. Here we have sought only to reveal the practical, ambitious, and often beautiful work of Hingham's many forgotten craftsmen—particularly the four hundred coopers and nearly seventy toymakers documented in this study—who, by equipping the country's kitchens, barns, boats, playrooms, and military storehouses, profoundly shaped local and national culture. It is our hope that this book will spur further interest in tracing Hingham's extraordinary material history and its legacy.

NOTES

1. William H. C. Walker and Willard Brewer, *A History of World's End* (Milton, Mass.: Trustees of Reservations, 1973).

2. For a history of Hingham's early settlement, see Lorena Laing Hart and Francis Russell Hart, *Not All Is Changed: A Life History of Hingham* (Hingham: Hingham Historical Commission, 1993), 17–32.

3. George Lincoln et al., *History of the Town of Hingham, Massachusetts, Historical* (Hingham: By the town, 1893), 1.1:201–4. Hart and Hart, *Not All Is Changed*, 19.

4. Edward Franklin Ripley, *Shepherd in the Wilderness: Peter Hobart, 1604–1679, A Founder of Hingham Plantation in Massachusetts* (Lanham, Md.: University Press of America, 2001), 59.

5. Hart and Hart, *Not All Is Changed*, 27–32.

6. Ibid., 19.

7. Ibid., 25.

8. John Coolidge, "Hingham Builds a Meetinghouse," *New England Quarterly* 34, no. 4 (December 1961): 435–61. See also Hart and Hart, *Not All Is Changed*, 41–42.

9. Hart and Hart, *Not All Is Changed*, 44–57.

10. The most complete history of Derby Academy is Theodore Swan Roscoe's *History of Derby Academy, 1784–1984* (Hingham: Trustees of Derby Academy, 1984). See also Alexander Macmillan, "The Hingham Heritage Museum at . . . Old Hersey?," *News and Notes out of the Ordinary* 46, no. 5 (June 2013).

11. For a discussion of needlework by Derby School students, see Betty Ring, *Girlhood Embroidery: American Samplers & Pictorial Needlework, 1650–1850*, 2 vols. (New York: Alfred A. Knopf, 1993), 1:152–55.

12. Roscoe, *History of Derby Academy*, 32–33.

13. Cap Vinal, *Joseph W. Lincoln, Accord, Massachusetts, 1/26/1859–2/16/1938* ([Hingham: Cap Vinal], 2002).

14. Gertrude Goldman, "200-Year-Old Cooper's Tools Found in Hingham," *Quincy Patriot Ledger*, January 28, 1946.

15. For more information on Tower and other Hingham clockmakers, see Brock Jobe, Jack O'Brien, and Gary Sullivan, *Harbor & Home: Furniture of Southeastern Massachusetts, 1710–1850* (Lebanon, N.H.: University Press of New England, 2009), 268–73, 297–302, 304–8, and 312–27.

16. George Lincoln et al., *History of the Town of Hingham Massachusetts, Genealogical* (Hingham: By the town, 1893), 3:266.

17. Sprague did not receive credit for his most prominent early work, including Audubon's book on quadrupeds; see Emanuel D. Rudolph, "Isaac Sprague, Delineator and Naturalist," *Journal of the History of Biology* 23, no. 1 (Spring 1990): 91–126.

18. Lincoln et al., *History of Hingham, Historical*, 1.2:163–64.

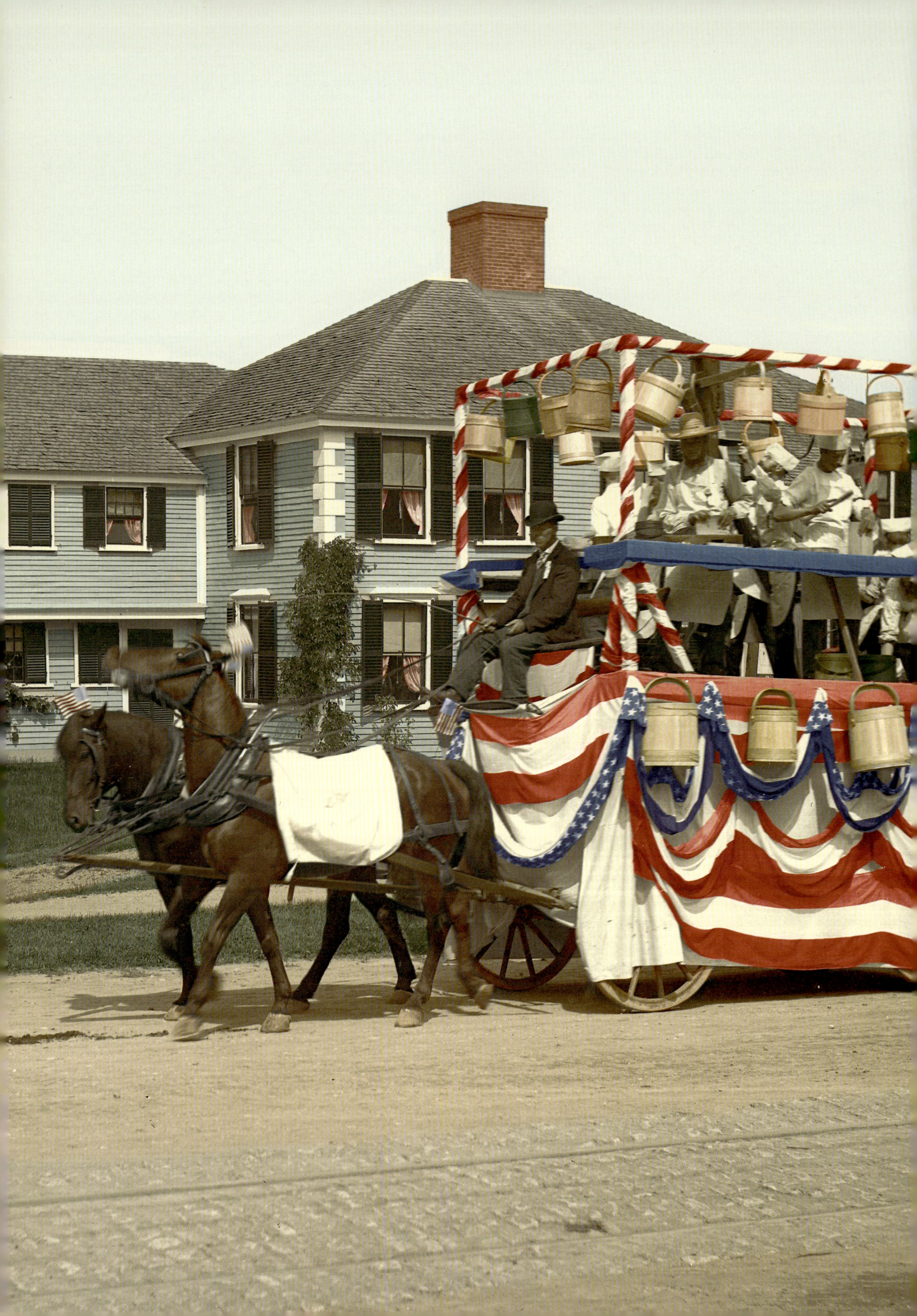

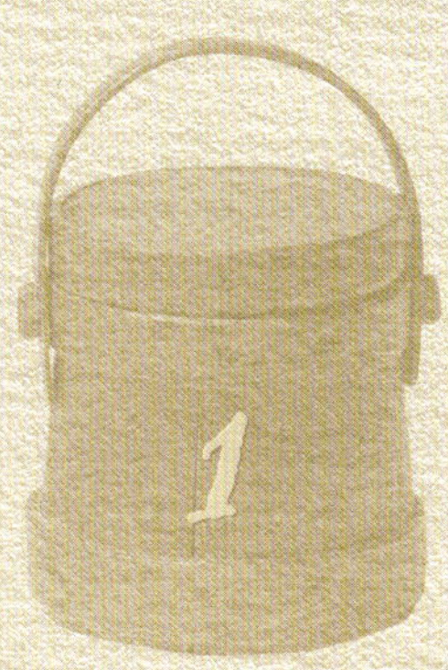

HINGHAM WOODENWARE:
WIDELY USED AND FAVORABLY KNOWN

The colony at Hingham, living at peace with their neighbors, soon became remarkable for manufacturing the most beautiful wooden buckets on the continent, from which it derives its present classical name. . . . Suffice it to say that it is . . . one of the loveliest spots the sun visits, that its ladies are unrivalled and its buckets unexcelled.

"History of Buckettown," 1816

FROM HINGHAM'S EARLIEST BEGINNINGS, THE RESOURCES TO DRIVE A vigorous coopering industry were present within the community. Like many of the small coastal settlements of the Massachusetts Bay Colony, the town devoted itself to three main industries: farming, fishing, and timbering. Early inhabitants busied themselves clearing fields—some land had already been opened by native Wampanoag people—and planting crops to feed their families and their livestock. They also took to the sea, where cod and mackerel abounded.

Wood, however, was the town's most valuable natural resource.[1] In *Wonder Working Providence* (1653), the first printed history of New England, the historian Edward Johnson noted that "the people [of Hingham] have much profited themselves by transporting Timber, Planke and Mast for Shipping to the Town of Boston, and also Ceder and Pine-board to supply the wants of other Townes, and also to remote parts, even as far as Barbardoes."[2] Records from the period confirm his observations. Hingham's first settlers harvested and milled trees to export framing members for boats and buildings; a few shipwrights even pursued this activity locally.[3] Tree cutting was so rampant that it threatened to deforest the town common. In 1655 town officials decreed: "The Town, taking into consideration the great detriment and damage that hath been and still is like[ly] to be, by the daily transporting of the Town's wood and timber. It is therefore ordered and agreed upon . . . That no person within this town shall . . . fall or cause to be falled any Wood or Timber upon the Town's Common to transport out of the Town upon the penalty of twenty shillings for every tree or part of a tree so fallen, or transported."[4]

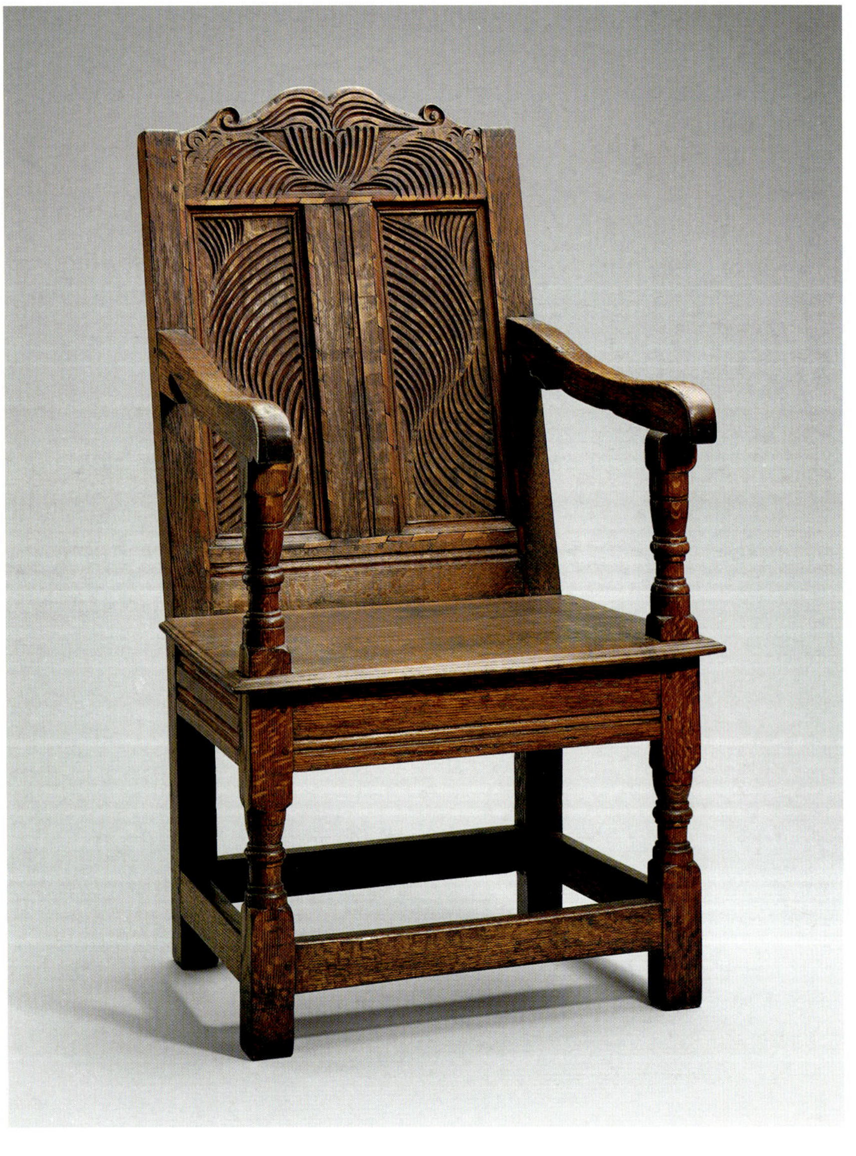

Figure 1.1
Joined chair, Hingham, 1635–85. Red oak, black walnut inlay; H 38 ¼, W (seat) 25, D (seat) 15 ⅝. Courtesy of First Parish, Hingham, Hingham Historical Commission, and Hingham Historical Society. Photograph © 2013 Museum of Fine Arts, Boston

The town's large stands of pine, oak, maple, and cedar trees attracted other skilled woodworkers—notably coopers, who produced a variety of wooden containers constructed of staves, hoops, and heads for both household and commercial use.[5] Thomas Lincoln (d. 1691), generally regarded as the town's first cooper, hailed from Beaminster in southwest England, where he probably learned the trade. In 1636 Lincoln was granted a home lot in Hingham; his descendants still occupy part of the original house.[6] Although few details of his career survive, Lincoln sometimes identified himself as a maltster and, on occasion, took on carpentry work: in 1671 he crafted partitions and a table for the schoolhouse.[7] According to tradition, he owned the finely carved and inlaid great chair (fig. 1.1) that descended through his family until about 1912, when it was donated to the Hingham Memorial Bell Tower beside Old Ship Church.[8] Lincoln's son, also named Thomas, operated a carpenter's shop on the town common, and he may have had a hand in its production.[9]

By the mid-seventeenth century coopering was a major industry in Hingham. Its increasing importance may account for a law passed in 1665 against removing hoop poles—withes for binding containers—from the common and transporting them out of town.[10] Legal measures to protect the interests of a growing class of craftsmen suggest their considerable economic and political influence; by 1700 an astonishing total of thirty coopers had joined Lincoln in the trade.[11] Perhaps the most intriguing of these newcomers was Edward Gould (ca. 1607–84/85), who in an early land deed listed his occupation as "pail-maker"—an indication that he specialized in making the ubiquitous wooden pails, or buckets, so essential to colonial life. Gould is thought to have first settled in Hingham in 1635, but he was not admitted as a townsman until a few years later.[12] He apparently conducted some business with Thomas Lincoln, the nature of which remains a mystery; in 1676 Lincoln settled an unspecified debt with the town on Gould's behalf.[13]

Coastal settlements with active harbors typically employed large numbers of coopers; they supplied the oak casks, kegs, and barrels the shipping trade depended on.[14] By 1648 coopers in the bustling seaport towns of Boston and Charlestown wielded enough power to acquire a charter to form their own guild. This closed company of craftsmen monitored the quality of coopered goods produced there and regulated entrance into the trade. Though its establishment in Boston was a sure indication of the coopers' standing in the community, the English guild system was not viable in colonial America; its protective, exclusionary rules may have exacerbated the fledgling colonies' chronic shortage of skilled labor. The Boston coopers' guild operated for just a few years, but its members continued to manufacture packaging for the region's robust farming, fishing, and brewing industries. They also exported huge quantities of shooks, or bundles of coopering parts, for assembly in distant ports.[15]

The coopers of Hingham, however, differed significantly from their Boston counterparts. They did not, by and large, construct barrels, casks, and kegs—that was the province of dry and wet coopers.[16] Instead, they engaged in "white coopering," manufacturing the lightweight household containers known as woodenware: pails (figs. 1.2, 1.3) to transport milk and water; tubs for curing meat, dyeing fabrics, and laundering; churns (fig. 1.4) for making butter; strainers and hoops for cheeses; boxes to store spices; and a variety of related goods (fig. 1.5).[17] Exactly why Hingham became the center of this activity is unknown. It is possible that an early, specialized craftsman like Thomas Lincoln or Edward Gould powerfully influenced the direction of the local industry, sparking a kind of communal distillation of practice. No matter the cause, however, the town of Hingham had, within a couple of generations, become synonymous with white coopering. When strapped for cash in 1687, the town was permitted to pay its tax bill to the General Court in Boston with surplus milk pails; by 1708 local boatmen like

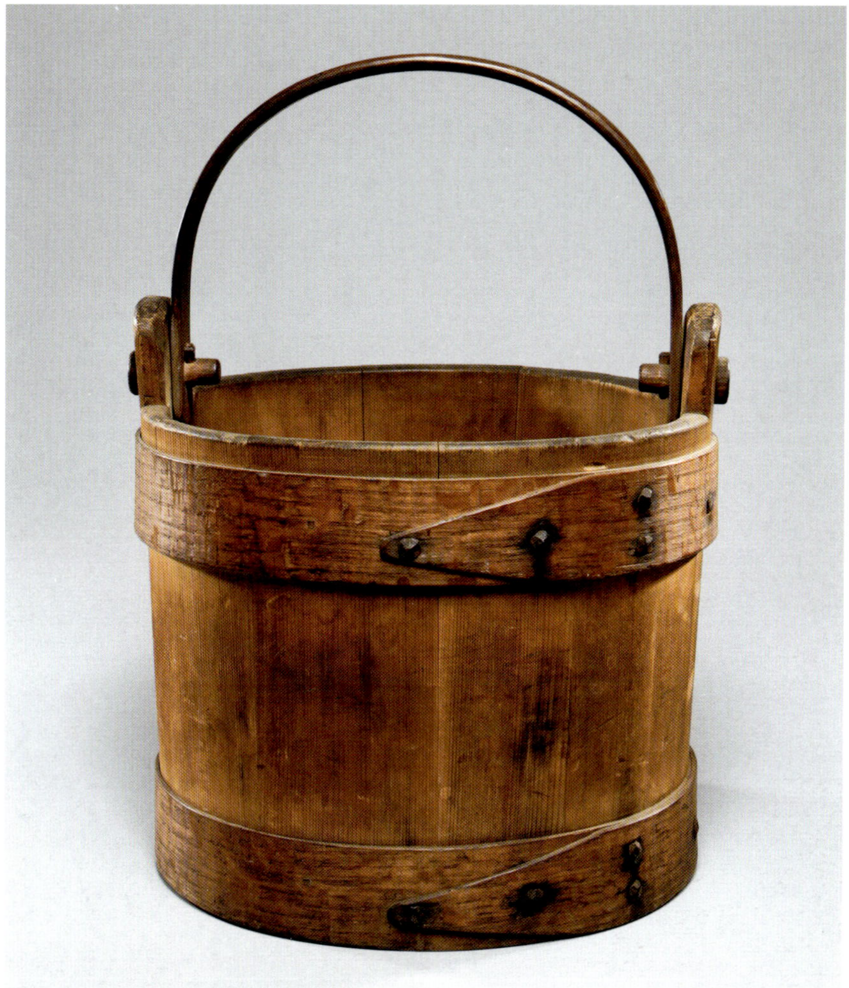

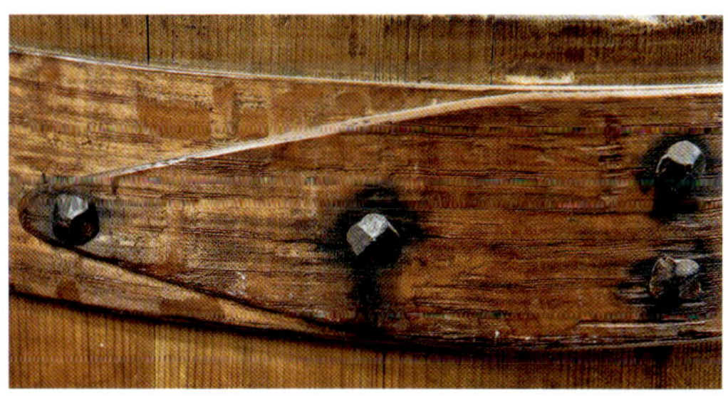

Figure 1.2
Pail, by Bela Sprague, Hingham, 1820–40. Stamped "B.S" on bottom. White pine, oak, iron nails; H 10 ¾, Diam 11. Courtesy of Hingham Historical Society

Figure 1.3
Detail of hoop on pail in figure 1.2

Figure 1.4
Churn, attributed to David Hersey, Hingham, 1820–30. Stamped "D. HERSEY" on lid. White pine, oak, ash, maple, iron nails; H 22 ¾, Diam 10 ⅞. Courtesy of Hanover Historical Society

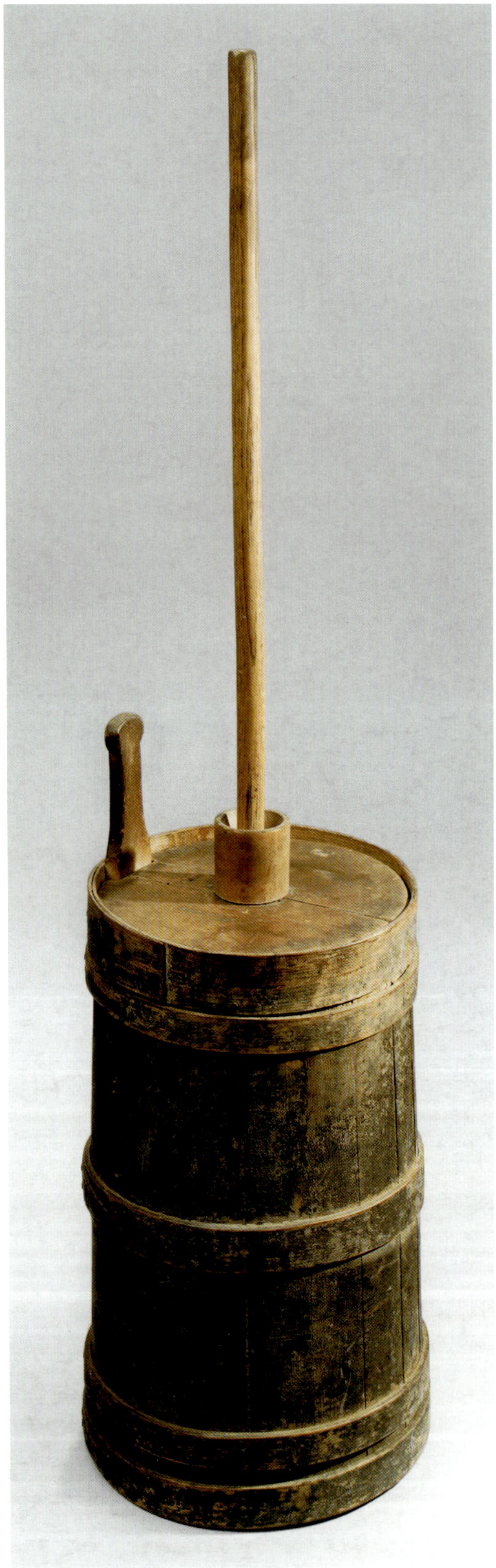

Lieutenant Jacob Beal were freighting hundreds of these goods and other "coopers ware" to the city.[18] Hingham was well on its way to earning the nickname it was known by for at least two centuries: Bucket Town.[19]

It was undoubtedly Hingham's innovative, entrepreneurial craftsmen who placed the town at the forefront of New England's white coopering trade. As the eighteenth century progressed, several family dynasties dominated local production of woodenware; generations of intermarriage further consolidated their prominence. Not surprisingly, they often clustered together along the same streets, where they operated small farms and independent coopers' shops. The Herseys settled on Austin's Lane, Ripleys on Main Street, Lincolns in West Hingham, and Leavitts on Leavitt Street. The same held true for other pockets of coopering established by the Beal, Whiton, and Tower families; the Groces populated a short stretch of Liberty Plain.[20]

In search of better opportunities—and usually more land—some coopers migrated west to Worcester County, north to New Hampshire and Maine, or south to Connecticut. Simon Burr was part of that emigration in the early 1700s, moving his family from Hingham to nearby Rehoboth, where he continued to operate a cooper's shop. His son, also named Simon (b. 1730), moved across the border to Rhode Island, where he adapted his skillset to local demands; in addition to buckets and tubs, he produced hundreds of hanks and mast hoops for a flourishing shipbuilding industry.[21] Despite dozens of coopers like the Burrs relocating with their trade, the industry remained centered in Hingham, the woodenware capital of the colonies.

Back in Hingham Simon Burr's cousin Elisha Burr (1703–78), offered boxes as part of his general line of woodenware; many members of the Burr family identified themselves as boxmakers.[22] When Elisha died in 1778, his shop contained "12 dozen boxes" and a quantity of "pail timber" for cutting staves. Indeed, after pails, boxes were Hingham's chief export (fig. 1.6).[23] Produced in round and oval shapes, these items were sold in gross lots, often in nests of four, which were packed in hogsheads for shipment.[24] Earlier examples (fig. 1.7) are typically constructed

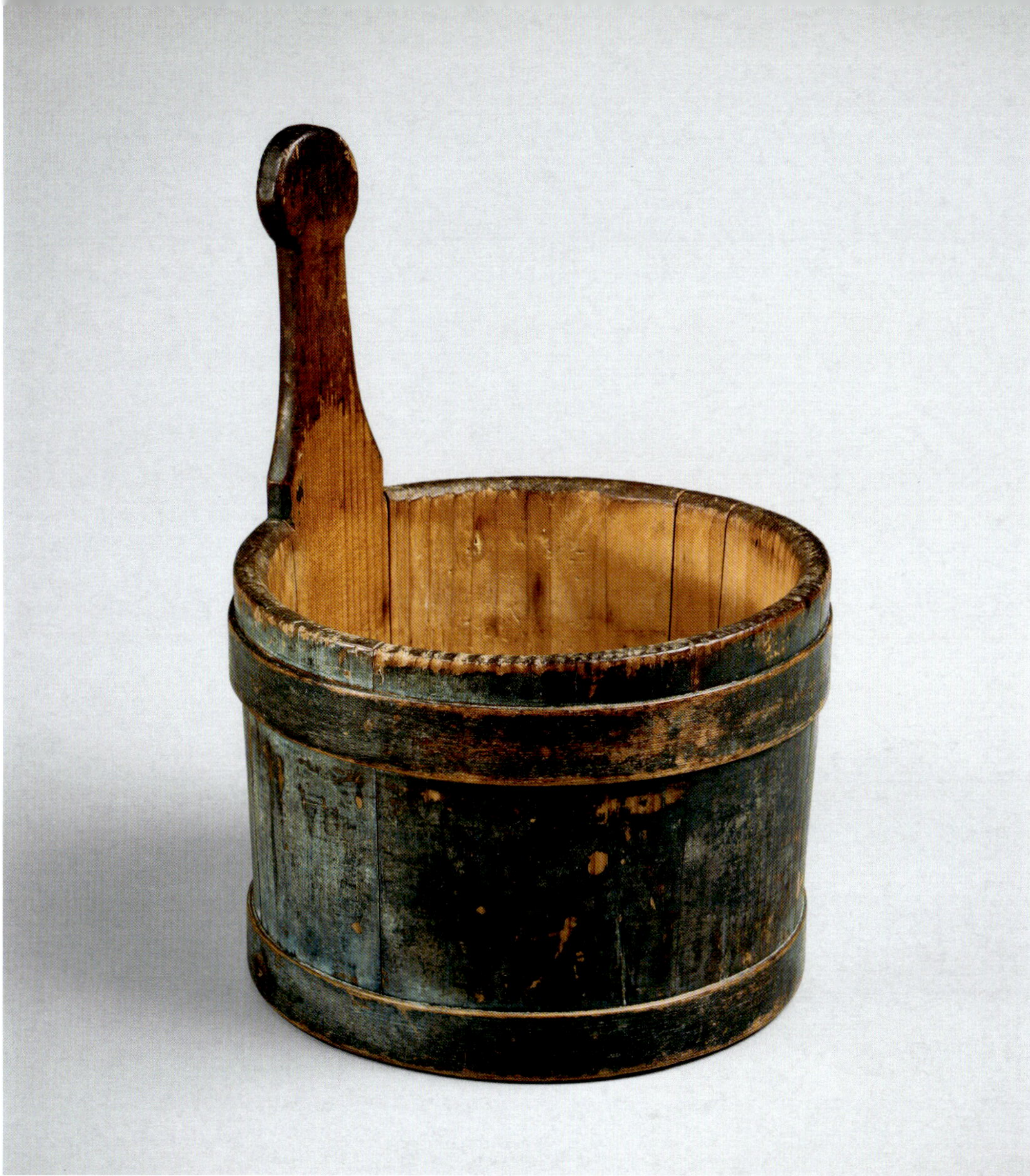

Figure 1.5
Piggin, by Caleb S. Hersey, Hingham, 1820–60. Stamped "C. S. HERSEY" on bottom. White pine, oak, iron nails; H 10 9⁄16, Diam 7 7⁄16. Courtesy of the Hersey family

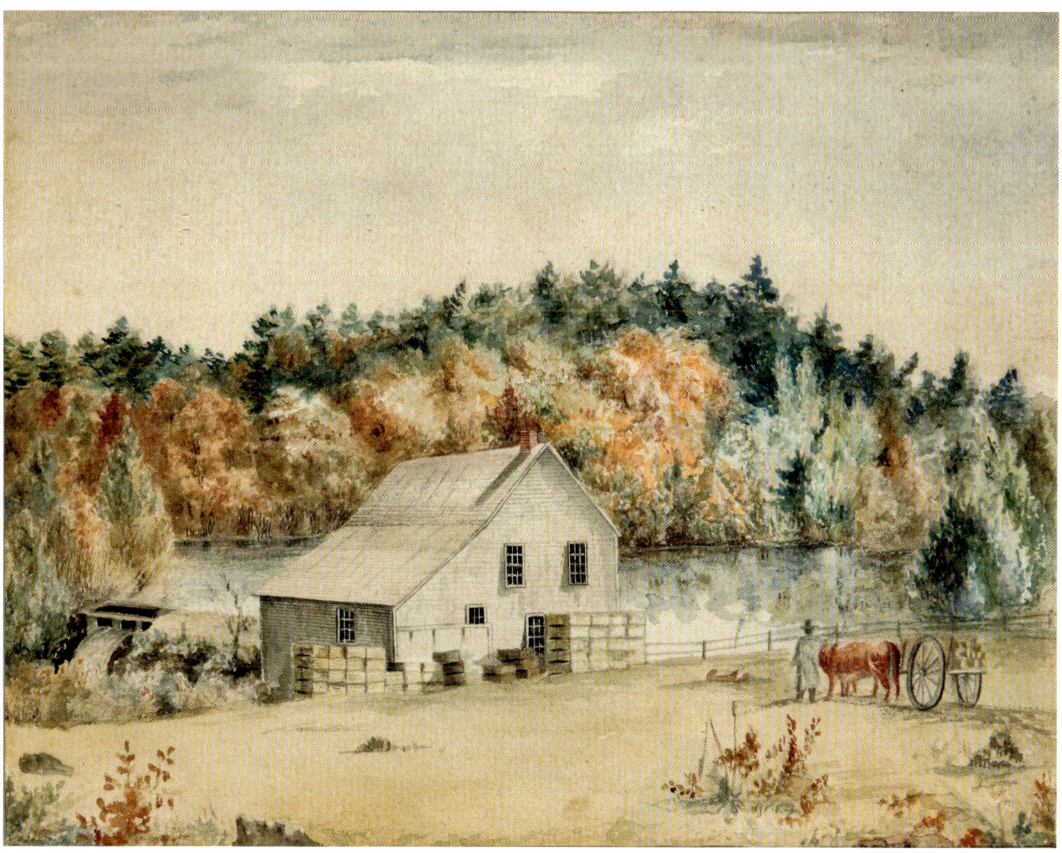

Figure 1.6
Leavitt's Shingle and Box Mill at Shingle Mill Pond, by Isaac Sprague, Hingham, 1835–45. Watercolor on paper; H 7 3⁄8, W 9 5⁄8. John Leavitt operated the mill in the early nineteenth century, sawing shingles and the heads for hoop boxes made by local coopers. Courtesy of the Hersey family

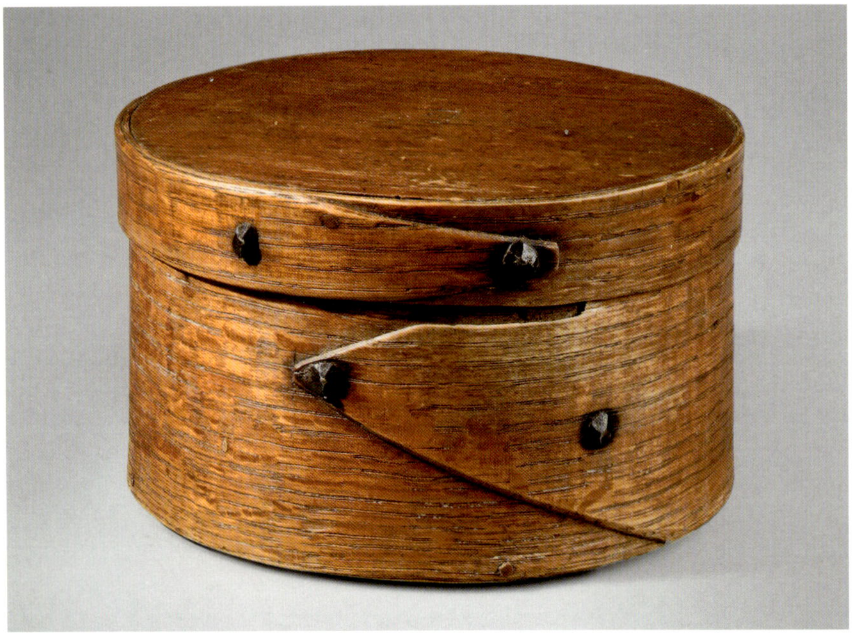

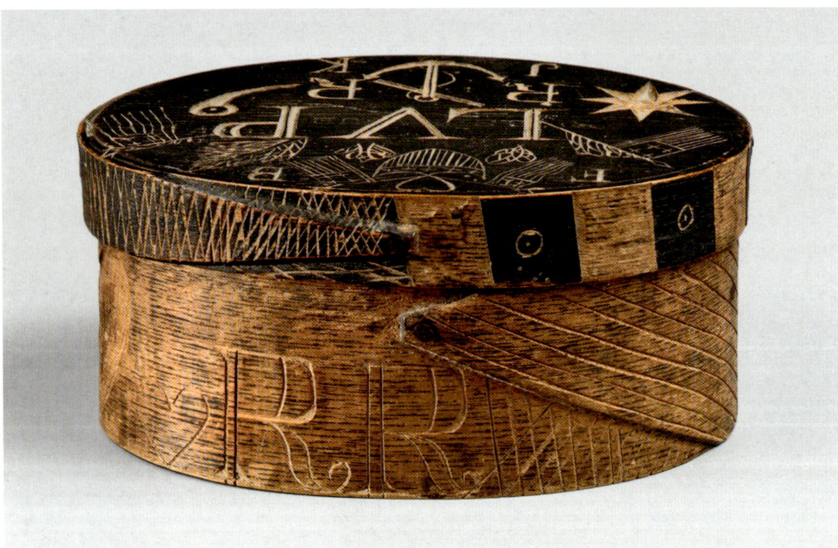

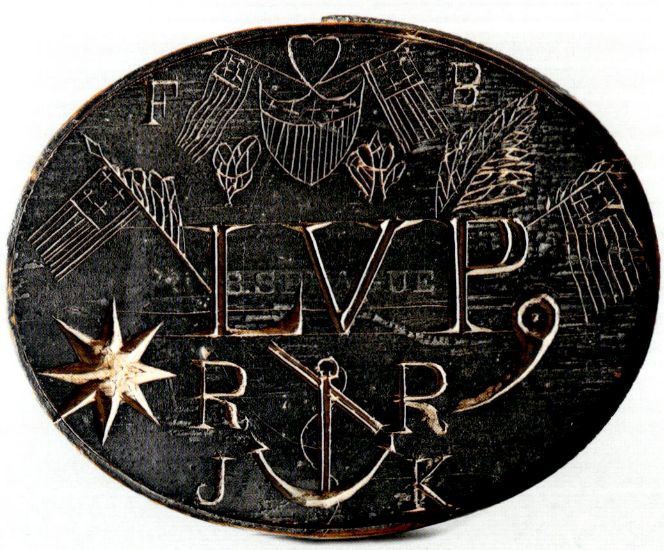

of ash or oak, but by the turn of the century coopers had begun making them from sugar maple—hence the name sugar boxes—and pine imported from nearby towns and from as far away as Maine; local supplies had possibly dwindled by this time.[25] As they were relatively inexpensive and easy to purchase, many served as a ready canvas for amateur artists. A carved and decorated example (figs. 1.8, 1.9) was probably the work of a sailor whiling away the hours at sea. Nearly a century later, these containers continued to be a staple of local coopers—and a much-coveted possession for many. In 1862 a terrible fire consumed the Hingham dry goods store of Royal Whiton. It was reported that "Mr. Whiton had in his second story several barrels filled with small wooden pill boxes . . . made in Hingham, which were carefully poured out of the window onto the sidewalks as one would throw cents to a crowd of boys to scramble for. Most of the boys went off with their pockets well filled with small boxes."[26]

Boston had solidified its position as New England's commercial center by the mid-eighteenth century. Ships from the colonies and ports all over the world crowded its wharves, loading and unloading a wealth of exotic cargoes and domestic goods such as fish, furniture, and timber.[27] Woodenware, customs records show, was also widely exported at this time. In 1745 Captain John Stephenson listed dozens of pieces in the bill of lading for the *Sea Flower,* a Hingham sloop destined for North Carolina.[28] Ships headed to Pennsylvania, Maryland, Virginia, South Carolina, and the West Indies also commonly carried such goods. An especially brisk trade developed with merchants in New York City. In 1744 and 1745 the sloops *Anne* and *Lydia* sailed from that city for Boston twice a month—a voyage that took eleven days. Before returning, they packed their vessels with hundreds of wooden pails, sieves, and sugar boxes, all probably made in Hingham and quickly dispersed among a burgeoning Manhattan population.[29]

For many buyers, packet boats docked in Boston were the first point of purchase for Hingham woodenware. Since the seventeenth century these small coastwise vessels had transported passengers and merchandise from town to city. In 1790 the *Hingham Packet*, captained by Jotham Lincoln, became the first to make regular trips—two hours each way. By the turn of the century several others had joined in.[30] Hosea Sprague, a Hingham native who worked as a library assistant at the Boston Athenaeum, relied on the boats to receive food, letters, and bundles of clothing from his family back home. In December 1812 he wrote to his brother Isaac that the "cranberry pies and some cheese" had arrived safely by water and "were very acceptable."[31] Indeed, Hingham's packet men prided themselves on prompt, swift passage. Competition ran high among the captains, who raced each morning to secure the head berth at Boston's Long Wharf near India Street; situated close to the Custom House, it was an ideal location at which to conduct business.[32] A photograph (fig. 1.10) taken around 1850 shows one of the boats in front of the Custom House, the ground beneath its bowsprit stacked chockablock with barrels of merchandise.

Station packets were yet another venue from which to market woodenware. Made from the hulls of retired ships, these "floating hotels," moored permanently at Long Wharf, were noted landmarks on the waterfront. Their cabins and kitchens gave guests—usually from Hingham—a place to stay when visiting Boston, while the boats' broad decks functioned as consignment shops for "buckets, eggs, knit woolen stockings, and other products" made by residents of the South Shore.[33]

By the end of the eighteenth century General Benjamin Lincoln (1733–1810) had emerged as Hingham's most prominent resident. The son of a wealthy farmer and politician, Lincoln spent most of his life in public service, including nearly three decades as town clerk and representative to the General Assembly. He acted on an even larger stage during the Revolution, when he commanded troops in several major battles and—as George Washington's second in command—formally accepted the surrender

Opposite page:
Figure 1.7
Round box, by Elisha Beal, Hingham, 1780–1830. Stamped "E. BEAL" on lid. White pine, oak, iron nails; H 3 5/16, Diam 5 7/8. Courtesy of the Hersey family

Figure 1.8
Oval box, by Bela Sprague, Hingham, 1820–60. Stamped "B. SPRAGUE" on lid. White pine, ash, iron nails; H 2 7/8, W 6 7/16, D 4 7/8. Courtesy of Tom and Sandra Ammann

Figure 1.9
Detail of lid on box in figure 1.8

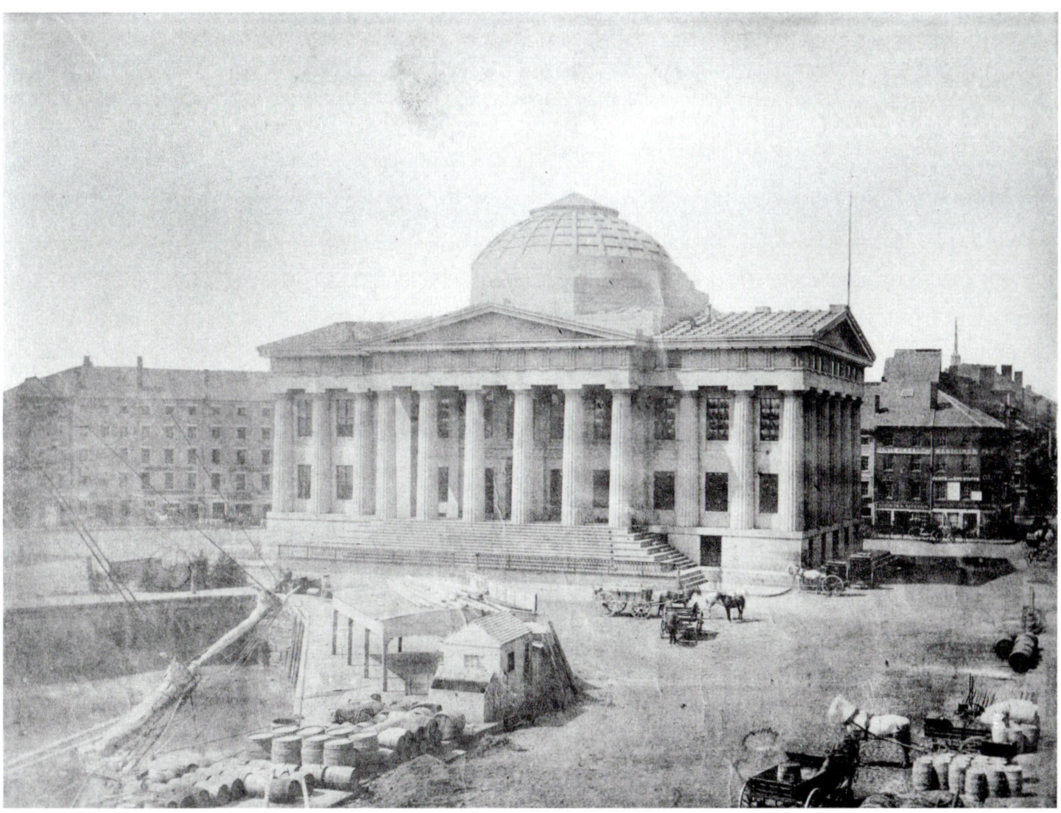

Figure 1.10
Boston Custom House and Hingham packet, photograph, 1840–50. Courtesy of Bostonian Society

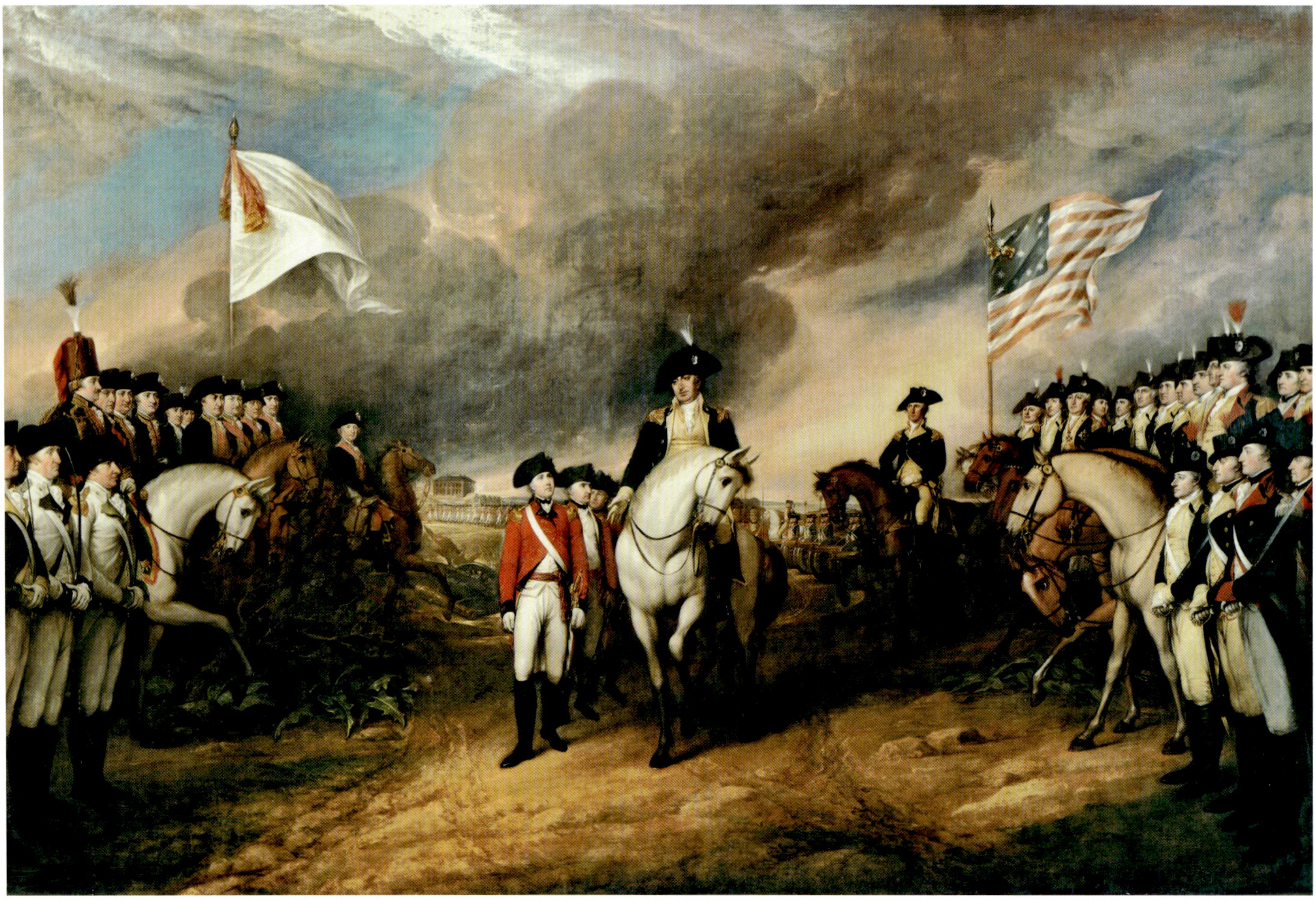

Figure 1.11
Surrender of Lord Cornwallis, by John Trumbull, 1820. Oil on canvas; H 12 feet, W 18 feet. General Lincoln is depicted at the center, mounted on a white horse, accepting the surrender of the British army at Yorktown, Virginia, in 1781. Courtesy of Architect of the Capitol

of the British army at Yorktown. John Trumbull later immortalized Lincoln's role in that event in a painting (fig. 1.11) commissioned by the U.S. Government for the newly built Capitol Rotunda, where it still hangs. Following the war, Lincoln held such distinguished positions as lieutenant governor of Massachusetts and collector of the Port of Boston. He was honored with a state funeral when he died in 1810.[34]

At the height of his career, Lincoln helped secure large military contracts for Hingham coopers. In October 1774, on the eve of the Revolution, he was appointed one of five commissaries to the Committee of Supplies, a group tasked with gathering arms, ammunition, and other necessary equipment for the colony's militia. A month later the five men voted unanimously to acquire 15,000 canteens for troops to take into the field. Under the influence of Lincoln, the committee looked to Hingham and its established network of coopers to produce the wooden drinking vessels quickly.[35] Almost immediately local traders such as Hawkes Fearing began stockpiling hundreds of canteens made by the Burr, Sprague, and Leavitt families. When he had collected a sufficient number, Fearing carted the goods—along with dozens of buckets—to a critical Patriot encampment at Roxbury.[36] One man who transported canteens reportedly hid them under piles of hay to avoid detection and arrest by the enemy.[37] They were probably moved in a similar manner a day before the attempted raids at Lexington and Concord, when the committee agreed to deposit the drinking vessels at storehouses in the countryside.[38]

Hingham's coopers found steady employment throughout the war. After the British evacuated Boston in March 1776, the Commissary General ordered 1,000 more canteens for a new regiment stationed at Noddle's Island in Boston harbor.[39] Just a few months later the Assembly requested an additional 5,000 for troops headed to New York and Canada to reinforce the Continental Army. Theophilus Cushing, a representative from Hingham, was elected to purchase the canteens, all of which came from his hometown.[40] He filled the order in less than twenty days—a testament to the skill and speed of the coopers—but he still fell under sharp criticism for delivering them too late.[41] Within a few weeks the Assembly directed him to purchase another 1,500 canteens for regiments headed west to Fort Ticonderoga. This time Deputy Secretary John Avery warned that "there are many complaints from different parts of this State for want of camp-kettles, canteens, &c., which you, as one of the Committee, were to provide. I need not press the necessity of forwarding them immediately, when I inform you that the men are now waiting on the road for those articles, and are very uneasy."[42]

Lincoln, Cushing, and other representatives rarely, if ever, purchased woodenware directly from craftsmen, relying instead on agents from Hingham to organize the coopers, broker the contracts, and oversee shipments. Between 1776 and 1778 Caleb Loring Jr. bought nearly 2,400 canteens from the workshop of Jonathan Lane. Samuel Norton (1744–1832), a wealthy merchant and half brother to Benjamin Lincoln, purchased another 470, all of which he paid for in cash. Lane credited both men roughly six shillings for "surveying" the goods, a turn of phrase that may indicate a commission for procuring him the work.[43] Loring maintained a similar relationship with the father-and-son duo Jacob and John Leavitt, coopers of Hingham Centre; during the same period they sold Loring nearly 2,000 canteens.[44] Undoubtedly many other coopers from the town turned to the manufacture of wooden drinking vessels during the war, but records of their contribution are lost or no longer survive.

In addition to provisioning the militia, Hingham made many of the canteens issued to soldiers in the Continental Army. In October 1782, shortly before peace talks with Great Britain began, Samuel Norton (fig. 1.12) wrote to Lincoln in Philadelphia, where he was engaged as secretary of War. Norton lamented that Hingham's coopers had not been paid for their work in over a year. He further reported that Colonel Jabez Hatch, deputy quartermaster of the army, was embarrassed by the shortage of cash, but he offered no help in settling the debt—nor did the continental treasurer. Norton implored his half brother for assistance, as he had already advanced a "scant pittance" from his own pocket and could no longer afford to do so.[45] Lincoln responded:

> *I feel too sensibly for the distresses of my neighbours the canteen makers not to view their situation with concern. I know many of them must suffer by the delay of payment of the money due from the public yet have it not in my power to give them that relief by procuring their money which they so much need & to which they have so just a claim. You say I was their employer, it is true I procured them the*

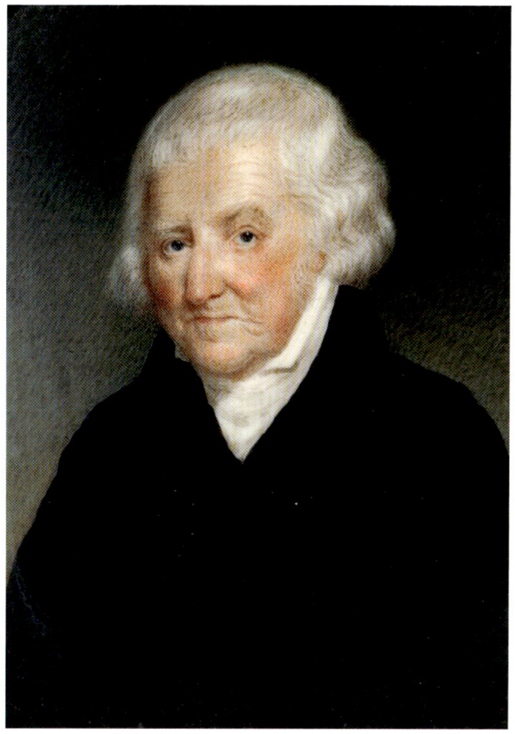

Figure 1.12
Portrait of Samuel Norton, by Sarah Goodridge, Boston, 1826. Oil on ivory; H 3 ½, W 3 ¼. Courtesy of the Hersey family

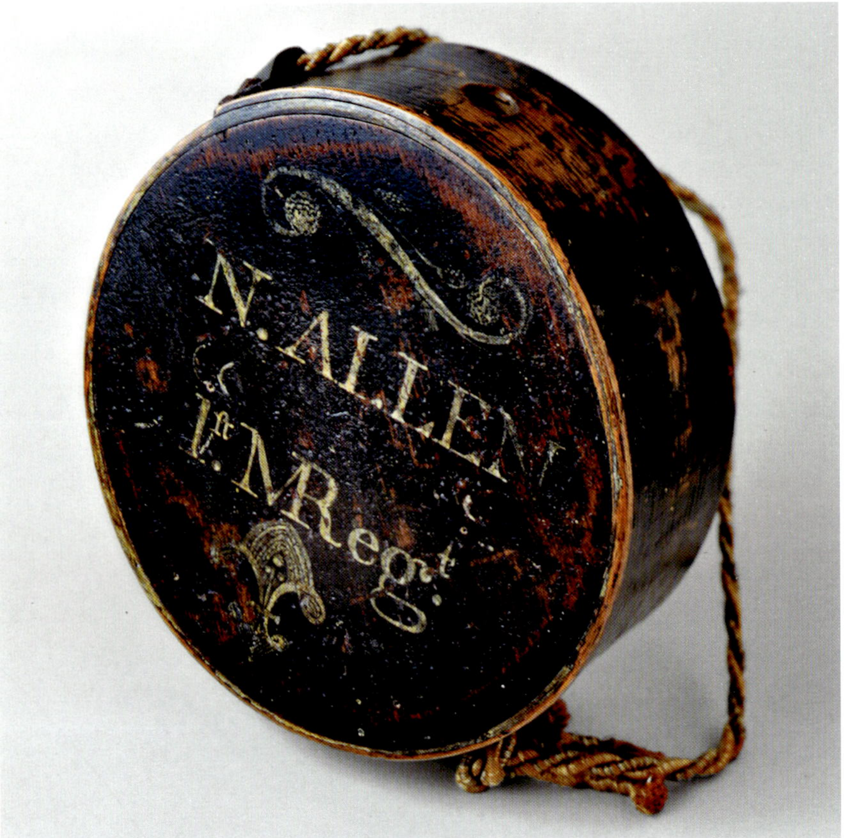

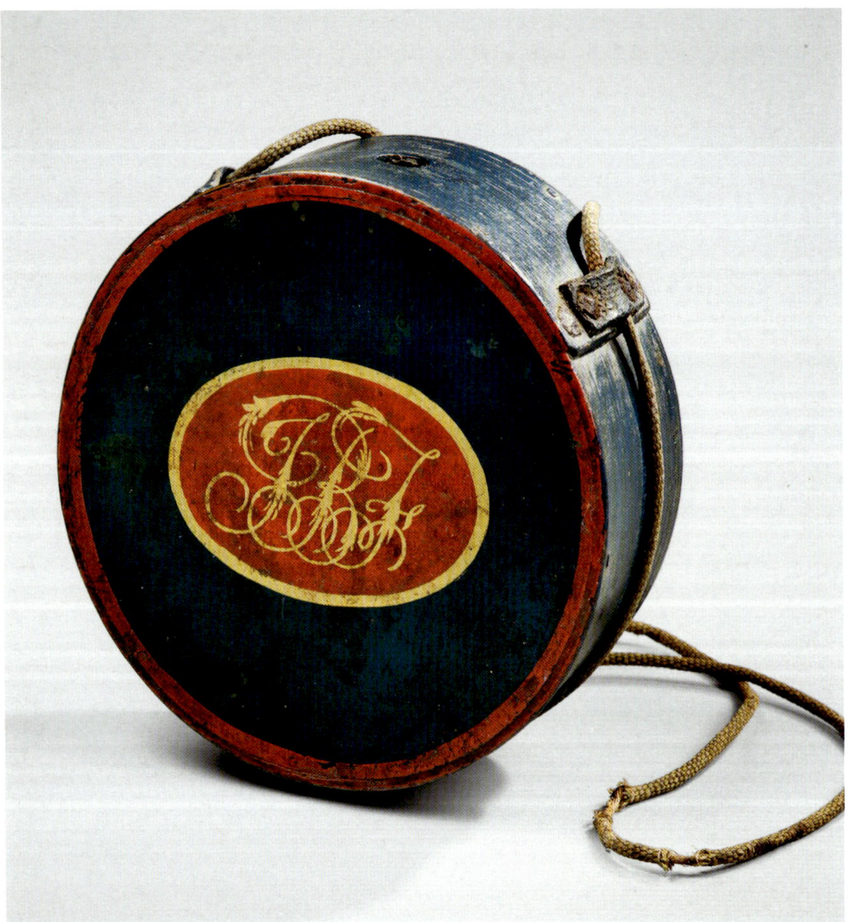

job, but I am sure it was not understood that I was personally responsible for the money.[46]

Lincoln added that he would do his best to repay the coopers when he returned home to Hingham in December. Whether they were ever compensated is unknown.

Few Revolutionary canteens survive with reliable histories of ownership; one is a decoratively painted example (fig. 1.13) used by Lieutenant Noah Allen of Sandisfield in Berkshire County. Although its place of manufacture is undocumented, Hingham is a strong possibility. In January 1781, after five years of service, Allen joined the 1st Regiment of Massachusetts in the Continental Army.[47] Around this time he was probably issued the round canteen later ornamented with his name and other identifying information. Collectors today commonly refer to this type of container as a "cheese box" canteen; it is similar in appearance and construction to the cheese or butter boxes also made by coopers.[48]

In the decades following the war canteens remained a staple of Hingham coopers, who continued to produce them in large numbers for the U.S. Army and volunteer militias throughout the region. A brightly painted example (fig. 1.14) made by Elisha Burr (1776–1863) of Hingham is among the best documented from the period.[49] Burr crafted it around 1822, and impressed his maker's mark, "E. BURR," on the reverse. It was among the twenty-six canteens purchased by the Independent Boston Fusiliers, a wealthy volunteer militia. According to a surviving bill of receipt, they paid Benjamin Curtis and Daniel Hubbard, fancy painters in Boston, $8.06 to embellish them with red and blue paint, $2.28 to gild the edges and varnish the surfaces, and $1.00 for stoppers and trimmings, or shoulder straps.[50] By the 1830s and 1840s such containers had begun to be replaced by canteens made of tin.

The traditional model of the farmer-craftsman, which had existed in Hingham since its founding, began to unravel in the late 1820s as a new breed of Yankee entrepreneur embraced advances in technology and methods of production.[51] Almost overnight small shops turned into large factories, seasonal labor became year-round, and powerful, innovative machinery replaced hand tools. This enormous change is embodied in Joseph Jacobs's career trajectory. A blacksmith by trade, Jacobs produced hammers in a modest workspace at the rear of his house on Main Street. With the purchase of a steam-powered engine and new equipment to grind and polish edge tools, his enterprise quickly expanded to a workforce of twenty-five to thirty-five men who turned out over 10,000 hatchets a year.[52] Other industries soon followed suit.

A young and ambitious cooper named Crocker Wilder (1802–75) led the shift to mechanization among Hingham's coopers. In 1830 he established C. Wilder & Sons, the first bucket factory (fig. 1.15) in town to use steam-powered saws and lathes, streamlining the manufacturing process with quick production and standardized parts.[53] Only a year later Hingham became an official port of delivery for the United States, which made it even easier for Wilder to export the thousands of buckets, pails, and churns he produced.[54] His success also coincided with the growth of the local fishing industry; Hingham's fleet of sixty-five fishing boats harvested 50,000 barrels of mackerel in 1831.[55] C. Wilder & Sons did not make the large barrels in which this bumper harvest was ultimately shipped, but it did craft the fish kits, or coopered tubs, for its temporary storage aboard ship.

In response to this growing threat to their livelihoods, Hingham's traditional coopers banded together in 1840 to produce a "cooper's price agreement." Printed as broadsides and probably posted in the tradesmen's shops, the agreement (fig. 1.16) attempted to shore up prices for handwork in the face of cheaper, mass-produced goods—a desperate effort that ultimately failed. Factory production surged, fed by low prices and an ever-widening market for woodenware; in 1845 C. Wilder & Sons relocated to a large, three-story building on Cushing Pond. The new site gave the company access to waterpower and

Opposite page:

Figure 1.13

Canteen carried by Noah Allen of the 1st Massachusetts Regiment, possibly made in Hingham, 1775–85. White pine, oak, iron nails; Diam 7 ¼, D 2 ⅞. Courtesy of Fort Ticonderoga Museum

Figure 1.14

Canteen, by Elisha Burr, Hingham, ca. 1822. Painted by Curtis & Hubbard for the Boston Independent Fusiliers, Boston, 1822. Stamped "E. BURR" on back. White pine, oak, iron nails; Diam 5 ⅝, D 2 5⁄16. Courtesy of Old Sturbridge Village

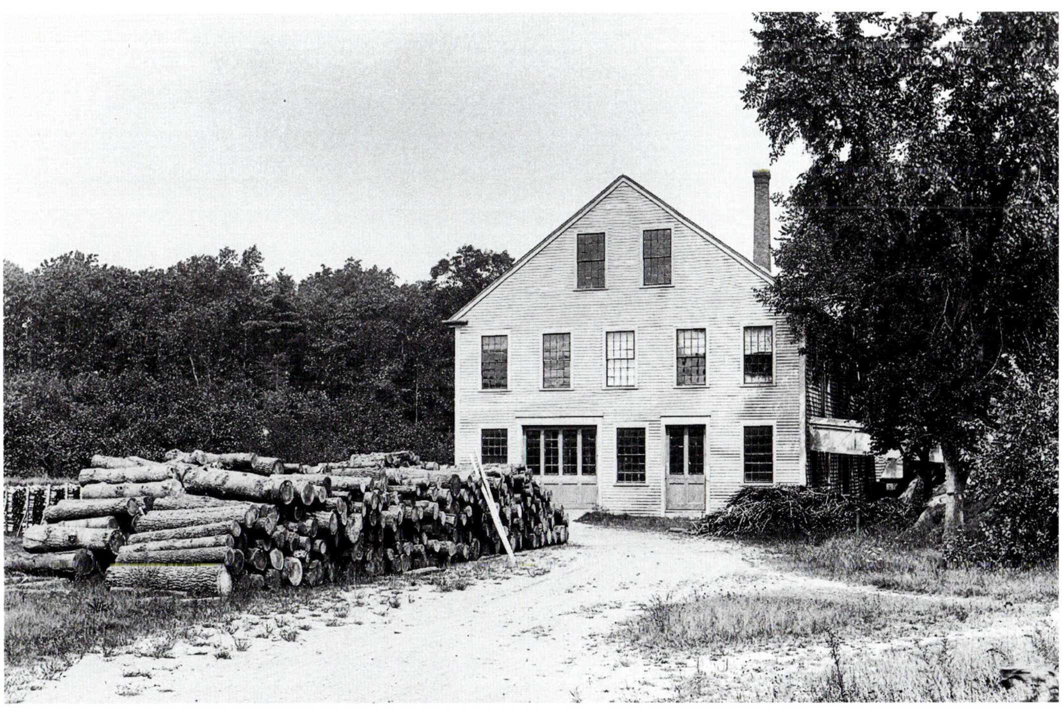

Figure 1.15

Wilder bucket factory, photograph by Frank Reed, Hingham, 1895. John P. Richardson Photograph Collection, courtesy of Hingham Historical Society

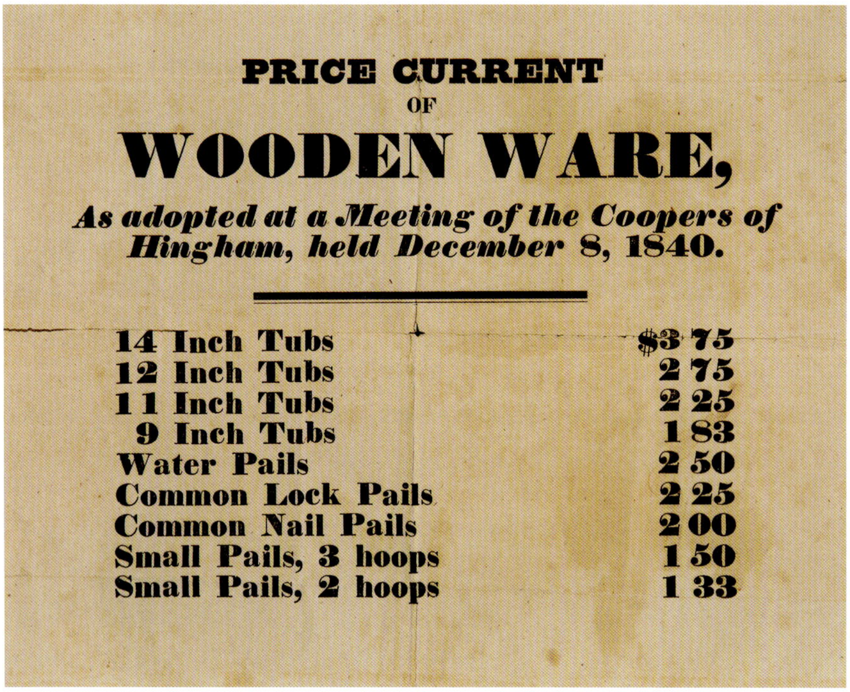

PRICE CURRENT
OF
WOODEN WARE,

As adopted at a Meeting of the Coopers of Hingham, held December 8, 1840.

14 Inch Tubs	$3 75
12 Inch Tubs	2 75
11 Inch Tubs	2 25
9 Inch Tubs	1 83
Water Pails	2 50
Common Lock Pails	2 25
Common Nail Pails	2 00
Small Pails, 3 hoops	1 50
Small Pails, 2 hoops	1 33

Figure 1.16
Price Current of Wooden Ware, Hingham, 1840. Broadside; H 7⅞, W 9⅞. Courtesy of the Hersey family

Opposite page:
Figure 1.17
Nest of buckets, by Nelson Corthell, Hingham, 1840–1900. Stamped "N. CORTHELL / HINGHAM" on each lid. White pine, oak, copper nails; largest: H 14¼, Diam 14⅝; smallest: H 9¾, Diam 9½. Courtesy of Peter and Diana Bennett

enough space for a lumberyard, drying house, and paint houses where unskilled laborers—usually young men—applied coats of blue, green, or gray paint to all types of woodenware.[56] At the height of production upwards of fifty people worked year-round in the heated factory. The buckets they assembled, tens of thousands annually, were shipped to markets throughout the country—including California mining operations—and as far away as Hawaii and the West Indies.[57] The Wilders remained Hingham's largest manufacturer of woodenware until the family sold the business in 1891.[58]

By midcentury several other craftsmen had achieved success with mechanized production, notably Elijah Whiton (1799–1871), Crocker Wilder's brother-in-law.[59] From the time of his youth until he died, Whiton kept a detailed journal chronicling his life and career.[60] A master of many trades, he excelled as a cooper, clockmaker, pumpmaker, surveyor, and manufacturer of scientific instruments. In the 1840s he developed a new scroll saw that cut nests of buckets, pails, and other containers from solid blocks of wood. To improve the saw's design, he enlisted the help of his nephew George B. Whiton, a member of the Shaker community at Harvard, Massachusetts—an ingenious mechanic and inventor in his own right who served as the community's resident carpenter, sievemaker, boxmaker, and manufacturer of seed presses.[61] The two men visited each other often. In May 1851 Brother Elijah Myrick, head of the Shakers' herb department, recorded in his journal that George had received by express mail "a box containing a modle of a saw mill for sawing pails & tubs for him to draft [plans for] his uncle Elijah."[62] A few months later Whiton received an official U.S. patent for his design.[63]

Elijah Whiton wasted no time, moving a one-and-a-half-story factory building to the entrance of Hersey Street. He later recalled: "I labored daily there to perfect the new process for making wooden ware. The mill was a decided success, and I began to realize the return for my labor. I was adding to my works a new method of making boxes by machinery, which begat some uneasiness among the hand workmen."[64] Tragically, in 1855 an arsonist burned the entire operation to the ground. After a tumultuous and expensive legal battle to recoup his losses, Whiton decided to leave the coopering business for good. He spent the rest of his days accumulating rental properties and speculating in land.[65]

Farther down Hersey Street stood Cedar Hedge, the experimental farm of Edmund Hersey (1823–1909), a pioneering agriculturalist and cooper.[66] In 1867 he revolutionized boxmaking with a machine that quickly and efficiently cut heads and bottoms for round and oval boxes. With the help of this tool, his new steam-powered factory reportedly produced over 1.5 million fig and fruit boxes in a single year.[67] At the peak of his success in 1871, Hersey sold the business for $6,000, negotiating a royalty of seventy-five cents for every one thousand boxes produced with his patented machinery.[68] Just a decade earlier his handcraft shop on the same site had earned him a modest annual salary of $240.[69] Hersey's sudden wealth allowed him to retire from manufacturing; he devoted the rest of his life to farming and other gentlemanly pursuits.

Hersey, Wilder, Whiton, and other industrialists pushed most of Hingham's traditional coopers to the brink of extinction. William Hobart, a young bucketmaker, probably trained in the small shop of his father. When he came of age in the 1850s, however, he sought employment at the very factory that had put his family out of business.[70] Countless others made the same transition or abandoned the trade altogether. For some, survival became a badge of honor; in 1894 Nelson Corthell proudly listed his business in the town directory as "the last of the many small shops"[71] (fig. 1.17). Without a new generation of handcraftsmen to occupy them, these modest workspaces—once ubiquitous in Hingham—fell into disuse or disrepair.

Hingham's major producers of woodenware, however, ultimately met a similar fate, facing stiff competition from outside the community. The town had enjoyed a near monopoly on the trade for almost two hundred years, but machine production eliminated the necessity—in Bucket Town and elsewhere—for skilled craftsmen. Laborers with relatively little training could now operate machines to cut parts and then hand-assemble the containers. At the same time, the advent of railroads expanded transportation far beyond the coastal waterways that had been so advantageous to Hingham's commerce. Finally, new factories in densely wooded areas of New Hampshire and western Massachusetts could manufacture their products more cheaply than Hingham's operations, which now depended on wood transported from those same forests.[72] By the 1850s and 1860s the town of Winchendon boasted over a dozen bucket and box factories. The firm of Murdock & Fairbanks alone employed seventy-five men and produced more than $80,000 in goods a year; in a few decades Murdock more than doubled his workforce.[73] Savvy manufacturers like Baxter Whitney, also of Winchendon, contributed to the demise of

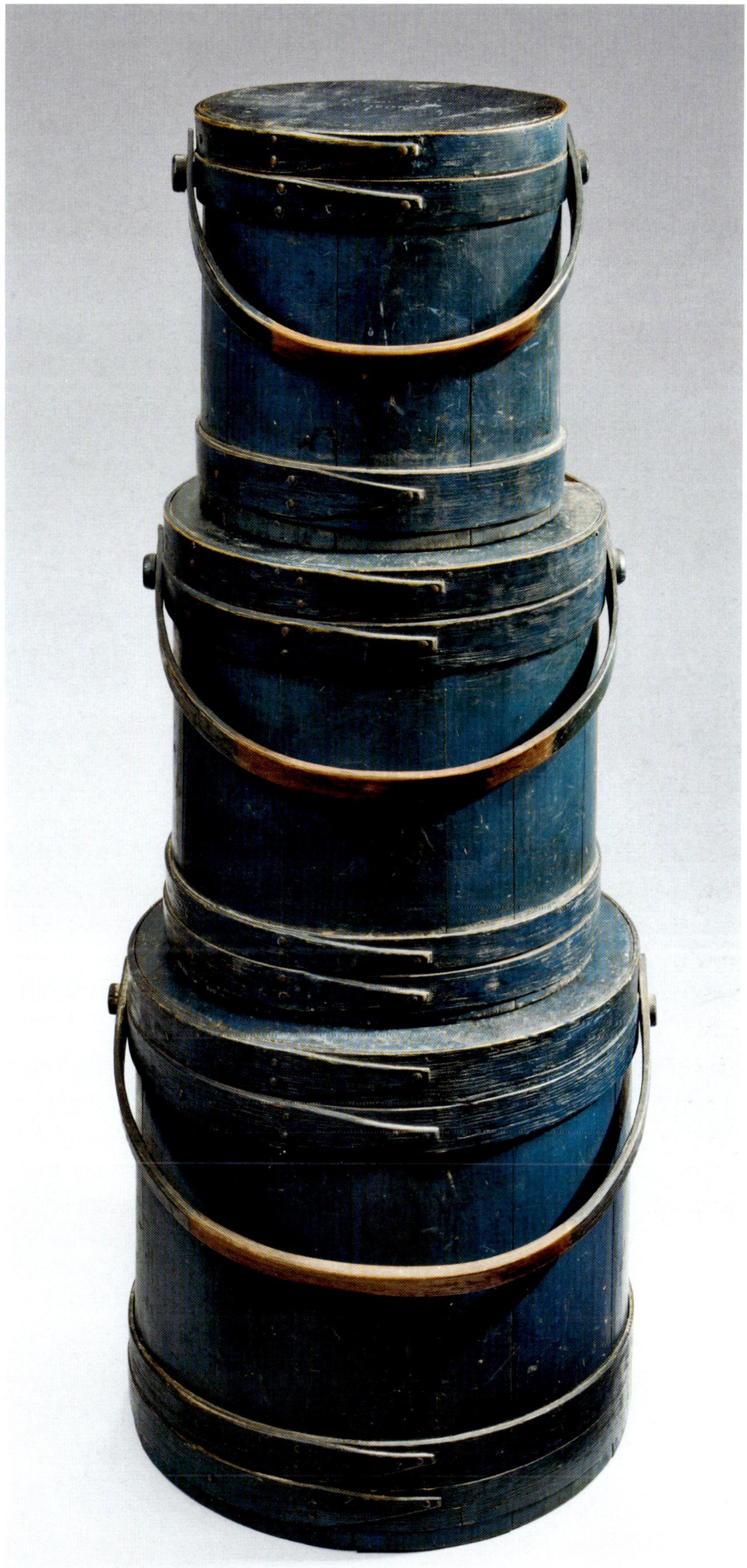

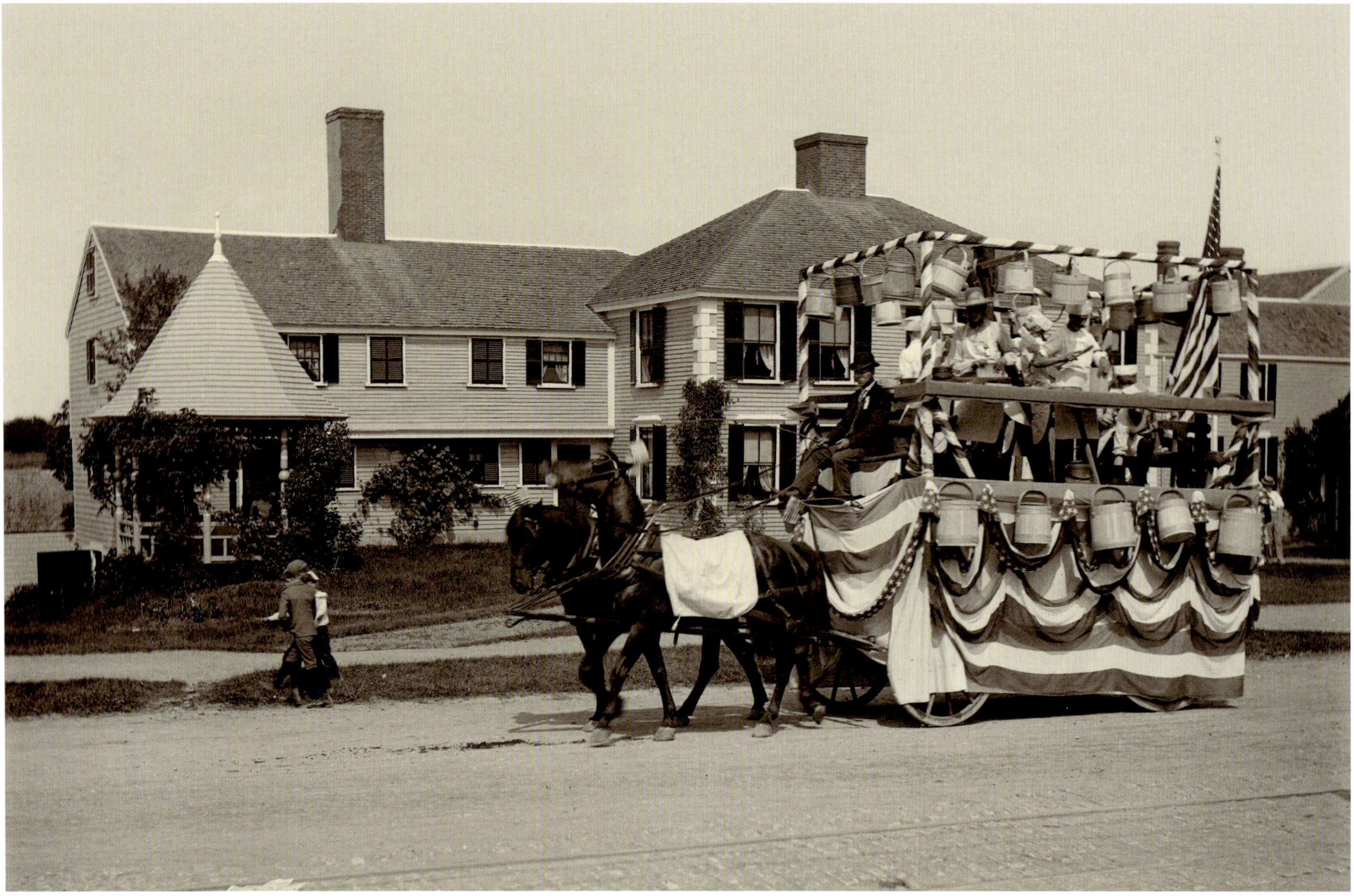

Figure 1.18
Anthony J. Sprague's Fourth of July parade wagon, photograph by William Blake Luce, Hingham, ca. 1900. John P. Richardson Photograph Collection, courtesy of Hingham Historical Society

Hingham's coopering industry too; he operated a large factory that developed and built the very machines used by established and upstart woodenware factories throughout the country.[74]

By the turn of the twentieth century Bucket Town could no longer claim a single manufacturer of commercial woodenware. As the small shops and factories vanished from the landscape, so did the many customs tied to their once-bustling industry—such as the appearance of coopers in Hingham's annual Fourth of July parade. A. J. Sprague, successor to C. Wilder & Son, was among the last of the town's beloved craftsmen to participate in the event. Around 1900 he outfitted a horse-drawn wagon with several coopers busy at benches making pails and buckets (fig. 1.18). Patriotic bunting festooned the exterior of the vehicle, as did the fruits of their labor—buckets and pails of different sizes and finishes, clattering against the wagon's sides. A local photographer captured the floating factory in its full glory—and documented one of the final celebrations of coopering in Hingham.

NOTES

1. For information on Hingham in the seventeenth century, see Lorena Laing Hart and Francis Russell Hart, *Not All Is Changed: A Life History of Hingham* (Hingham: Hingham Historical Commission, 1993), 14–43.

2. Edward Johnson, *Johnson's Wonder-Working Providence, 1628–1651*, ed. J. Franklin Jameson (New York: Charles Scribner's Sons, 1910), 115–16.

3. For references related to timbering in seventeenth-century Hingham, see Thomas Lechford, *Note-Book Kept by Thomas Lechford, Esq., Lawyer, in Boston, Massachusetts Bay, from June 27, 1638 to July 29, 1641*, ed. Edward Everett Hale Jr. (Cambridge, Mass., 1885), 304; John Winthrop, *The Journal of John Winthrop, 1630–1649*, ed. Richard Dunn, James Savage, and Laetitia Yaendle (Cambridge: Belknap Press of Harvard University Press), 391, 685; William Aspinwall, *A Volume Relating to the Early History of Boston, Containing the Aspinwall Notarial Records from 1644 to 1651* (Boston: Municipal Printing Office, 1903), 100–101, 124, 235; and *Records of the Suffolk County Court, 1671–1680, Part II* (Boston: Colonial Society of Massachusetts, 1933), 528.

4. Hosea Sprague, "Hingham Records from 1635, to 1700, copied (verbatim) line for line and page for page, from the beginning of the first volume to the 168th page of the 2nd volume" (Hingham, Mass.), 55, Town Clerk's Office, Hingham.

5. For a history of coopering, see Kenneth Kilby, *The Cooper and His Trade* (London: John Baker, 1971); see also Kenneth Kilby, *Coopers and Coopering* (Princes Risborough, Buckinghamshire, U.K.: Shire Publications, 2004).

6. George Lincoln et al., *History of the Town of Hingham, Massachusetts, Genealogical* (Hingham: By the town, 1893), 3:3. See also F. J. Nicholson, "A Clue to the English Origins of Thomas Lincoln the Cooper and William Lane of Hingham MA," *American Genealogist* 64 (1989): 214–15.

7. Lincoln is identified in town records as "cooper Lincoln" and "Thomas Lincoln, cooper"; Selectmen's Book of Records, Hingham, Mass., 1661–1717, Town Clerk's Office, Hingham.

8. The chair descended through the Lincoln family until it was anonymously donated to the Hingham Memorial Bell Tower around 1912, the year the tower was dedicated to the town's first settlers. A brass plate on the chair is inscribed, "This chair belonged to Thomas Lincoln The Cooper who came from England in 1635. He was the great, great grandfather of General Benjamin Lincoln." A fragmentary paper label on the chair is inscribed, "Chair brought by one of first settlers of Hingham Thomas Lincoln, the cooper." For a discussion of furniture practices in seventeenth-century Hingham, see Robert Blair St. George, *The Wrought Covenant: Source Material for the Study of Craftsmen and Community in Southeastern New England, 1620–1700* (Brockton, Mass.: Brockton Art Center / Fuller Memorial, 1979), 60–64.

9. On August 16, 1665, the town voted to allow Thomas Lincoln, carpenter, to "set a shop upon the common over against his house lot"; Sprague, "Hingham Records," 139.

10. Throughout the seventeenth century the town enacted laws to limit the harvesting of hoop poles and timber from the town common; Selectmen's Book of Records, 1661–1717. For an excellent description of hoop poles and their use in cooperage, see "The Hoop-Pole Industry: Talk with a Man Who Has Made Money Following It," *New York Times*, July 30, 1893.

11. For a list of coopers working in Hingham in the seventeenth century, see appendix C.

12. Gould was admitted to the town February 10, 1641, but did not purchase land until 1645; Sprague, "Hingham Records," 46, 96. For additional information on Gould, see Robert Charles Anderson, *The Great Migration Begins: Immigrants to New England, 1620–1633*, 3 vols. (Boston: New England Historic Genealogical Society, 2003), 3:121–23.

13. Selectmen's Book of Records, 1661–1717.

14. In the seventeenth century coopers were the second most common woodworkers in Ipswich, Mass.; see Robert Tarule, *The Artisan of Ipswich: Craftsmanship and Community in Colonial New England* (Baltimore: Johns Hopkins University Press, 2007), 71–77.

15. Stephen Innes, *Creating the Commonwealth: The Economic Culture of Puritan New England* (New York: W. W. Norton, 1995), 222–24.

16. Large-scale cooperage, like barrelmaking, existed at the harbor, where there was a steady demand for containers to ship locally harvested foodstuffs. This branch of coopering was overshadowed by the many white coopers who specialized in smaller woodenware for export.

17. Kilby, *The Cooper and His Trade*, 43–45. For early nineteenth-century depictions of woodenware in domestic and commercial settings, see Harris Myers, *William Henry Pyne and His Microcosm* (Stroud, Gloucestershire, U.K.: Sutton Publishing, 1996), 182, 222–23

18. Joseph B. Felt, *An Historical Account of Massachusetts Currency* (Boston: Perkins & Marving, 1839), 47. Beal also freighted timber to Boston and surrounding towns; Jacob Beal, account book, 1708, box FP17, Hingham Historical Society, Hingham.

19. The earliest known use of the nickname Bucket Town is in "History of Buckettown," a satirical essay in a book of handwritten poems and stories by students of the Derby School in Hingham (it has often and erroneously been credited to *Two Years before the Mast* [1840] by Richard Henry Dana Jr.); Salmagundi Society, "Salmagundi the Younger; or, The Whim Whams of ye Lads & Lasses of Buckettown," Hingham, Mass., 1816, 43–44, Ms. qAm.

2295, Special Collections, Rare Books and Manuscripts Department, Boston Public Library, Boston.

20. Austin's Lane was renamed Hersey Street in 1827. When Obadiah Groce (1708–49), a second-generation craftsman, died in his prime, he left behind a valuable selection of farmer's and cooper's tools, as well as a shop containing six dozen finished pails and a large quantity of "bottoms, staves, hoops, & buttons" for dozens more; Obadiah Groce, inventory, filed September 15, 1753, Suffolk County Probate Records, 49:52.

21. Simon Burr, account book, 1769–98, Hersey Family Collection.

22. Lincoln et al., *History of Hingham, Genealogical,* 2:102.

23. Elisha Burr, inventory, filed November 6, 1778, Suffolk County Probate Records, 77:628.

24. In 1775 the Hingham cooper John Burr was paid by the trader Hawkes Fearing for "7 nests of boxes in hogshead"; see Hawkes Fearing, account book, 1775–78, box FP56:1, Hingham Historical Society, Hingham.

25. John S. Barry, *A Historical Sketch of the Town of Hanover, Mass., with Family Genealogies* (Boston: Samuel G. Drake, 1853), 32; see also Duke de La Rochefoucault Liancourt, *Travels through the United States of North America, the Country of the Iroquois, and Upper Canada, in the Years 1795, 1796, and 1797,* 4 vols. (London: T. Davison, 1799), 1:482.

26. Katharine Lincoln, "Old Hingham and the New," Hingham Histories, microfilm reel 80.7, Hingham Public Library, Hingham.

27. James A. Henretta, "Economic Development and Social Structure in Colonial Boston," *William and Mary Quarterly* 22 (January 1965): 81–88.

28. Boston clearance records, 1744–1784, Mss. L19, Boston Athenaeum, Boston. This was probably the same sloop that served in the Crown Point Expedition in 1756; see George Lincoln et al., *History of the Town of Hingham, Massachusetts, Historical* (Hingham, Mass: By the town, 1893), 1.1:262.

29. Boston clearance records, 1744–1784, Mss. L19.

30. Lincoln et al., *History of Hingham, Historical,* 1.2:242–43.

31. Hosea Sprague, Boston, to Isaac Sprague, Hingham, December 14, 1812, Mss. .B1 4, Boston Athenaeum, Boston.

32. In preparation for the 1893 *History of Hingham,* the historian George Lincoln interviewed, through written correspondence, many of the town's older residents. Captain Luther Stephenson shared his memories of the Hingham packets; see Luther Stephenson, Rogers, Maine, to George Lincoln, Hingham, March 27, 1890, Hingham Records, 1637–1931, Ms. N-1404, box 13, folder May 1884, Massachusetts Historical Society, Boston.

33. Lincoln et al., *History of Hingham, Historical,* 1.2:243. For another description of the Hingham Station packet, see Wendell O. Hunt, "Sixty Years Married: Wedded Life of Mr. and Mrs. Dyer of Hingham," *Boston Daily Globe,* January 31, 1888, 4.

34. David B. Mattern, *Benjamin Lincoln and the American Revolution* (Columbia: University of South Carolina Press, 1998).

35. William Lincoln, ed., *The Journals of Each Provincial Congress of Massachusetts in 1774 and 1775, and of the Committee of Safety* (Boston: Dutton and Wentworth, 1838), 35, 509.

36. Hawkes Fearing, account book, 1775–78, box FP65, Hingham Historical Society, Hingham; see also John Leavitt, account book, 1771–92, Hersey Family Collection.

37. Henry G. Pickering, *Nathaniel Goddard: A Boston Merchant, 1767–1853* (Cambridge, Mass.: Riverside Press, 1906), 59.

38. The canteens were to be deposited in equal numbers in Sudbury, Concord, and Worcester; see Lincoln, *The Journals of Each Provincial Congress,* 517.

39. On April 19, 1775, the committee resolved to purchase and transport 1,000 canteens and 1,000 pails, among other supplies, to a storehouse near Noddle's Island; see *The Acts and Resolves, Public and Private, of the Province of the Massachusetts Bay,* vol. 19 (Boston: Wright & Potter, 1918), 337.

40. On June 25, 1776, Cushing was appointed to the committee to purchase "canteens or wooden bottles" for 5,000 men. On June 27 the Assembly resolved to deposit 448 canteens in Medfield, 601 in Haverhill, 1,070 in Westford, 742 in Rehoboth, 1,136 in Brookfield, 261 in Pittsfield, 374 in Northampton, and 374 in Hadley; see Peter Force, ed., *American Archives: Consisting of a Collection of Authentick Records, State Papers, Debates, and Letters and Other Notices of Publick Affairs,* 5th ser. (Washington, D.C.: U.S. Congress, 1843), 287–89, 294–95.

41. On July 26 Timothy Newell, another member of the committee to purchase canteens and other supplies, reported to the Assembly, "The delay is wholly owing to the canteens not coming from Hingham." He further warned of the need for "more workmen to be put upon the canteens supplied by Theophilus Cushing, Esq., at Hingham," probably to outfit them with stoppers and carrying straps; ibid., 608–9.

42. Cushing was appointed to the committee to purchase the canteens on July 13, 1776. John Avery reprimanded Cushing in a letter dated July 30, 1776; ibid., 322, 551, 684.

43. Lincoln et al., *History of Hingham, Genealogical,* 3:94. Jonathan Lane, account book, 1776–98, John P. Richardson Manuscript Collection, Hingham Historical Society, Hingham.

44. Leavitt also sold canteens to Hawkes Fearing and canteen bottoms to Amos Sprague, a cooper; see John Leavitt, account book, 1771–92, Hersey Family Collection.

45. Samuel Norton, Hingham, Mass., to Gen. Benjamin Lincoln, Philadelphia, August 21, 1782, Wigglesworth Family

Papers, Ms. N-114, box 1, Massachusetts Historical Society, Boston.

46. Gen. Benjamin Lincoln, Philadelphia, to Samuel Norton, Hingham, Mass., September 11, 1782, Wigglesworth Family Papers, Ms. N-114, box 1, Massachusetts Historical Society, Boston.

47. F. A. Gardner, M.D., "Colonel John Fellows's Regiment," *Massachusetts Magazine: A Quarterly Magazine Devoted to History, Genealogy and Biography* 2, no. 3 (July 1909): 155. For an excellent survey of canteens used in America, see Mike O'Donnell, *U.S. Army & Militia Canteens, 1775–1910* (Alexandria, Va.: O'Donnell Publications, 2008).

48. Hingham-type canteens were being produced by 1758, when the Massachusetts Assembly resolved to purchase "One wooden bottle made with one hoop (if to be procured) for each man or a canteen." Hingham canteens and boxes, often called hoop boxes in the eighteenth and nineteenth centuries, were made with single-hoop construction; see *The Acts and Resolves, Public and Private, of the Province of the Massachusetts Bay,* vol. 16 (Boston: Wright & Potter, 1909), 312.

49. For information on Burr, see cat. 5.

50. Curtis & Hubbard to Boston Fusiliers, bill of receipt, November 19, 1822, 1968.40.3.16, Old Sturbridge Village, Sturbridge.

51. For a comparable history of the rise of mechanized coopering factories in Richmond, N.H., see Richard A. Martin, *The Only Mill in Town: The Story of the Pail-Making Industry in Richmond, New Hampshire* (Richmond: Friends of Historic Richmond, 1995).

52. Lincoln et al., *History of Hingham, Historical,* 1.2:159–60.

53. "History of the Bucket Mill," *Hingham Journal,* April 4, 1902, 1. For more information on Wilder, see cat. 8.

54. The Hingham customhouse remained open for forty-five years; Lincoln et al., *History of Hingham, Historical,* 1.2:385.

55. Ibid., 172.

56. For a description of the factory, see "C. & A. Wilder and the Elijah Whiton Factories," *Hingham Journal,* May 6, 1853, 2. Isaiah F. Tower and Oliver W. Cushing are listed as pail painters in the 1855 Massachusetts State Census; see appendix C.

57. "C. & A. Wilder and the Elijah Whiton Factories," 2.

58. The dissolution of C. Wilder & Son was advertised in the *Hingham Journal,* June 5, 1891.

59. Lincoln et al., *History of Hingham, Genealogical,* 3:305.

60. Elijah Whiton, journal, 1802–70, John P. Richardson Manuscript Collection, Hingham Historical Society, Hingham.

61. Elijah Myrick, daybooks, 1850–52, Shaker Manuscripts, Hancock Shaker Village, roll 4, vols. 15–17.

62. Ibid. George visited his uncle in Groton on January 25, 1850. In later years Elijah traveled from Hingham to see his nephew.

63. Elijah Whiton, "Machine for Sawing Volutes," U.S. Patent 8400, issued September 30, 1851.

64. Entry for January 1852, Elijah Whiton, journal. Whiton's factory was located at the entrance of Hersey Street near Hobart Street; Lincoln et al., *History of Hingham, Historical,* 1.2:164.

65. Ibid.

66. Lincoln et al., *History of Hingham, Genealogical,* 2:324; Massachusetts Vital Records, 1841–1910, 47:102.

67. Edmund Hersey, "Improvement in Machines for Cutting the Heads and Bottoms of Wooden Boxes," U.S. Patent 62,846, issued March 12, 1867; Edmund Hersey, "Machines for Making Heads of Wooden Boxes," U.S. Patent 169,700, issued November 9, 1875. Hersey's machinery reportedly made 160 strawberry-box bottoms per minute and 260 hoops per minute; "The Box and Toy Business," *Hingham Journal,* August 11, 1876, 1. Lincoln et al., *History of Hingham, Historical,* 1.2:164.

68. In 1871 Edmund Hersey sold the company to the partnership of Seymour & Cain, but he remained president and principal stockholder; see Seymour & Cain, account book, 1871–72, Hingham Bicentennial Microfilm Collection, Hingham Public Library.

69. U.S. Federal Census, Non-Population Schedule, 1860, Hingham, Plymouth County, Massachusetts.

70. Lincoln et al., *History of Hingham, Historical,* 1.2:39. Hobart's occupation is listed as "bucket factory" in the 1855 Massachusetts State Census.

71. *Directory of Hingham and Hull, Mass., 1894* (Quincy, Mass.: J. H. Hogan, 1894), 119.

72. For information on the development of white coopering in Richmond, New Hampshire, see Martin, *The Only Mill in Town.*

73. Glen Wheeler, *Winchendon: Images of America* (Charleston, S.C.: Arcadia Publishing, 1997), 47–49.

74. Alvin Streeter, a noted manufacturer of woodenware machinery, was the last to operate the factory before it burned in 1902; see "History of the Bucket Mill." See also *Inland Massachusetts Illustrated* (Worcester, Mass.: Elstner, 1891), 150.

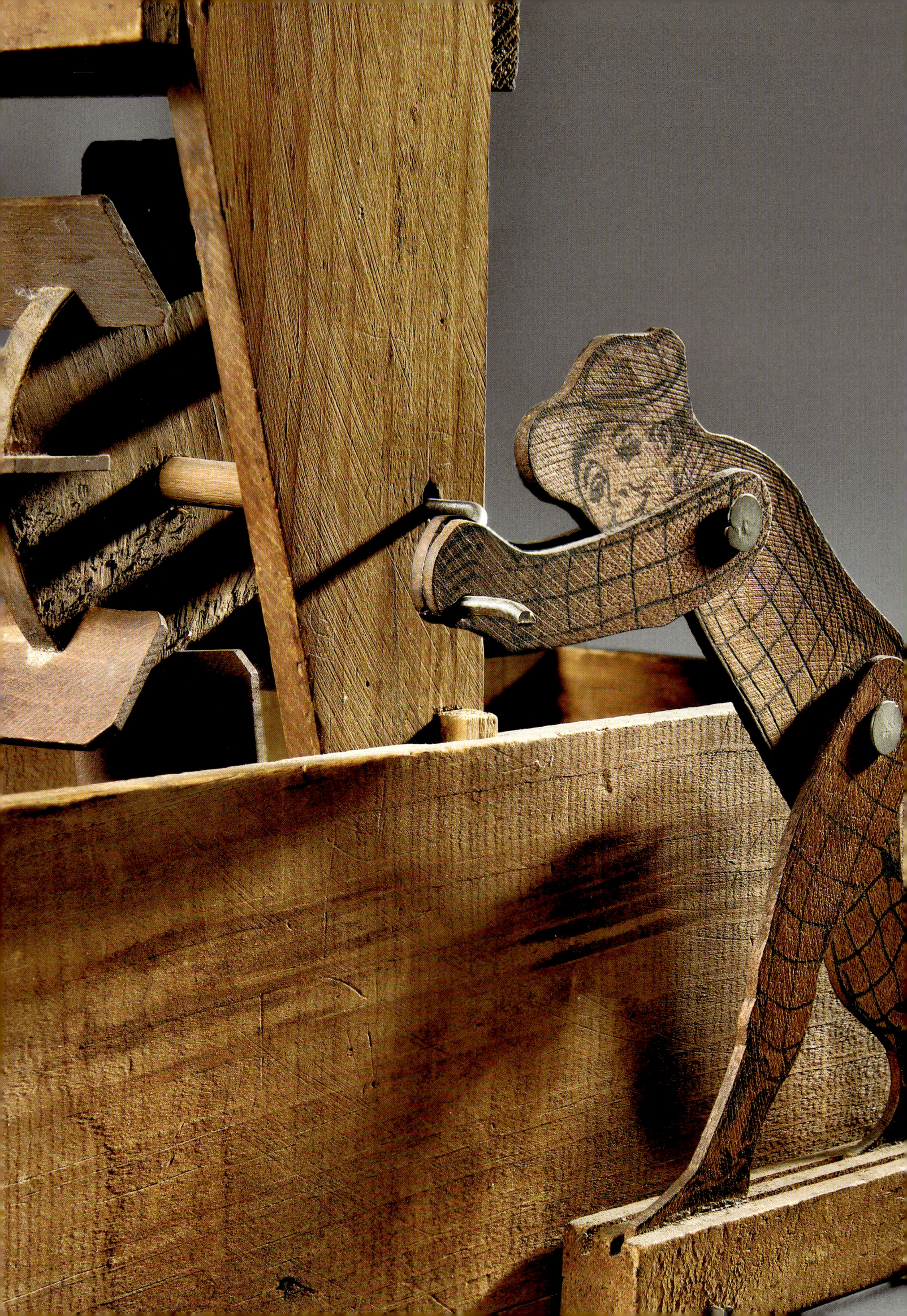

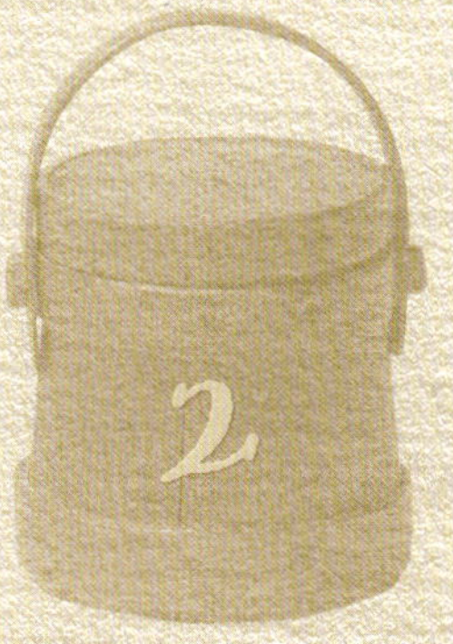

HINGHAM TOYS: LITTLE THINGS FOR SMALL FOLKS

Hingham has been the centre and source of . . . toys for nearly or quite one hundred years . . . the children of the country would sorely miss their Christmas gifts if any calamity should overtake the workshops of this quiet but industrious and ingenious "Yankee Town."

Hingham Journal, *August 11, 1876*

CHILDREN'S TOYS WERE AN IMPORTANT COMMODITY IN EIGHTEENTH-CENTURY Boston. By 1733 William Price, a print and map seller, advertised Dutch and English playthings at his corner store near the old brick meetinghouse.[1] Other storekeepers offered similar goods, but, like Price, none dealt exclusively in toys. That changed in the 1790s when several enterprising men and women established the city's first toy shops and even a toy warehouse.[2] Jane Salter was among these early pioneers. For nearly a decade she sold retail and wholesale clients a "fresh assortment" of drum sets, wax dolls with moving eyes, alphabet cards, and other imports especially suited for boys and girls.[3] Domestic toys, however, were harder to come by. For

Page 28:
Detail of gristmill sand toy by William S. Tower, courtesy of the Hersey family

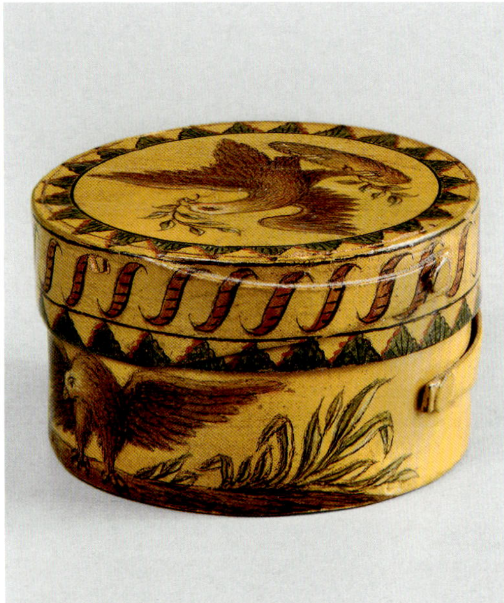

Figure 2.1
Miniature round box, attributed to Hingham, 1820–60. White pine, maple, copper tacks; H 1, Diam 2. Although the small box was likely made in Hingham, its vibrant paint and eagle decoration are the work of an unidentified hand. Courtesy of Jane and Gerald Katcher

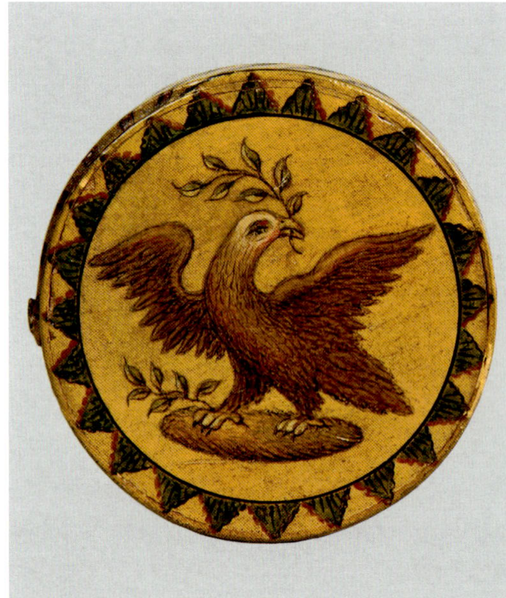

Figure 2.2
Detail of lid on box in fig. 2.1

those items customers often turned to neighborhood craftsmen like tinsmiths and cabinetmakers who occasionally made toys as a sideline to their regular trade.[4] This practice flourished in nearby Hingham, where coopers fashioned miniature buckets and other novelty woodenware (figs. 2.1, 2.2). From their shops emerged the first and largest community of professional toymakers in America.[5]

Hingham's toy industry originated on Hersey Street. Beginning in the 1730s, dozens of members of the Hersey family established modest farms there with small workshops for coopering.[6] Around the time of the Revolution, if not earlier, they began to craft "small ware," in addition to the full-size buckets, tubs, and other wooden containers that had made the town famous (fig. 2.3). The homestead of Reuben Hersey (1757–1845), Hingham's earliest known toymaker, was a major production hub and training ground for future generations of craftsmen.[7] At least five of Reuben's sons and grandsons would become prominent toymakers; his son Cotton Hersey (1792–1881) was the first. In 1814 Cotton built a small Cape beside his father's house and a shop across the street.[8] He too identified himself primarily as a cooper, but by the 1830s he had begun to focus on specialty items, including finely made boxes and toys. Though Cotton and his shop were undoubtedly a popular attraction for both children and adults, accounts suggest that he was a shy man who rarely ventured away from his farm.[9]

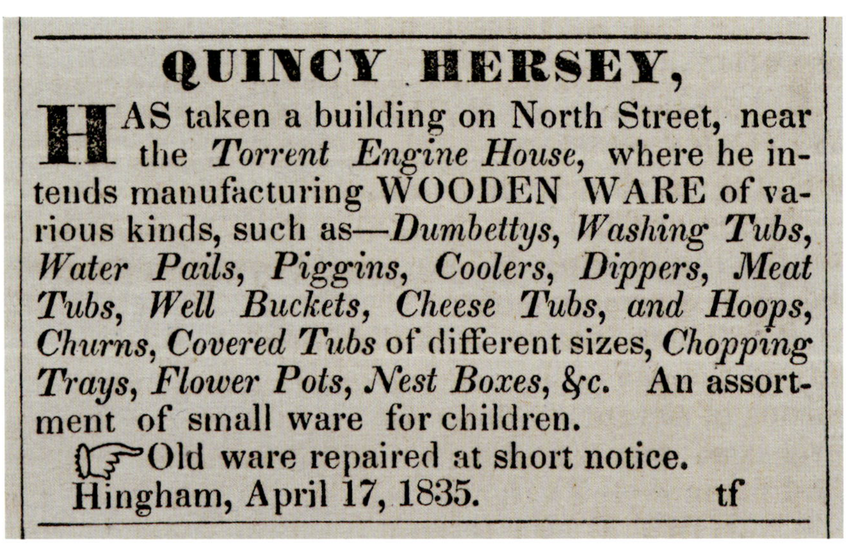

QUINCY HERSEY,

HAS taken a building on North Street, near the *Torrent Engine House*, where he intends manufacturing WOODEN WARE of various kinds, such as—*Dumbettys, Washing Tubs, Water Pails, Piggins, Coolers, Dippers, Meat Tubs, Well Buckets, Cheese Tubs, and Hoops, Churns, Covered Tubs* of different sizes, *Chopping Trays, Flower Pots, Nest Boxes, &c.* An assortment of small ware for children.

☞Old ware repaired at short notice.
Hingham, April 17, 1835. tf

Figure 2.3
Advertisement for woodenware and small ware for children, by Quincy Hersey, *Hingham Gazette,* April 17, 1835. Courtesy of Hingham Historical Society

Technology and tourism drove Hersey's business. In 1830 Crocker Wilder established the town's first steam-powered bucket factory, manufacturing tubs, pails, and churns more quickly and inexpensively than his local competitors, who continued to labor with hand tools.[10] While Hingham's traditional coopers struggled to keep pace, Cotton Hersey's workshop prospered. His small boxes and toys had found a new and growing audience: affluent summer visitors. In 1819 the steamship *Eagle* began regular trips between Boston Harbor and Hingham, carrying upwards of

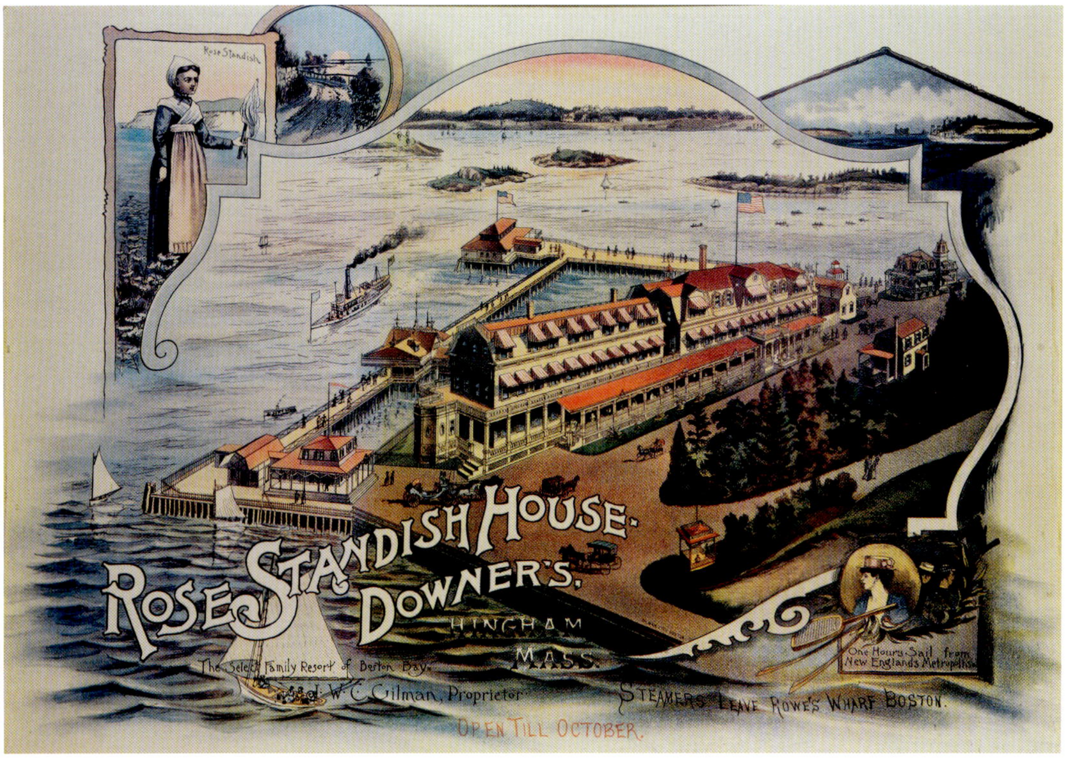

Figure 2.4
Rose Standish House: Downer's, Hingham, Mass. The Select Family Resort of Boston Bay, printed in Boston, 1880–90. Chromolithograph on paper; H 20, W 28 ½. Courtesy of the Hersey family

two hundred passengers—many in search of recreation and entertainment.[11] The Sportsman Hotel, built in 1826, stood at the head of nearby Nantasket Beach, described by one contemporary as a "fashionable place of resort."[12] Vacationers could also stay near the wharf at Hingham's swank Old Colony House, where the seaside was just a short carriage ride away.[13] In the decades that followed, faster transportation and a burgeoning hotel scene (fig. 2.4) made it even easier for throngs of beachgoers to descend on the region—a steady and well-heeled market for Hingham's toymakers.

Cotton Hersey typically stamped his name or initials into the soft pine tops and bottoms of his toys and boxes, a practice Hingham coopers had adopted around the turn of the nineteenth century. Although many of his items were left undecorated, dozens of others feature brightly painted surfaces, some of which he probably applied in his shop. Examples include a charming miniature bucket painted blue (fig. 2.5) and a diminutive green painted washtub with dark green hoops and a white interior (fig. 2.6). He crafted these containers in the same manner as their full-size counterparts, with hand-cut staves and hoops held in place with copper tacks and tiny wooden pegs.

Children and adults lucky enough to own Hersey's toys cherished them. In 1953 eighty-six-year-old Bessie Lincoln, granddaughter of Hingham's first historian, donated her

Figure 2.5
Miniature bucket, by Cotton Hersey, Hingham, 1820–60. Stamped "C.H" on lid. White pine, maple, copper tacks; H 2 ⅜, Diam 2 ⅜. Courtesy of the Hersey family

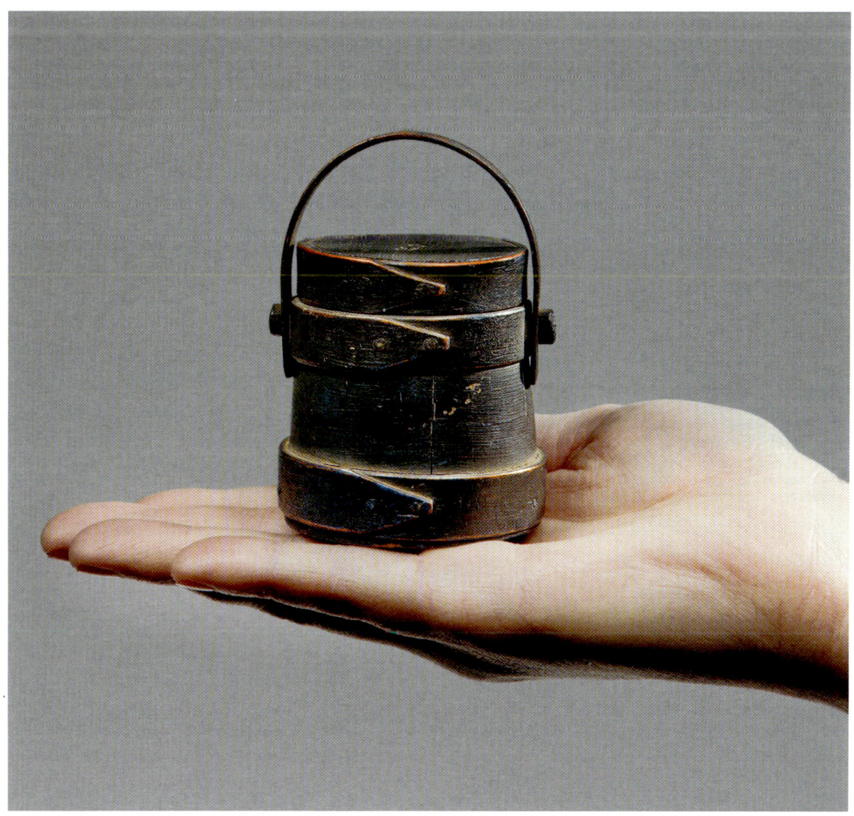

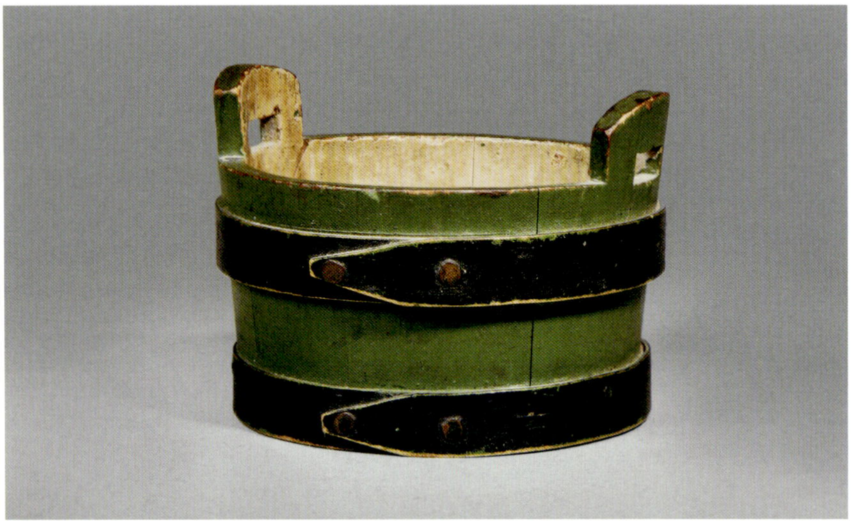

Figure 2.6
Miniature tub, by Cotton Hersey, Hingham, 1820–60. Stamped "C.H" on bottom. White pine, maple, copper tacks; H 2, Diam 2 9⁄16. Courtesy of Peter and Diana Bennett

childhood dollhouse to the Essex Institute—now the Peabody Essex Museum—in Salem, where she had spent her youth. Her mother, with the assistance of the "ladies" of the town, had crafted most of the house's furnishings. Lincoln noted, however, that the small pail in the kitchen was "made in the old Hingham bucket shop" of Cotton Hersey, which she may have visited as a young girl.[14] Dorothy Durgin, a member of the Shaker community in Canterbury, New Hampshire, probably felt a similar fondness for the miniature bucket by Hersey (see pl. 11.6) that she owned. Around midcentury Brother Elijah Brown (1772–1851), a noted woodworker, gave her the plaything as a token of his affection.[15] The thin yellow stain on the exterior of the bucket is reminiscent of the finish found on Shaker woodenware made about the same time.

By 1850 Samuel Hersey (1820–1909) had eclipsed his uncle Cotton as Hingham's most recognized and beloved toymaker.[16] He produced many of the same goods—from tiny tubs to painted boxes—but dropped the title of cooper and began advertising himself as a "Box and Toy Ware Manufacturer," an occupation mostly unheard of at the time.[17] Several other coopers in the family also rebranded themselves, but none achieved Samuel's level of success.[18] He distinguished himself from these makers by marking his goods with printed paper labels, some of which describe him as a retailer and wholesaler. A typical example bearing his name and location on Hersey Street is found on a stack of miniature oval nesting boxes (figs. 2.7, 2.8) formerly owned by the pioneering folk art collector Nina Fletcher Little.[19] Over the years the maple exterior of the largest box—measuring a mere three inches in length—has oxidized to a golden brown color, whereas the smaller boxes hidden inside are as pale as the day they were made. Such neatly designed and constructed containers earned Hersey great praise from his peers and, occasionally, cash prizes at local fairs.[20]

Samuel Hersey further outstripped his Hingham brethren by introducing furniture to his line of miniature wooden toys—arguably the first professionally made dollhouse furnishings in America. Around 1864 Catherine C. Biddle, a socialite from Philadelphia, purchased a rectangular table and chairs by Samuel for the kitchen of her extraordinary model town house. She exhibited this small architectural wonder at the Great Central Fair for the U.S. Sanitary Commission and raffled it to raise money for Civil War hospitals.[21] A similar set of furniture (fig. 2.9) belonged to Mary Eliza Low (1819–1878), the daughter of a prominent Gloucester, Massachusetts, ship's captain.[22] Hersey patterned the tall-post bed, chest of drawers, washstand, and side chairs after the simple cottage furniture popular in New England at the time. Someone—possibly Mary—painted each piece with colorful fruit vines, floral bouquets, and green pinstriping. The furniture remained a prized possession in her family and was passed down through at least three generations of women.[23] Annie Low, the third owner, recalled playing with it as a child in the early 1900s: "I remember this set in a large white box on the top shelf of a closet in my grandmother's 'best bedroom.' Only on rainy days when time hung heavy for a small girl was I allowed to take down the box, unwrap each piece from its layers of tissue paper, and once more taste the joy that comes when one plays with a treasure."[24]

Despite a long paper trail and the survival of dozens of Samuel's wooden toys, relatively little is known of his day-to-day activities. In 1940 Nellie Corthell Cobb (b. 1855), his niece and a former neighbor, reminisced: "He certainly was a master craftsman. He was my uncle and when I was a child I enjoyed going in his shop and watching him make toys which he kept displayed for sale on shelves behind glass doors. The shop, like himself, was always immaculate."[25] Cobb added that Samuel sold goods only from his small building on Hersey Street, and that "during the summer a great many people came to see and buy them."[26] Manufacturing records confirm the shop's success. In 1860 Hersey estimated annual sales of $1,200—almost double what a traditional cooper made.[27] Having no wife or children and only a modest parcel of land to farm, he devoted over half a century to the toymaking trade.

Before long, however, Hersey faced stiff competition from an upstart toymaker in Hingham's industrial center. In August 1869 the local newspaper reported:

> *The Messrs. C. Wilder & Son have just completed a large building in connection with their factory to be occupied by William S. and Charles S. Tower for the manufacture of toy ware of all kinds and description. Their machinery is of the best quality and nearly all in place and will be put in operation soon. The Messrs. Towers are thorough mechanics. Specimens of their work have been seen at our Agricultural Fairs. We hope they will be successful in the undertaking. We understand that they will employ quite a number of hands, mostly girls, in the manufacture of their ware.*[28]

The son of a regionally famous clockmaker, William Sewall Tower (1826–98) possessed a strong mechanical aptitude, but wood was his medium; he soon excelled at coopering.[29] By 1859 he was working at the Hingham factory of Crocker Wilder, where he mastered industrial tools and earned a reputation for the "fancy buckets" he produced during off hours.[30] The most impressive of these, a lidded specimen with an accompanying

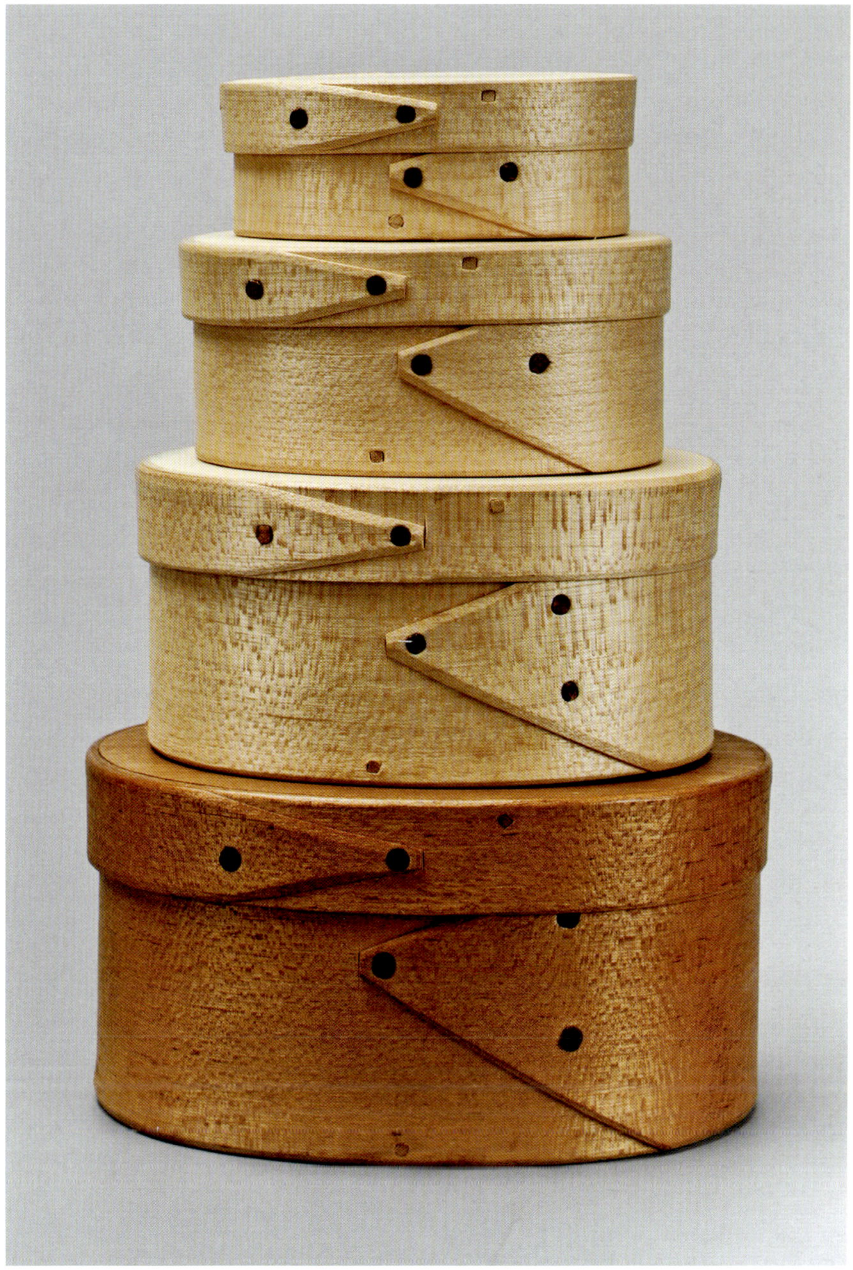

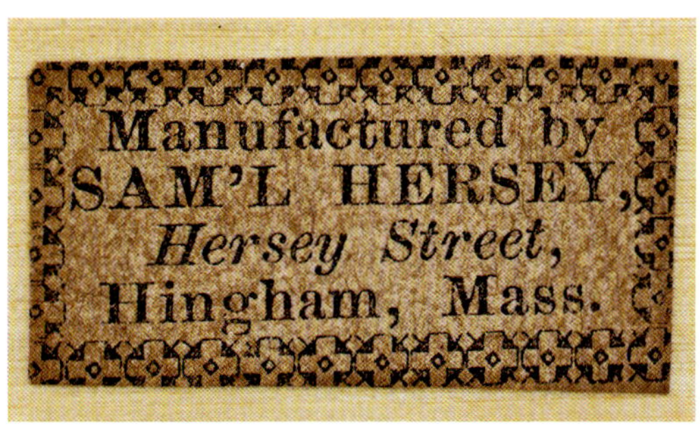

Figure 2.7
Nest of miniature oval boxes, by Samuel Hersey, Hingham, 1850–70. White pine, maple, birch, copper tacks; largest: H 1 ⅞, W 3 9/16, D 2 ⅝; smallest: H 1 3/16, W 2 ⅛, D 1 ½. Courtesy of Historic New England, gift of Bertram K. and Nina Fletcher Little

Figure 2.8
Detail of label on box in figure 2.7

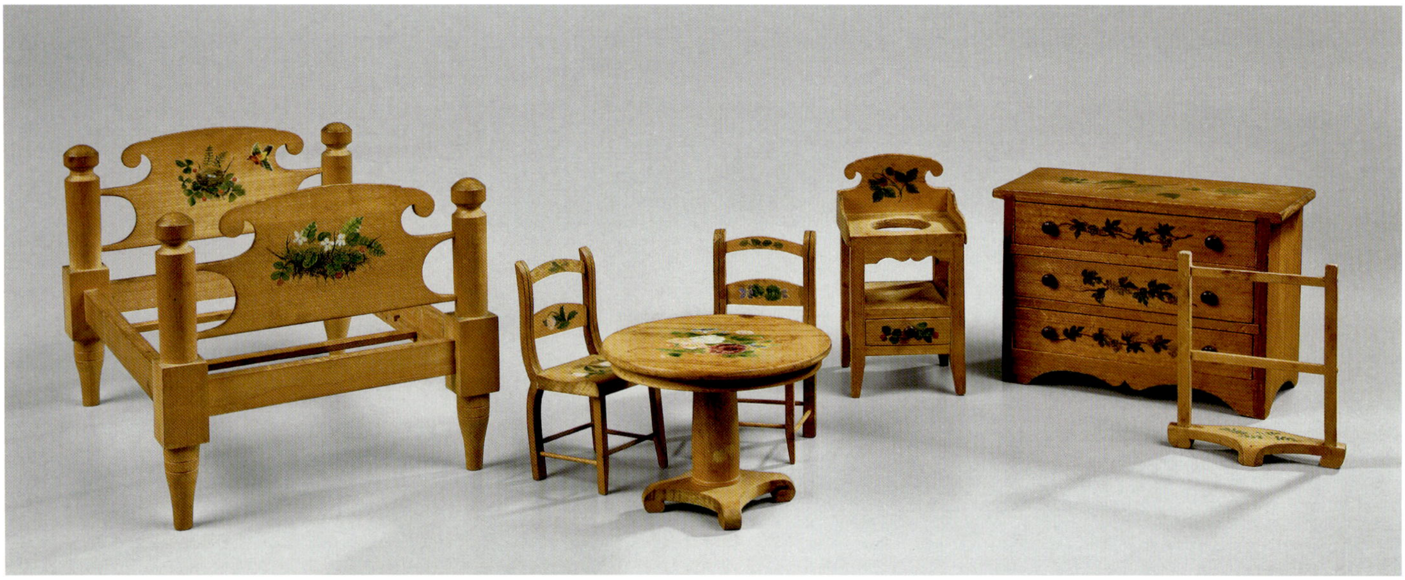

Figure 2.9
Suite of miniature furniture, by Samuel Hersey, Hingham, 1850–70. Labeled "Manufactured by / SAM'L HERSEY / *Hersey Street,* Hingham, Mass." White pine, maple, iron nails; bed: H 6 ¼, W 6 ⅞, D 10 ⅛; chair: H 4 ½, W (seat) 6 ⅞, D (seat) 3 ⅛. Courtesy of the Hersey family

calling card tray, were made around 1865 and exhibited throughout Massachusetts (fig. 2.10).[31] A remarkable feat of cooperage, it consists of 47,568 separate pieces of red and white cedar organized in a herringbone pattern and decorated with inlaid compass stars. The staves, like those in the buckets produced by Wilder, are bound together with riveted metal hoops.

In 1869 Tower left the bucket business to open William S. Tower & Company, one of the country's first factories dedicated entirely to the production of wooden toys.[32] His brothers, George and Charles, joined him in the enterprise, though little is known about their involvement. Tower first established himself in a two-story building adjacent to the bucket mill of his former employers, from whom he leased the property.[33] An early view of their side-by-side operations shows his clapboarded workshop with a large painted sign hung on the front (fig. 2.11). He outfitted the space with mechanized saws, lathes, planes, and other coopering tools purchased from the recently defunct Hingham Woodenware Company.[34] The machinery, only a couple of years old, was powered by a steam engine and water from nearby Cushing Pond.[35]

The new William S. Tower & Company factory employed approximately ten full-time laborers.[36] Some were skilled mechanics, like George Fearing and Loring Cushing, who would go on to become master toymakers in their own right. Women—who could be paid less than men—made up the rest of the workforce and were probably valued for their experience in fine handwork, useful for the finishing of small wooden toys.[37] When Jenny Cook worked for the company in the early 1890s, she used a diminutive tack hammer (fig. 2.12) to assemble a variety of playthings, most of which were held together with tiny iron fasteners. The tool and its history remained in her family as a valued keepsake until her niece donated it to the Hingham Historical Society years later.

By 1881 William S. Tower & Company boasted a line of 125 toys ranging from miniature woodenware and dollhouse furniture to croquet sets and puzzle boards.[38] Tower illustrated a sampling of these goods in a trade card (fig. 2.13) printed around this time. It assured buyers "All Toys made in the most thorough manner." Surviving examples of his work, however, bear the hallmarks of fast and thrifty production. Virtually all the parts for Tower toys were cut, turned, or shaped with the aid of machines; further painstaking and precise methods of craftsmanship were eliminated by opting for simple designs and assembly that required less skill. This is perhaps most evident in the octagonal wash sets that Tower sold in large quantities. The nested tubs and wire-handled pails (fig. 2.14) are held together only with flexible tin bands, many of which came undone over time. Tower expanded his line of toyware to include "instructive amusements" like Tower's Log Cabin—an old-fashioned homestead complete with timber fences and a well sweep (see pl. 14.5).[39] The parts were delivered—and could then be stored—in nailed wooden boxes marked with the company's printed paper labels. In some cases the packaging became an integral part of the toy. For example, the base of Tower's gristmill (fig. 2.15) is formed by the box in which the toy was originally packaged. An accompanying shovel allowed children to scoop sand or other material into a hopper, which passed through a paddle wheel and into a bin. When it was in motion, a stenciled figure of a man wearing a hat and checkered clothes appeared to run the mill by way of a metal hand crank.[40]

The firm distributed its toys through a robust trade network. Many were sold locally from a salesroom in Hingham, or nearby at summer destinations such as Downer Landing and the Pavilion at Nantasket Beach; Tower's trade cards even advertised hand-delivery to "persons on the South Shore from Hull to Scituate."[41] Items could also be ordered directly through mail catalogues found in the back pages of popular magazines like the *Youth's Companion*. Agents in Boston and New York City further extended Tower & Company's reach to the western United States and Canada, as well as "Liverpool,

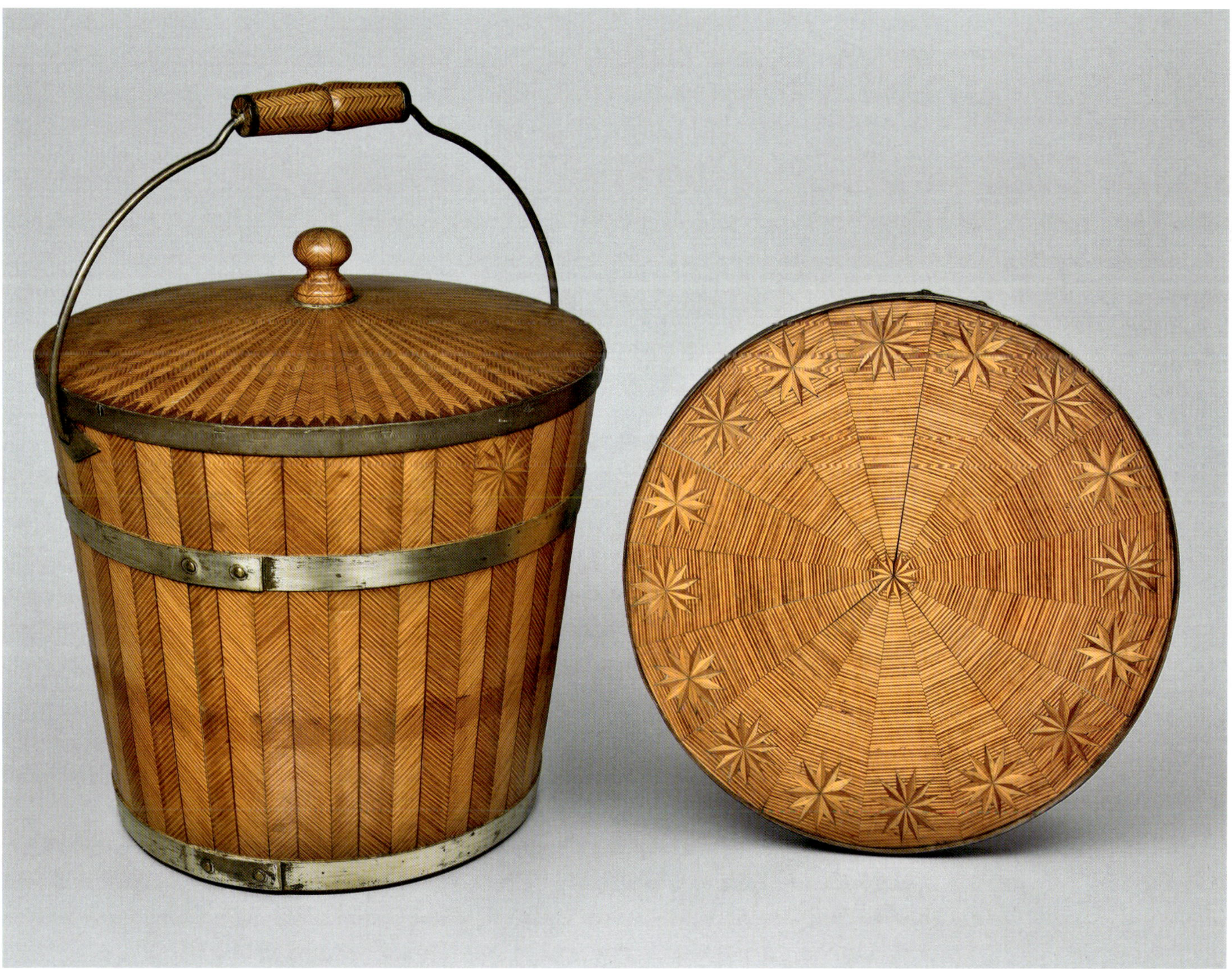

Figure 2.10
Fancy bucket and calling card tray, by William S. Tower, Hingham, ca. 1865. Red and white cedar, tinned iron hoops, wire handle; bucket: H 12 ¾, Diam 11 ⅞; tray: H 1 ½, Diam 11 13⁄16. Courtesy of Hingham Historical Society

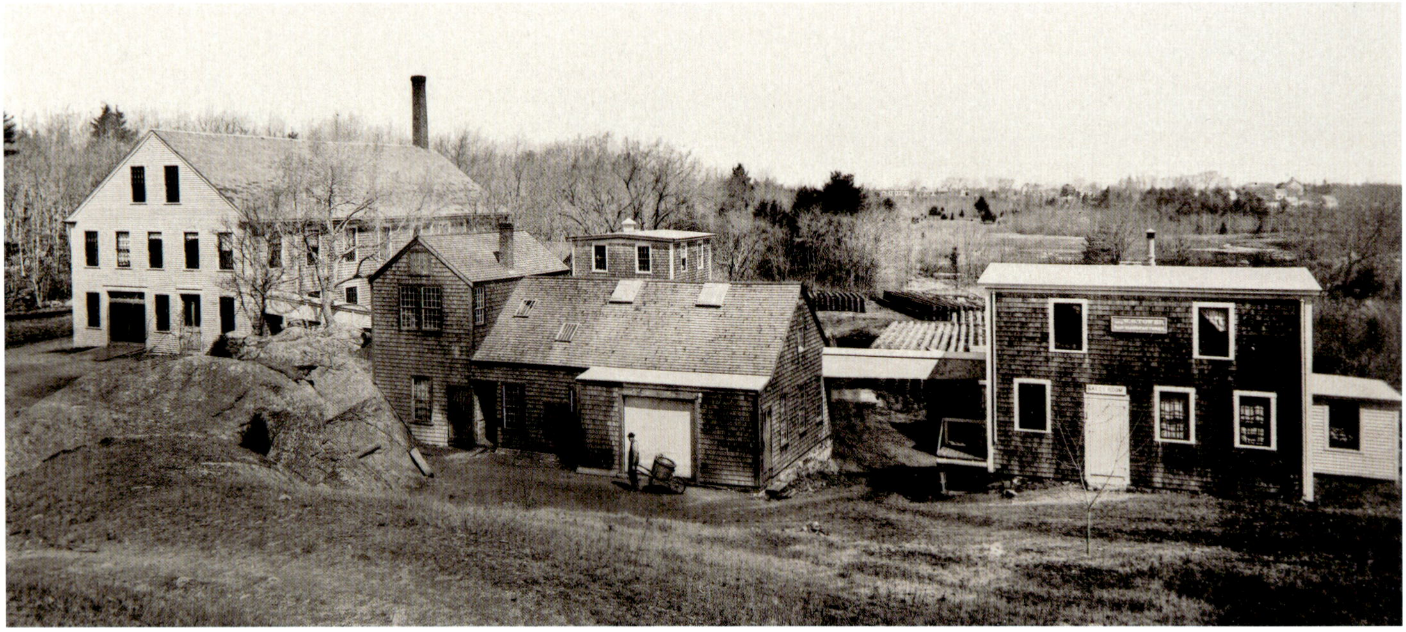

London and other places in Europe."[42] In an effort to gain a foothold in such distant markets, Tower exhibited his toys at major national and international fairs—even winning a prize at the Exposition Universelle in Paris in 1878.[43] Although he almost certainly marketed these goods through trade catalogues, none is known to have survived. Later, in the 1890s, he supplied retailers with cabinet cards of his toys produced by local photographers.[44]

The last decade of the century marked a period of boom and bust for the firm. Around 1892 control of the factory passed to Tower's son-in-law, Frank Burrell, who renamed it the Tower Toy Company.[45] A year later the *Hingham Journal* announced tremendous growth for the company, which added a 12,000-square-foot wareroom in anticipation of doubling its workforce from twenty to forty people. The factory reportedly used "1000 pounds of tacks and nails, three tons of metals for hoops for toy pails, twenty-five miles of wire for pails, etc. and cut up over one quarter of a million of lumber" annually.[46] Within a few years Burrell would relocate the operation to Hersey Street, where he added a "steam job shop and planing mill."[47] By 1899 "Mr. F. W. Burrell's shrill little whistle morning, noon, and night was a pleasant reminder that something in the way of business was being done in Hingham."[48]

Despite an outward appearance of prosperity, the Tower Toy Company failed to keep up with the increasingly mechanized

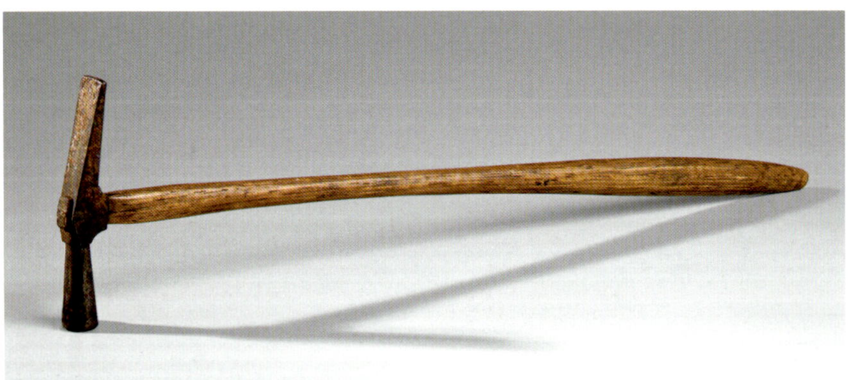

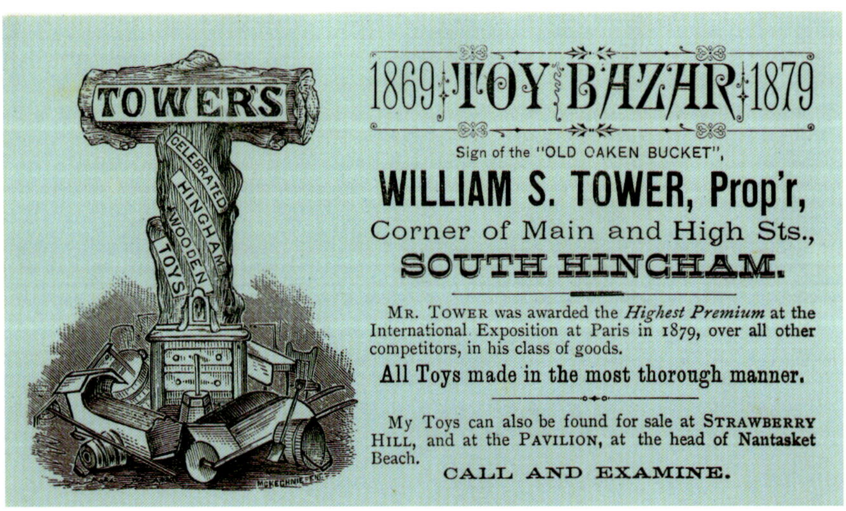

world of toy production. By 1880 Mason & Converse in Winchendon, Massachusetts, already employed three times as many workers.[49] That firm's specialty, also "light woodenware" and wooden toys, grew swiftly under the leadership of Morton E. Converse. Within a decade his operation ballooned to over three hundred workers and six acres of factory space, where he manufactured a huge variety of wooden and metal playthings. Winchendon became known as Toy Town; Hingham, once a major player, faded in prominence.[50] In 1900, without much warning, Burrell sold the business to George R. Healy Manufacturing Company of New York.[51] Tower Toys would never again be produced in Hingham—or anywhere else.

Nearing extinction, Hingham's toy trade experienced a stunning resurgence at the turn of the century. In 1901, only months after Tower Toys shuttered its doors, a group of fifteen wealthy and civic-minded men and women, including the philanthropist Susan B. Willard, established the Society of Arts and Crafts, Hingham.[52] The society sought to "promote artistic work in all branches of handicraft," to "bring designers and workmen into mutually helpful relations, & to encourage workmen," especially toymakers, "to execute designs of their own."[53] Despite the town's relatively small population, the group successfully recruited more than a dozen artists and skilled craftsmen to submit their goods. Items that passed inspection were labeled with the society's official trademark, the iconic Hingham Bucket, and sold at a permanent salesroom in Hingham and at handicraft shows throughout New England. According to one reporter, the society's annual exhibitions attracted crowds of "visitors within the entire cruising radius of the modern automobile."[54]

George W. Fearing (1838–1918), an early member of the society, became emblematic of its efforts.[55] He was featured regularly in the group's promotional material (fig. 2.16), often described as the "sole survivor" of Hingham's coopering and toymaking trades.[56] As a former employee of the Wilder bucket mill and Tower toy factory, Fearing was, in fact, among the few remaining craftsmen to have worked in both industries at their height.[57] As a tradesman he had spent most of his career working with mechanized tools. It was not until the late 1880s that Fearing opened a small toymaking shop behind his house on Water Street, where he specialized for the

Opposite page:
Figure 2.11
William S. Tower toy factory and salesroom at Cushing Pond, photograph by Charles W. Eddy, 1885. Tower worked in the flat-roofed building in the far right of the image. From *Hingham: Illustrated with Pen and Camera* (1885)

Figure 2.12
Tack hammer used by Jenny Cook at the Tower toy factory, 1890–95. L. 9 ½, W 2 ¼. Courtesy of Hingham Historical Society

Figure 2.13
William S. Tower Toy Bazar trade card, 1879. Engraving on paper; H 2 ¾, W 4 ¾. Courtesy of American Antiquarian Society

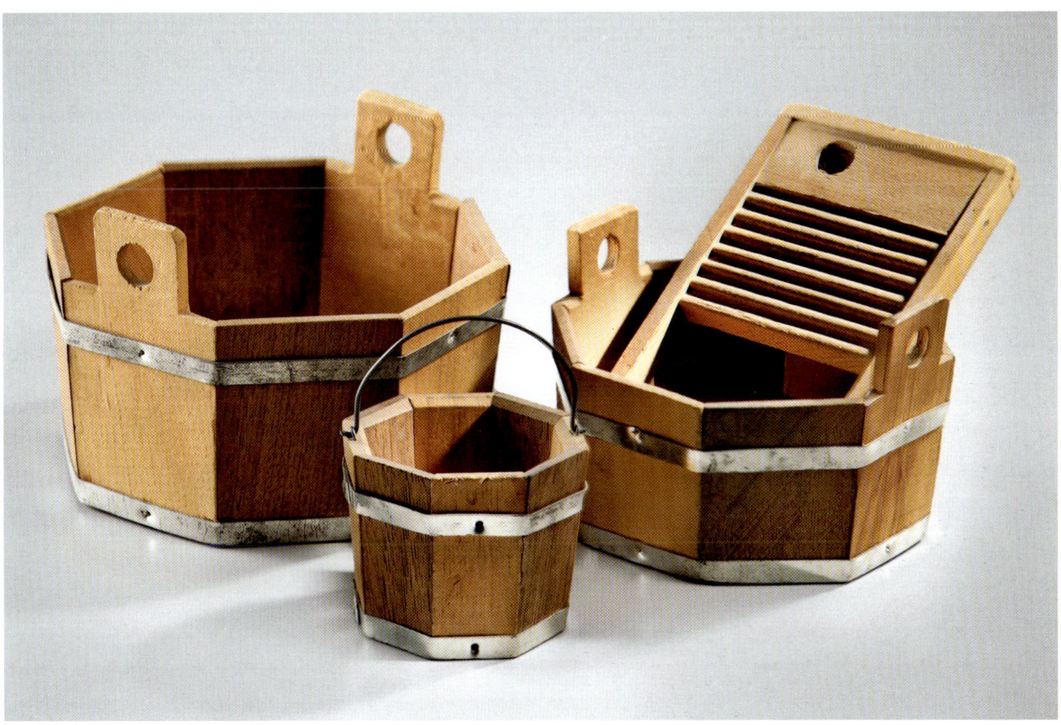

Figure 2.14
Miniature nesting washtubs, washboard, and pail, by William S. Tower, Hingham, 1869–85. Stamped "WM. S. TOWER / SO. HINGHAM, MASS." on each bottom. White pine, walnut, tinned iron bands, wire handles; largest tub: H 4 ¼, W 4 ¼; pail: H 2 1/16, W 1 ¾. Courtesy of Boston Children's Museum

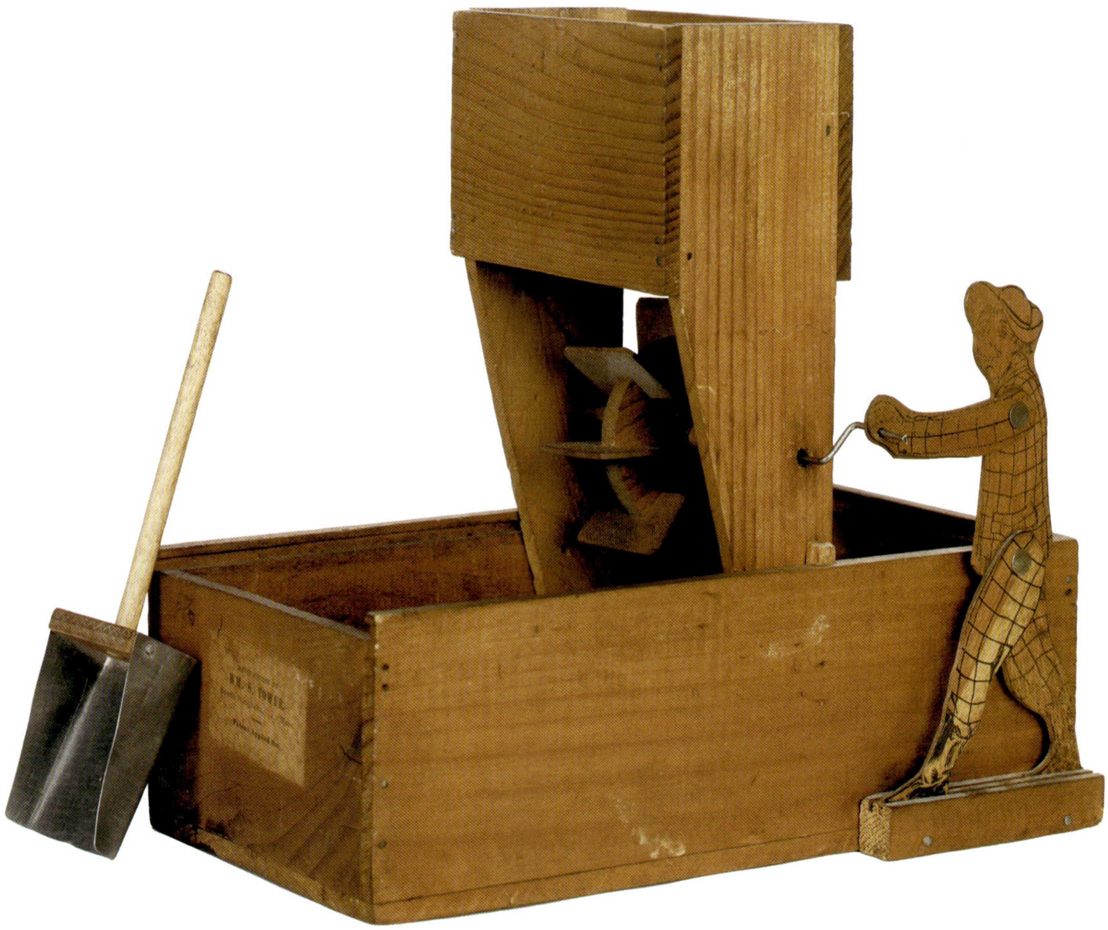

Figure 2.15
Gristmill sand toy, by William S. Tower, Hingham, 1869–85. Labeled "MANUFACTURED BY / WM. S. TOWER, / South Hingham, Mass. / Patent Applied for." White pine, tin shovel, iron nails; H 9, W 5, D 9 ½. Courtesy of Shelburne Museum

first time in handcrafted items like miniature woodenware, furniture, and lawn ornaments.[58] Before joining the society, he had kept a relatively low profile in the community, doing little to advertise his wares.

Children and adults alike held dear his masterfully coopered items; these diminutive, handcrafted containers (see pl. 16.2) formed a tangible link to Hingham's preindustrial past. The society emphasized this association by celebrating Fearing as "the son of a Hingham bucket-maker" who used a set of ancient tools that "could no longer be duplicated" and who marked his goods with a carved hickory brand passed down by his father.[59] The picture painted was not entirely accurate, but it helped tie Fearing to a bygone era that fascinated consumers. Building on this marketing scheme and the growing enthusiasm for the Colonial Revival, he began to craft miniature furniture (see pls. 16.3–16.5) modeled after colonial examples, some of which had been made famous through the work of antiquarians such as Wallace Nutting, who published a series of illustrated books celebrating the merits of furniture made in early America.[60] Affluent families could purchase reproductions of these forms from Nutting or collect the real thing from local estate auctions and antiques shops. Their children now had the option of playing with tasteful miniatures that espoused the same virtues of heritage and handcraftsmanship.

The career of Loring H. Cushing closely paralleled that of his good friend George Fearing. The men reportedly labored at the same bench at Wilder's bucket mill and again at the Tower toy factory, where Cushing worked as a general manager for two decades before leaving the toy business in the early 1890s.[61] They even served in the same company during the Civil War, when "on many a night they slept beneath the same blanket or they tramped side by side through many a long weary march."[62] It is not surprising, then, that Cushing also became an early member of the Society of Arts and Crafts. Though he had never operated a shop of his own, he converted the loft of his carriage house to a workroom complete with a motorized lathe and saw. The site became a popular destination for tourists who hoped to catch a glimpse of the old toymaker at work. Like Fearing, Cushing looked to full-sized examples for inspiration when producing his miniature colonial furniture (see pls. 17.4–17.6). He described the process to a reporter for the *Boston Globe* in 1910:

> *I have spent days working on some small detail that had to be made absolutely correct before it met with my approval. And I know it is the same with Mr. Fearing. Take one of my low-boys [fig. 2.17], for an example, for I had to make a dozen and more before I had made one which was entirely satisfactory to me. The low-boy in question I copied from one of 1760 belonging to a neighbor. Even now that I have been able to show a perfect reproduction in miniature form I find it necessary to spend any amount of time and painstaking labor on each of the legs; whereas but for that I could make this piece of furniture in a day.*[63]

The success of Cushing and Fearing prompted a renaissance in Hingham toymaking. Between 1900 and 1920 the town was home to over a dozen men and women, old and young, plying the trade of toymaker for the first time. Though most of these craftspeople joined the Society of Arts and Crafts, others seemed content to benefit from its proximity and the traffic it generated through its active publicity campaign. Ezra Wilder Jr. (1843–1929), a retired shoemaker, spent the last decade of his life crafting miniature furniture (figs. 2.18, 2.19), possibly at the clock and watch shop formerly occupied by his father and grandfather on Friend Street.[64] Despite a fairly long career in the toy trade, his name and goods, of which several examples survive, lack any association with the society, which suggests that he was never admitted to its ranks, and that his work was never officially sanctioned. Daniel Litchfield (1862–1943), on the other hand, was a proud member.[65] A custodian at a local schoolhouse, Litchfield spent his summer vacations and evening hours at the trade. The quaintness of his operation was described in an early newspaper article profiling the society:

> *The Litchfield cottage, neat and New Englandy, just off Main Street and near Frank Rogers's studio, shows nothing*

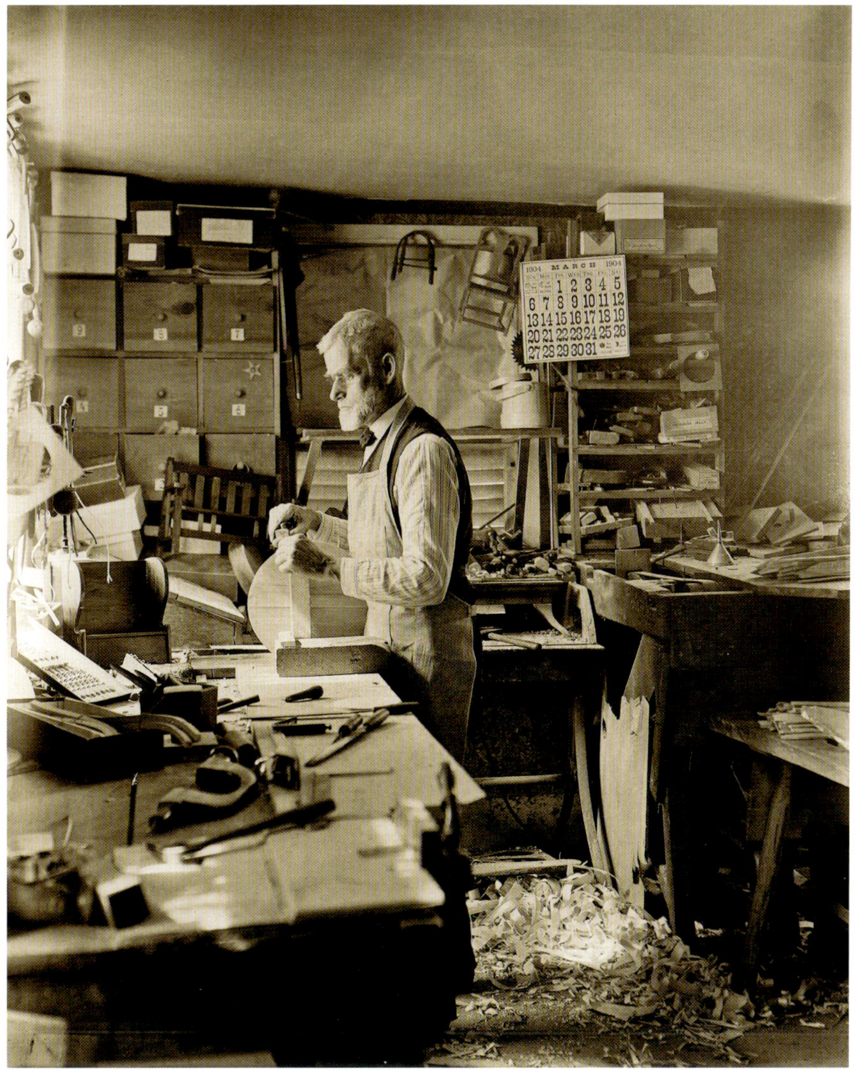

Figure 2.16
Toymaker George Fearing at work in his shop on Water Street, photograph by William Blake Luce, Hingham, 1904. John P. Richardson Photograph Collection, courtesy of Hingham Historical Society

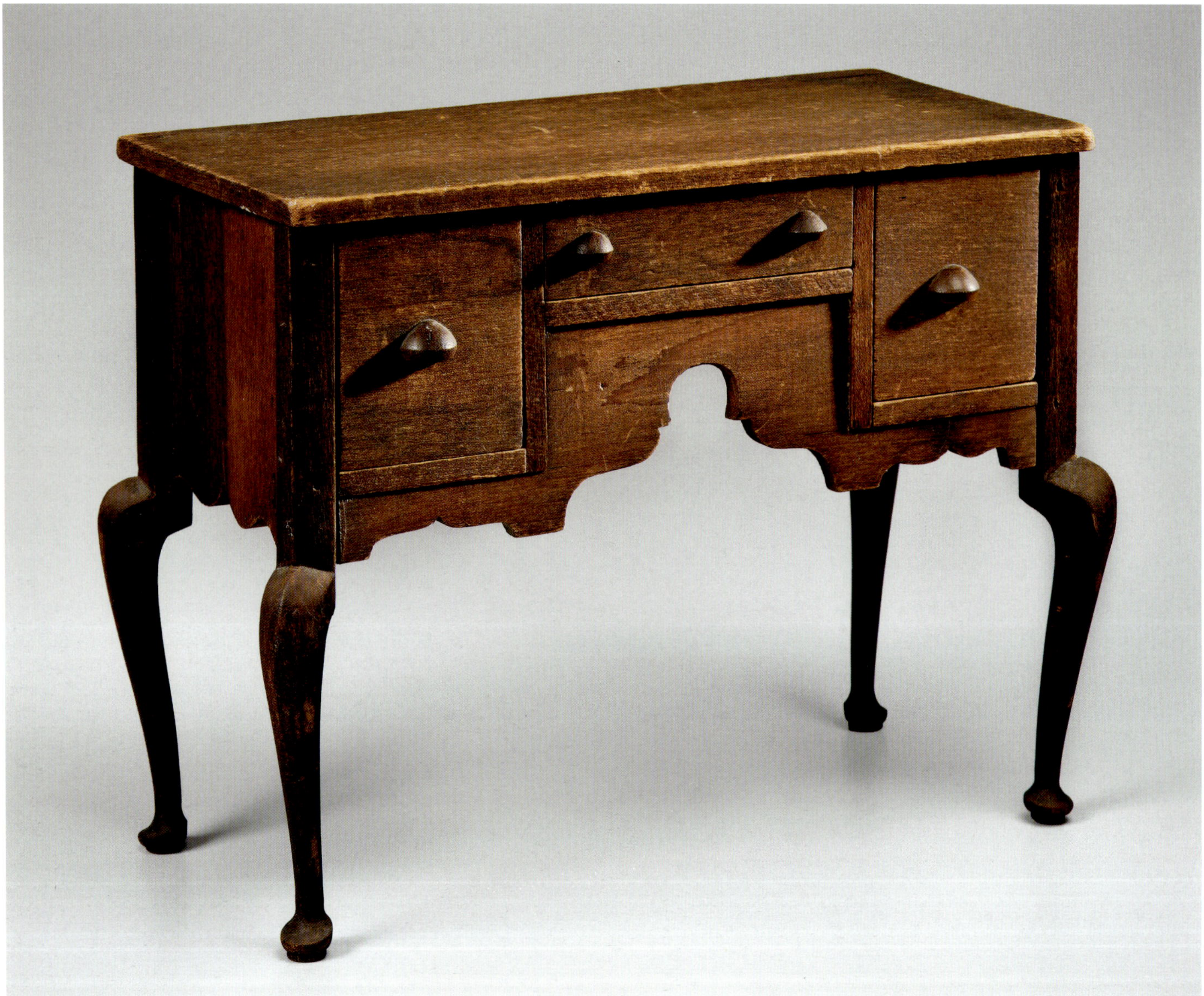

Figure 2.17
Miniature lowboy, by Loring H. Cushing, Hingham, 1901–15. Labeled "MADE BY / LORING H. CUSHING / So. Hingham, Mass." Tulip-poplar; H 7 ½, W (legs) 10, D (legs) 4 ¾. Courtesy of Chad and Dianne Forman

exceptional on the outside, but in the shed behind the house you find a most amazing panoply of paraphernalia. Here is a one-man furniture factory, its power supplied by a gasoline motor in the cellar. Mr. Litchfield is an ingenious mechanic who will put you together anything from a Jacobean settle to a Georgian tall clock, always on a scale suitable for the nursery. It is, indeed, a pretty side line he has built up. His doll's furniture is in homes from Eastport to San Diego.[66]

Official records for the Society of Arts and Crafts, Hingham, no longer survive. The group probably disbanded in the early 1920s, around the time their advertisements in local directories ceased. Hingham's fame as a center of toymaking, however, continued to drive business to the few shops that remained open, but without the support and publicity of the society, demand for wooden toys and colonial-style miniatures dwindled.

Remarkably, one toymaker—Ralph Jones (1886–1977)—continued to produce wooden toys long after other craftsmen had left the business.[67] In 1910, at age twenty-four, Jones was a newcomer to the trade, specializing in fine miniature furniture that rivaled the work of his peers. By the 1930s and 1940s, however, he had transitioned to a larger wholesale operation. His goods became less expensive—and

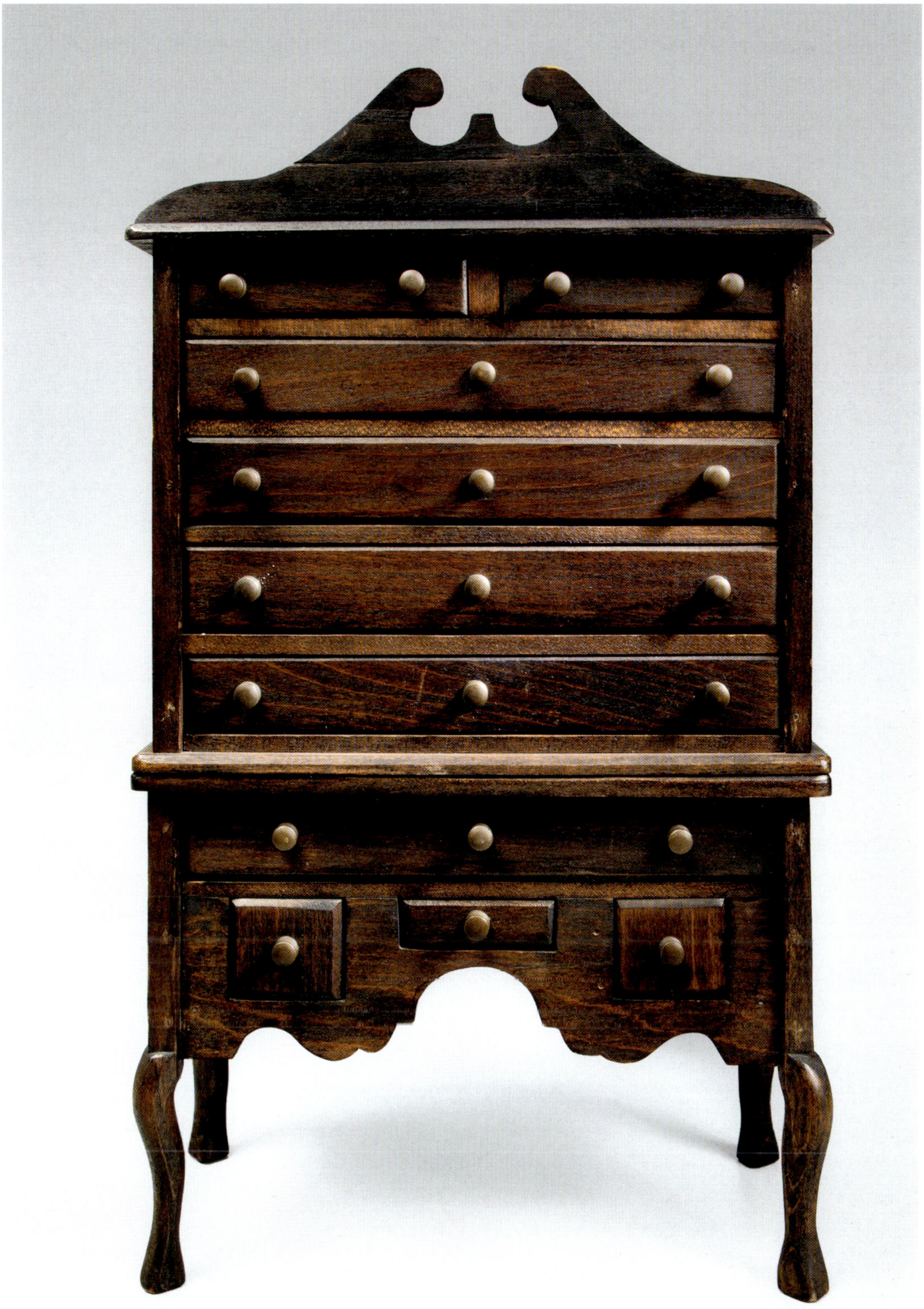

Figure 2.18
Miniature highboy, by Ezra Wilder, Hingham, 1901–20. Tulip-poplar; H 15, W 8 11⁄16, D 3 7⁄8. Courtesy of Hingham Historical Society

no longer showed the attention to detail and materials that had so elevated his earlier work. Jones continued on this path, broadening his catalogue to include items such as miniature cranberry scoops and sea chests that appealed to the thousands of tourists who visited Cape Cod during the summer. They purchased his

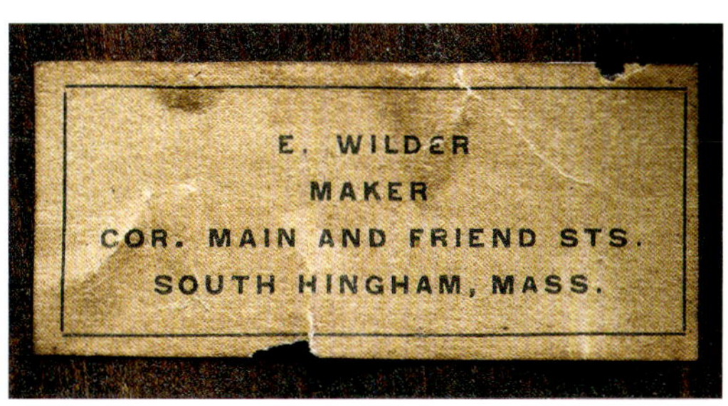

Figure 2.19
Detail of label on highboy in figure 2.18

wares from the trinket shops lining the main commercial highway. Although his business had shrunk to a one-man operation by the 1960s, Jones still produced many of these goods from a workshop in Hingham—a place of refuge and industry late in his life—until he passed away at the age of ninety-one. His departure from the trade marked the end of a toymaking tradition that had continued uninterrupted in Hingham since the late eighteenth century.

NOTES

1. *Boston Gazette,* January 8, 1733, 4. For more information on Price, see Marshall McClintock and Inez McClintock, *Toys in America: A Profusely Illustrated History* (Washington, D.C.: Public Affairs Press, 1961), 29–30; see also Emma Forbes Waite, "William Price of Boston: Map-Maker, Merchant, and Churchman," *Old Time New England* 46 (October–December 1955): 52–56, iv.

2. Little research has been done on Boston's early toy trade. The *Boston Directory* for 1796 lists Mary Butler and Abigail Woodman as proprietors of toy shops. Butler had advertised toys as early as 1791; see *Columbian Centinel,* Boston, June 18, 1791, 111. For more information on Woodman, see Jacqueline Barbara Carr, *After the Siege: A Social History of Boston, 1775–1800* (Boston: Northeastern University Press, 2005), 176. The *Boston Directory* for 1805 lists Mary Ruggles and Jane Salter as proprietors of toy shops; Kenter, Von Harten, & Co. is listed as proprietor of a toy warehouse.

3. The *Boston Directory* for 1800 lists Richard Salter as proprietor of a toy shop at 14 Union Street. Two years earlier he advertised a "fresh supply" of children's toys; see *Russell's Gazette,* Boston, July 23, 1798, 4. Following his death in 1803, Salter's daughter, Jane Russell, took control of the shop; see *Columbian Centinel,* Boston, October 15, 1803, 3. Following Salter's death a description of her stock in trade was listed for sale in the *Boston Commercial Gazette,* November 23, 1815, 2. For a detailed inventory of her store, see Jane Salter, inventory, recorded October 16, 1812, Suffolk County Probate Records, 110:508. For information on the robust toy trade between Germany and America during the early nineteenth century, see Mary Audrey Apple, "German Toys in Antebellum America," *The Magazine Antiques* 162, no. 6 (December 2002): 92.

4. For a discussion of craftsmen who also made toys, see McClintock and McClintock, *Toys in America,* 71–73.

5. For an overview of the Hersey family of toymakers in Hingham, see David A. Schorsch, "The Herseys: Toy Ware and Box Manufacturers in Bucket Town," *Maine Antique Digest* (August 1995): 8b. Few professional toymakers worked in eighteenth-century America. The Lutheran minister Henry Muhlenberg of Pennsylvania noted the presence of a toymaker from Nuremberg in Philadelphia during the eighteenth century; this information comes courtesy of Lisa Minardi. In 1758 George Adam Wiedner of Berks County, Pennsylvania, advertised a runaway servant, John Henry Uncle-bock, "by trade a toy-maker"; *Pennsylvania Gazette,* Philadelphia, February 16, 1758, 4.

6. Daniel Hersey first settled in the area around Austin's Lane, later known as Hersey Street, in 1730; George Lincoln et al., *History of the Town of Hingham, Massachusetts, Genealogical* (Hingham: By the town, 1893), 2:300.

7. For a biography of Reuben Hersey, see "The Box and Toy Business," *Hingham Journal,* August 11, 1876, 1.

8. For a history of Cotton Hersey's house at 168–170 Hersey Street, see Julian Loring's notes on Hingham houses, Hingham Historical Society, Hingham. For more information on Hersey, see cat. 11.

9. As Hersey advanced in age, he became a frequent topic of local conversation. Numerous articles about him appeared: "Pleasant Gathering," *Hingham Journal,* January 14, 1876, 2; *Boston Journal,* January 11, 1878; "Old Folks at Home," *Hingham Journal,* August 9, 1878, 2; "Not Exactly So," *Hingham Journal,* August 30, 1878, 2; and *Hingham Journal,* "Obituary," December 23, 1881, 1.

10. "History of the Bucket Mill," *Hingham Journal,* April 4, 1902, 1.

11. George Lincoln et al., *History of the Town of Hingham Massachusetts, Historical* (Hingham: By the town, 1893), 1.2:244.

12. Solomon Lincoln, *Sketch of Nantasket in the County of Plymouth* (Hingham: Gazette Press, 1830), 14. For more information on the early hotels of Nantasket and Hingham, see *Tourists' Guide Down the Harbor: Hull and Nantasket* (Boston: John F. Murphy, 1897).

13. Lincoln et al., *History of Hingham, Historical,* 1.2:245. For more information on Hingham as a resort town, see Lorena Laing Hart and Francis Russell Hart, *Not All Is Changed: A Life History of Hingham* (Hingham: Hingham Historical Commission, 1993), 112, 161–68.

14. Bessie Lincoln to the Trustees of the Essex Institute, May 1, 1931, Acc. no. 128,637.

15. June Sprigg, *Shaker Design* (New York: Whitney Museum of American Art, 1986), 115.

16. For a genealogy of the Hersey family, see Stephen E. Hersey, *The Hersey Family: Tracing the Descendants of William Hersey*

of Hingham, Massachusetts, 1635–1998, 7th ed. (Westford, Mass.: Stephen E. Hersey, 1998), 37. See also Hersey family tree, compiled by Francis C. Hersey, South Boston, Massachusetts, 1895, Hersey family collection.

17. Samuel Hersey first advertised himself as a "Toy Ware & Box Manufacturer" in the "Hingham Business Directory," *Hingham Journal*, February 1, 1850, 1. For more information on Hersey, see cat. 12.

18. Caleb Strong Hersey, a distant cousin, spent his career vacillating between toymaking and commercial coopering; see appendix A. Edward Lewis Hersey showed exceptional talent as a toymaker, but he died young of tuberculosis; see cat. 13. Joseph Hersey Jr. abandoned the toy business altogether to specialize in boxes. Perhaps the most consistent of these later Hersey craftsmen was Reuben Hersey III. He operated a small shop for most of his life, but earnings from his real estate ventures greatly enhanced his income; see chap. 3. Edward, Joseph Jr., and Reuben III were Samuel's first cousins; all four Herseys were grandsons of Reuben.

19. Nina Fletcher Little, *Neat & Tidy: Boxes and Their Contents Used in Early American Households* (1980; repr., Haverhill, Mass.: Society for the Preservation of New England Antiquities, 2001), fig. 154 and pp. 157–58.

20. In 1873 Hersey was awarded a gratuity of two dollars for toys he exhibited at the Hingham Agricultural and Horticultural Society Fair; see *Transactions of the Hingham Agricultural and Horticultural Society for the Year 1873* (Hingham: Hingham Agricultural and Horticultural Society, 1873), 49.

21. Deborah Waters, *Delaware Collections in the Museum of the Historical Society of Delaware* (Wilmington: Historical Society of Delaware, 1984), 146–47.

22. Her father was Captain Frederick Gilman Low. For a family genealogy, see Mary Eliza Low's 1878 obituary, newspaper clipping, Hersey Family Collection.

23. A history of owners is written on the lid of a white box in which the set of furniture has been stored for nearly a century. The line of descent is as follows: Mary Eliza Low (1819–78), to her sister-in-law Amanda Low (1833–1920), Gloucester, Mass.; to her granddaughter Annie Low (Lufkin) Herrick (b. 1898).

24. "Doll's Furniture," *Boston Evening Transcript*, February 24, 1940.

25. Mrs. William C. Cobb to Mrs. Arthur R. Herrick, postmarked April 22, 1940, Hersey Family Collection. Hersey's shop was overturned during a violent storm in 1879; *Hingham Journal*, July 18, 1879.

26. Mrs. William Cobb to Mrs. Arthur R. Herrick, postmarked April 28, 1940, Hersey Family Collection.

27. For manufacturing records for Hersey and several Hingham coopers, see U.S. Federal Census, Non-Population Schedule, 1860, Hingham, Plymouth County, Mass.

28. "Letter from South Hingham," *Hingham Journal*, July 23, 1869, 2.

29. George Lincoln et al., *History of the Town of Hingham, Massachusetts, Genealogical* (Hingham: By the town, 1893), 3:264. Massachusetts Vital Records, 1841–1910, 482:656. For more information on the clockmaker Reuben Tower, see Brock Jobe, Jack O'Brien, and Gary Sullivan, *Harbor & Home: Furniture of Southeastern Massachusetts, 1710–1850* (Lebanon, N.H.: University Press of New England, 2009), 313–19, 323–27. Tower is identified as a turner in the 1855 Massachusetts State Census; this probably refers to his role within a bucket factory, probably that of C. & A. Wilder, where turners made a variety of parts, including rods and lids for churns.

30. According to the *Transactions of the Hingham Agricultural and Horticultural Society, from October, 1858 to March, 1861* (Hingham: Hingham Agricultural and Horticultural Society, 1861), 69, "the fancy buckets made by W. S. Tower, at the factory of C. & A. Wilder, were very neatly executed."

31. *The Tenth Exhibition of the Massachusetts Charitable Mechanic Association at Faneuil and Quincy Halls, in the City of Boston, September and October 1865* (Boston: Wright & Potter, 1865), 118; judges described it as "a very curious pail" and "a very elaborate piece of workmanship." Tower probably also exhibited the bucket at the annual exhibition of the Hingham Agricultural and Horticultural Society. A piece of the society's letterhead, now at the Hingham Historical Society, is inscribed with a description of the bucket and "Manufactured for the Hingham Agricultural & Horticultural Society by William S. Tower of Hingham, Mass." Tower also crafted a "Fancy Top Card Table," thought to be the calling card tray that now accompanies the bucket; see the *Eleventh Exhibition of the Massachusetts Charitable Mechanic Association at Faneuil and Quincy Halls, in the City of Boston, September and October 1869* (Boston: Wright & Potter, 1869), 126.

32. *Hingham Journal*, August 20, 1869, 2.

33. C. Wilder & Son owned the building that Tower used as a toy factory; see Appraisal of Partnership Property, C. Wilder & Son, May 5, 1891, in John P. Richardson Papers, "Hingham Woodenware Chronology, Individual Coopers, Toy, and Box Makers," Hersey Family Collection.

34. The Hingham Woodenware Company announced the sale of its engine and coopering machinery after only a couple of years in business; see "Valuable Manufacturing Property for Sale," *Hingham Journal*, July 30, 1869. Tower purchased the machinery a few weeks later; see "Toy Ware, etc.," *Hingham Journal*, August 27, 1869, 2.

35. U.S. Federal Census, Non-Population Schedules, 1870 and 1880, Hingham, Plymouth County, Mass. Tower's factory relied on steam power when it moved to Hersey Street in 1896.

36. In his first year Tower employed eight men and two women. Skilled mechanics made $1.75 a day, whereas ordinary laborers made $1.00 a day; see U.S.

Federal Census, Non-Population Schedule, 1870, Hingham, Plymouth County, Mass.

37. For information on women factory workers in New England, see Benita Eisler, ed., *The Lowell Offering* (Philadelphia: J. P. Lippincott, 1977). See also Thomas Dublin's *Transforming Women's Work: New England Lives in the Industrial Revolution* (Ithaca: Cornell University Press, 1994); and Sarah M. Evans's *Born for Liberty: A History of Women in America* (New York: Free Press, 1989).

38. "Hingham Toy Ware," *Hingham Journal,* October 21, 1881, 1.

39. A description of Tower's Log Cabin appeared in an advertisement in the *Youth's Companion,* July 3, 1879, 228.

40. A second example of a gristmill sand toy and an original shop sign used by Tower are illustrated in Bernard Barenholtz and Inez McClintock, *American Antique Toys, 1830–1900* (New York: Harry N. Abrams, 1980), 276.

41. In 1878 Tower advertised toys "not only at the High at Main Street shop, but also at Hull, Strawberry Hill & the Pavilion at the head of Nantasket Beach"; see *Hingham Journal,* June 28, 1878, 3. A few years later he advertised goods for sale at "Tower's Bazaar at Rockland Café"; see *Hingham Journal,* June 9, 1882, 3. By 1894 the company also sold toys at Downer Landing; see *Directory of Hingham and Hull, Mass., 1894* (Quincy, Mass.: J. H. Hogan, 1894), 25.

42. For a detailed description of Tower's factory, see "Manufacturers at South Hingham," *Hingham Journal,* June 2, 1876, 2.

43. Tower was awarded the bronze medal for his exhibition of toys; see U.S. Congress, House, *Reports of the United States Commissioners to the Paris Universal Exposition, 1878,* 46th Congress, 3rd sess., 1880, H. Ex. Doc. 42, part 1:347, 369, and 437.

44. In the 1890s the Tower Toy Co. produced a series of cabinet cards with pictures and descriptions of toys and order numbers for wholesale purchasers; several cabinet cards for toy wash sets are in the Hersey Family Collection. The Tower Toy Co. also sent prospective buyers postcards with wholesale information; see James Pierotti, *Hingham: Postcard History Series* (Charleston, S.C.: Arcadia, 2005), 109.

45. Burrell is first listed as proprietor of the Tower Toy Co. in 1894; *Directory of Hingham and Hull, Mass., 1894,* 25.

46. In 1893 the Tower Toy Co. occupied three factories in South Hingham, including the recently acquired Whitman-Whitcomb factory on the corner of Main and Friend streets; see *Hingham Journal,* March 31, 1893, 1.

47. *Hingham Journal,* June 12, 1896, 1.

48. Hart and Hart, *Not All Is Changed,* 125.

49. For manufacturing records for Morton & Converse, see U.S. Federal Census, Non-Population Schedule, 1880, Winchendon, Worcester County, Mass.

50. Ellery Bicknell Crane, ed., *History of Worcester County, Massachusetts,* 3 vols. (New York and Chicago: Lewis Historical Pub. Co., 1924), 2:674.

51. *Annual Statistics of Manufactures, 1900: Fifteenth Report* (Boston: Wright & Potter, 1901), 18. Healy purchased a number of factories that produced "iron and wood specialties, wood turnings, toys and novelties"; see *Iron Age* 65 (March 29, 1900): 42. According to McClintock and McClintock, *Toys in America,* by 1905 Tower was run by the Griswold Co. of New York; no record of Griswold is known, though it may have purchased the business from Healy.

52. Susan B. Willard, "The Hingham Society of Arts and Crafts," in *Hingham: A Story of Its Early Settlement and Life, Its Ancient Landmarks, Its Historic Sites and Buildings* (Hingham: Daughters of the American Revolution, Old Colony Chapter, 1911), 120–23. For more information on the Society of Arts and Crafts, Hingham, see C. Chester Lane, "Hingham Arts and Crafts, Their Aims and Objects," *Craftsman* 3 (December 1903): 276–81. The Society of Arts and Crafts, Boston, founded in 1897, was the first such organization incorporated in the country. It played a large role in introducing the Arts and Crafts Movement to America and provided a successful model and support system for towns like Hingham that sought to do the same. For more information, see Edward S. Cooke et al., *Inspiring Reform: Boston's Arts and Crafts Movement* (Wellesley, Mass.: Davis Museum and Cultural Center, 1997).

53. For pamphlets issued by the Society of Arts and Crafts, Hingham, in 1904, see Circulars and Leaflets, 1901–8, box L 1901, Printed Material, Massachusetts Historical Society, Boston.

54. F. W. Coburn, "Fifteen Years of Arts and Crafts at Hingham," *Boston Herald,* July 23, 1916, D3.

55. Lincoln et al., *History of Hingham, Genealogical,* 2:228. Fearing died in 1918; *Report of the Financial Affairs of the Town of Hingham for the Year Ending December 31, 1918* (Hingham: Hingham Journal Job Print, 1919), 85.

56. Officers of the Society of Arts and Crafts, Hingham, "Hingham Village Industries," *Southern Workman* 32, no. 9 (September 1903): 537–41.

57. "Toy Makers of Hingham," *Boston Daily Globe,* March 13, 1910, SM11.

58. For a description of Fearing's toy business, see "South Hingham," *Hingham Journal,* April 4, 1888, 1.

59. "Arts and Crafts Exhibit at Hingham, Mass.," *Drawing and Manual Training Journal* 2, no. 3 (December 1903): 123–24.

60. For more information on Nutting and the commercialization of the Colonial Revival movement in America, see Thomas Denenberg, *Wallace Nutting and the Invention of Old America* (Hartford, Conn.: Wadsworth Atheneum, 2003).

61. For a contemporary account of Cushing's and Fearing's careers, see "Toy Makers of Hingham"; see also "Artist Toy Makers of America: Reappearance of Handmade

Playthings in Nursery," *Boston Daily Globe,* December 24, 1911, SM3.

62. "Toy Makers of Hingham."

63. Ibid.

64. Lincoln et al., *History of Hingham, Genealogical,* 3:325. Wilder died in 1929; *Report of the Town of Hingham for the Year Ending December 31, 1929* (Hingham: Journal Print, 1929), 76. Wilder is listed as a toymaker in town directories and *Hingham Annual Reports* from 1912 to 1920.

65. U.S. Federal Census, 1900, Hingham, Plymouth County, Mass. Litchfield died in Weymouth in 1943; see *Report of the Town of Hingham for the Year Ending December 31, 1943* (East Weymouth: Weymouth Gazette Press), 65.

66. F. W. Coburn, "Fifteen Years of Arts and Crafts at Hingham," *Boston Herald,* July 23, 1916, D3.

67. Lincoln et al., *History of Hingham, Genealogical,* 2:143. Social Security Death Index entry for Ralph T. Jones, 1886–1977, SS# 030-18-1259. For more information on Jones, see cat. 20.

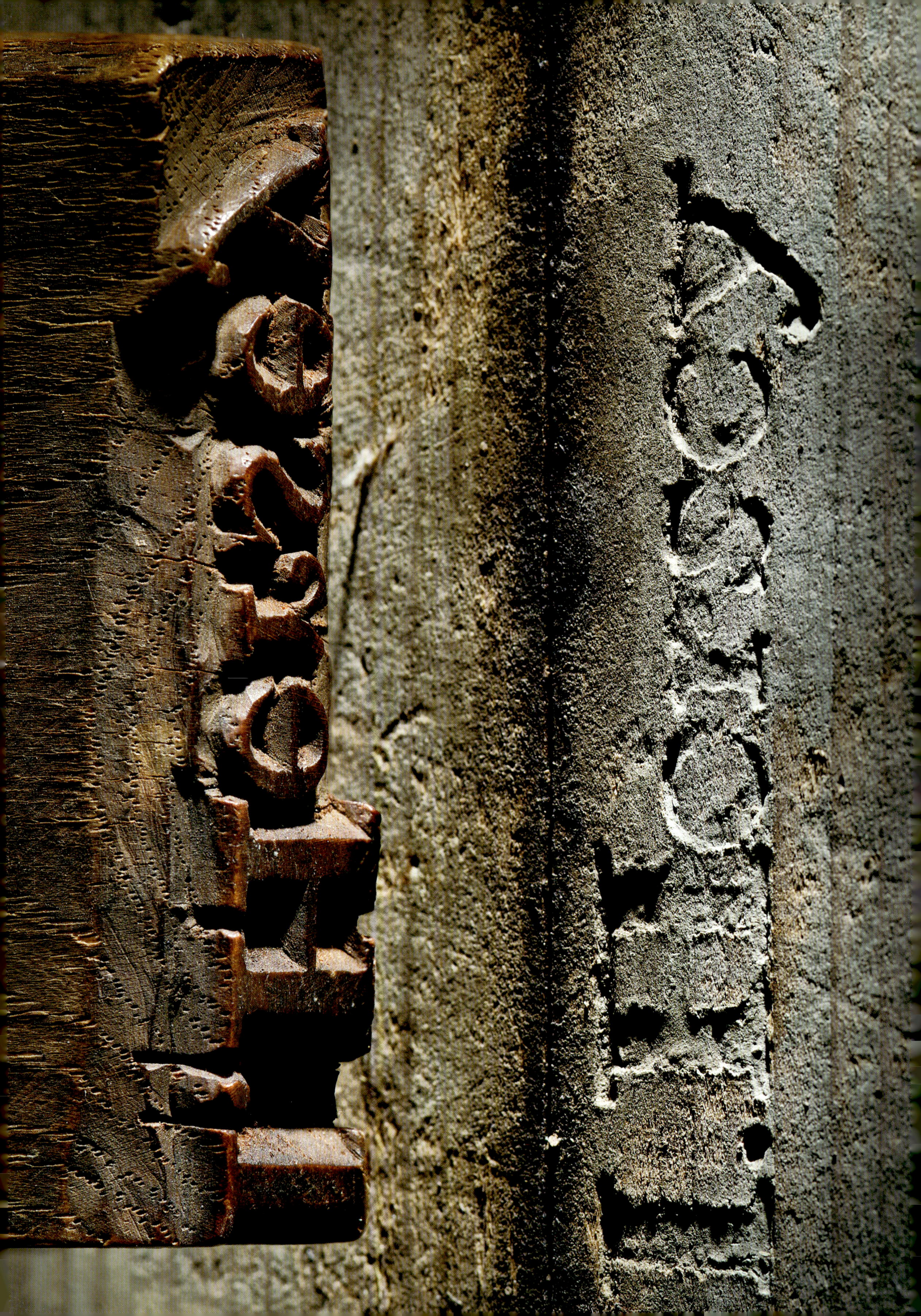

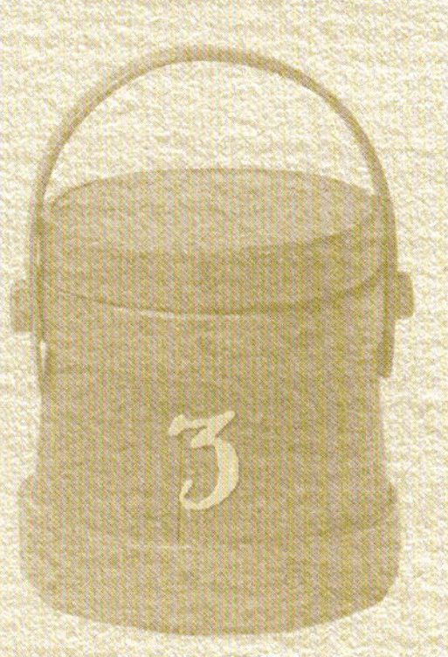

HERSEY FARM: THE FIRST FAMILY OF TOYMAKERS

This old farm shed, which stood in deteriorating condition for decades, held in its bosom incredible treasures of a time long ago.

John P. Richardson, 2008

AT THE CENTER OF HISTORIC HINGHAM STANDS HERSEY FARM—A PICTURESQUE eighteen-acre estate set on a rocky hillside (fig. 3.1). To the east of the property stretch large pastures and vestiges of old orchards, enclosed by woods and cedar swamps; to the south, the farm has a long view of the many saltbox homes clustered along Hersey Street. The early residents of Hingham would probably still recognize this landscape. Hersey Farm, however, is unique among the town's remarkably preserved sites: it has passed through six generations of Hersey husbandmen and master craftsmen—coopers, farmers, and entrepreneurs—whose activities helped shape local culture and commerce for more than two centuries.[1]

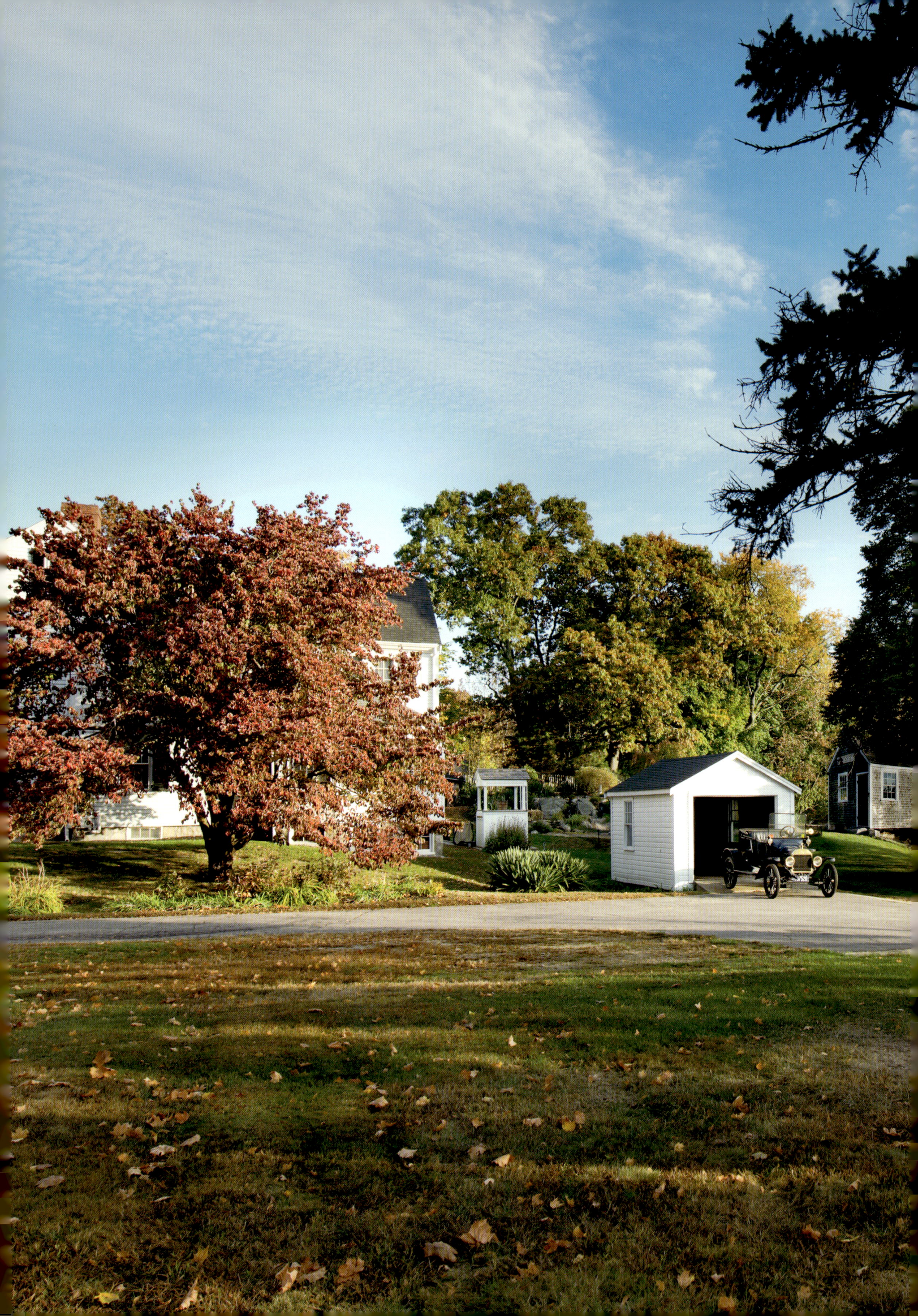

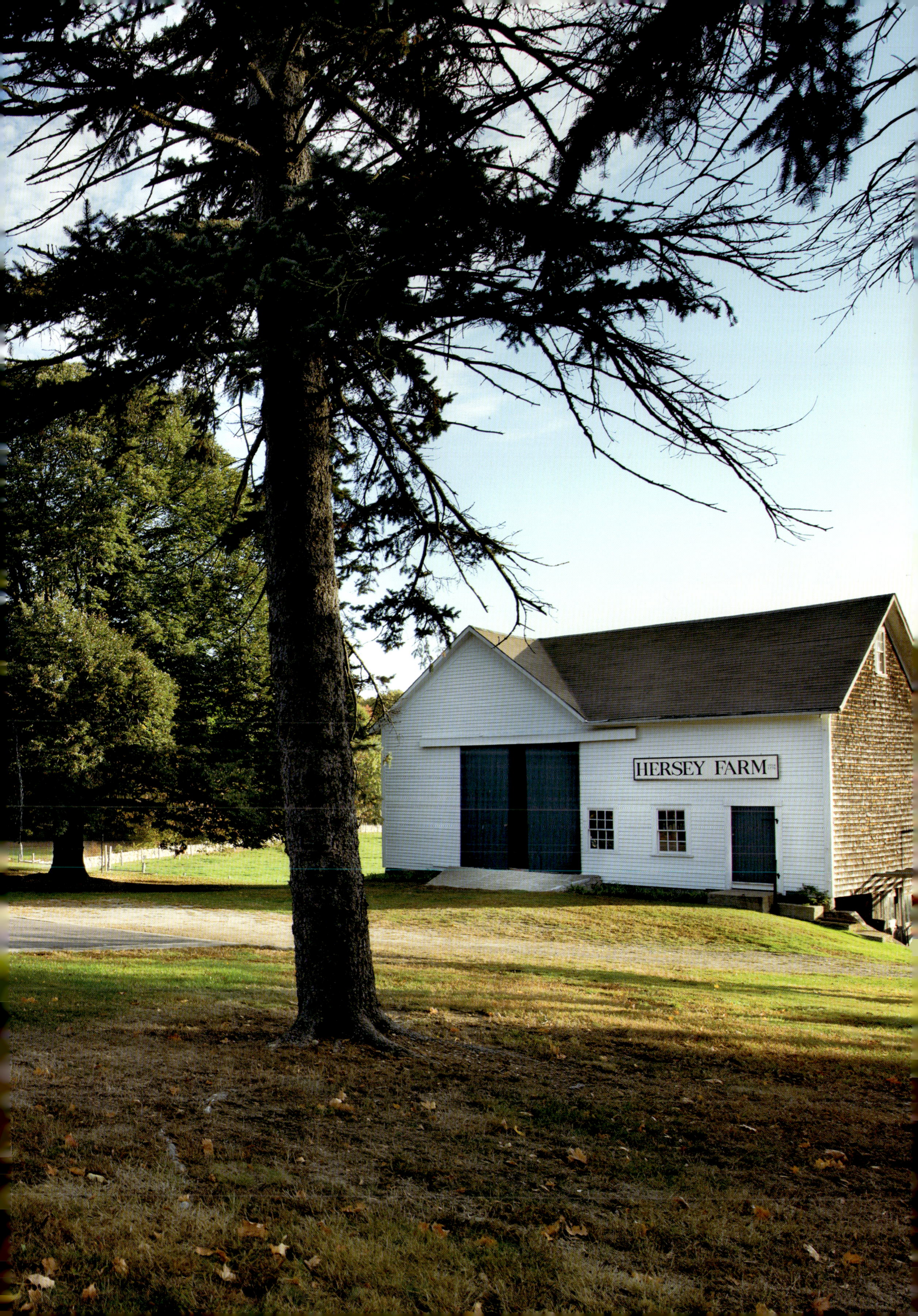
HERSEY FARM

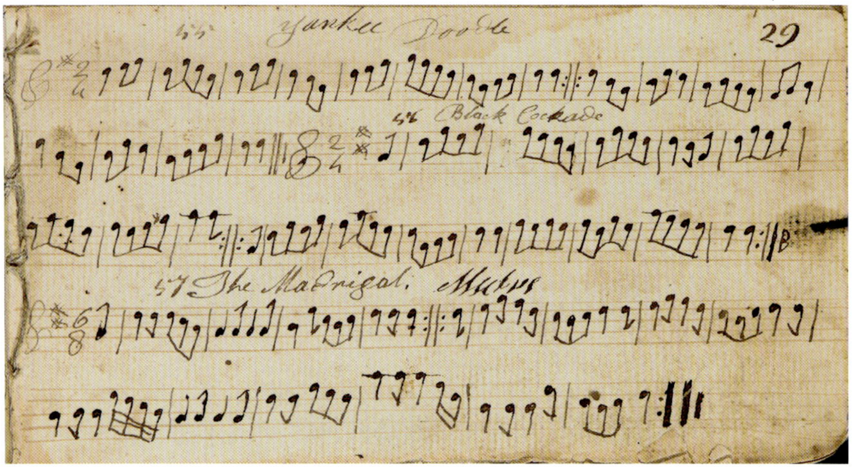

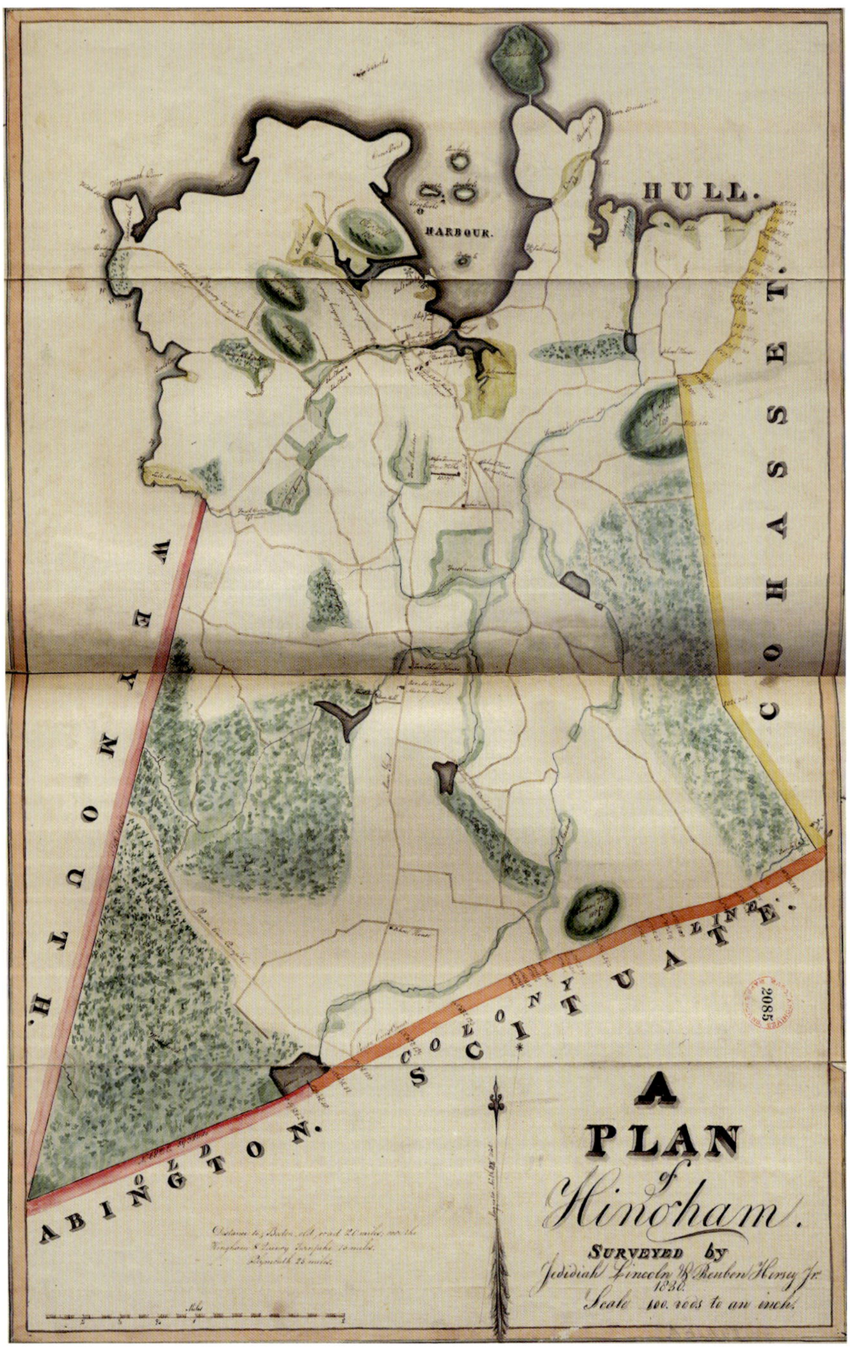

In the summer of 2007 descendants of the original proprietor began the long, difficult task of emptying the old homestead of its contents. Centuries of accumulated possessions filled the farm's two houses, a bank barn built into a ledgy slope, a poultry shed, and other outbuildings—an extraordinary record of life in Hingham from the time of the Revolution onward. Treasures lay everywhere. The handwritten music book (fig. 3.2) of Reuben Hersey (1757–1845), a fifer in the Continental Army, was found in a crowded attic.[2] It included such politically charged tunes as "Yankee Doodle"—its brash title scrawled at the top of a page—and "On Our Way to Boston." A surveying journal kept by his eldest son and namesake, Reuben Hersey Jr. (1780–1844), turned up in a wooden trunk. Hundreds of measurements were recorded in its pages—the basis for Reuben Jr.'s seminal 1830 map of Hingham (fig. 3.3).[3] Eighty years later Reuben Leavitt Hersey (1885–1938) erected the family's first garage. This modest structure, probably ordered from a Sears, Roebuck & Company catalogue, still housed the Model T coupe he purchased in 1915 from a local Ford dealer.

The biggest surprise, however, was discovered in a derelict outbuilding perched on a hillside in the backyard. Wild vines had covered the exterior and a lock, long rusted shut, had for decades prevented curious family members from entering.[4] The old lock was struck off; inside rested a jumble of tools, cast-off farm implements, and other objects, many of which had remained undisturbed for nearly a century. As the space was carefully cleared, a remarkable scene began to take shape. The little "shop," as it had always been called, was in fact an intact coopering and toymaking shop used by multiple generations of Hersey craftsmen from the 1830s through the early 1900s. It still held their original hand tools, strips of ash for hoops, a red-painted chest filled with tiny copper tacks to fasten toy woodenware, patterns large and small, molds, and finished and unfinished products—a true time capsule of the town's most important industries.

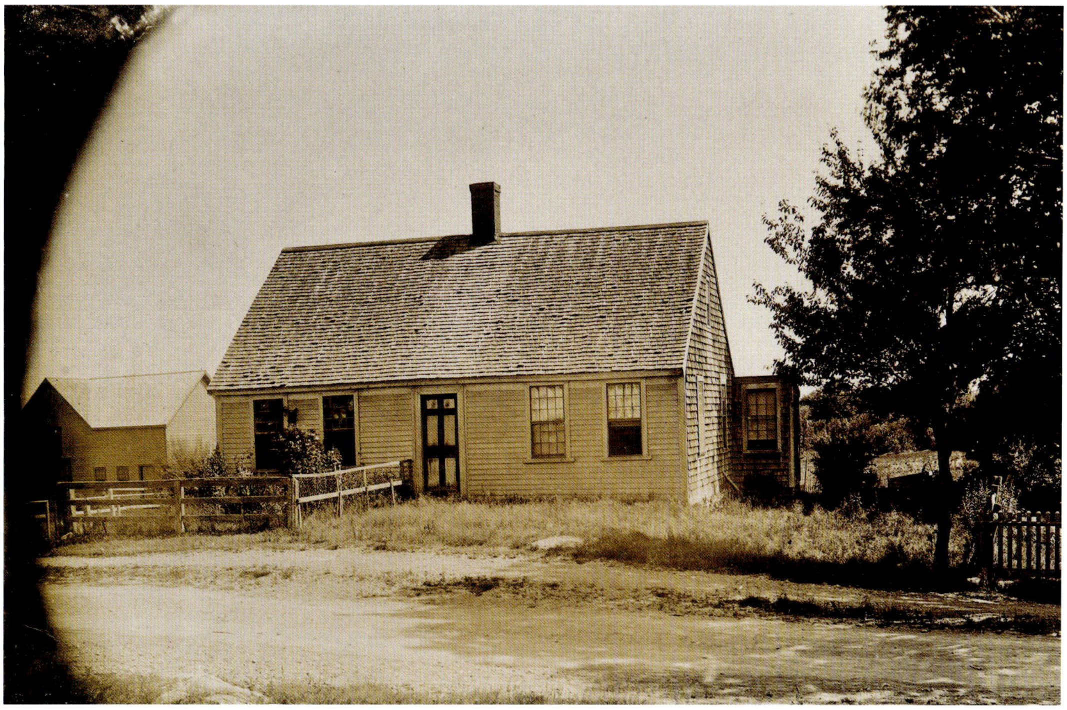

Pages 48–49:
Figure 3.1
Hersey Farm, established ca. 1776

Opposite page:
Figure 3.2
Music book of Reuben Hersey, 1774–1818. H 3 ⅝, W 6 ⅜. Hersey Farm Collection

Figure 3.3
A Plan of Hingham, surveyed by Jedidiah Lincoln and Reuben Hersey Jr., 1830. Watercolor on paper; H 27 ⅞, W 18 ⅞. Courtesy of Massachusetts State Archives

Above left:
Figure 3.4
House of Reuben Hersey, built ca. 1776, photograph, 1880–1920. Hersey Farm Collection

The Hersey shop is the last furnished cooper's shop standing in Hingham—and, more significantly, the only extant preindustrial toymaker's shop in America. This small structure and the original, largely intact colonial farmstead it sits on form the lone surviving example of the mixed-use agricultural and industrial properties that once dominated Hingham's landscape.[5]

The line of Hersey craftsmen began with Daniel Hersey (1682–1766).[6] The grandson of a prominent Hingham settler, it is likely that he apprenticed to Matthew Whiton (1635–1725), a cooper and neighbor who served as a constable with Daniel's father.[7] In 1730 Daniel built a small house on Austin's Lane—a winding cart path linking two larger roads.[8] The surrounding land offered several advantages to the craftsman-farmer: access to major thoroughfares, lush pasture for cattle, fertile soil for crops and orchards, and dense swamps of Atlantic white cedar, a desirable wood for the manufacture of pails. Most important, the site provided ample space for Daniel's offspring to establish homesteads of their own. In 1753 his son Isaiah built a home on the opposite side of the lane, where he, too, operated a small cooperage and farm.[9] This pattern repeated itself many times over, until both sides of the lane—by then renamed Hersey Street—were populated with the small farms and coopers' shops of Daniel's descendants. Between 1730 and 1900 more than thirty members of the Hersey family plied the coopering trade there—an extraordinary network of craftsmen whose bonds of industry and kinship were expressed in the very landscape itself.[10]

Reuben Hersey (1757–1845), a grandson of Daniel Hersey and founder of Hersey Farm, built his small clapboarded Cape (fig. 3.4) on Hersey land around 1776.[11] According to tradition, the unfinished house sat boarded up during his military service in James Lincoln's company of minutemen.[12] Reuben marched with Lincoln's men to Lexington and Concord on April 19, 1775—as a fifer, in the front lines—probably with his well-worn music book in his pocket. Though the company arrived late to this action, Reuben was part of the contingent that took possession of Dorchester Heights in March 1776.[13] The original front door of his house (figs. 3.5, 3.6) still survives, built of massive, reinforced boards with large strap hinges and an old coat of blue paint.

An accomplished cooper and ambitious businessman, Reuben further expanded the

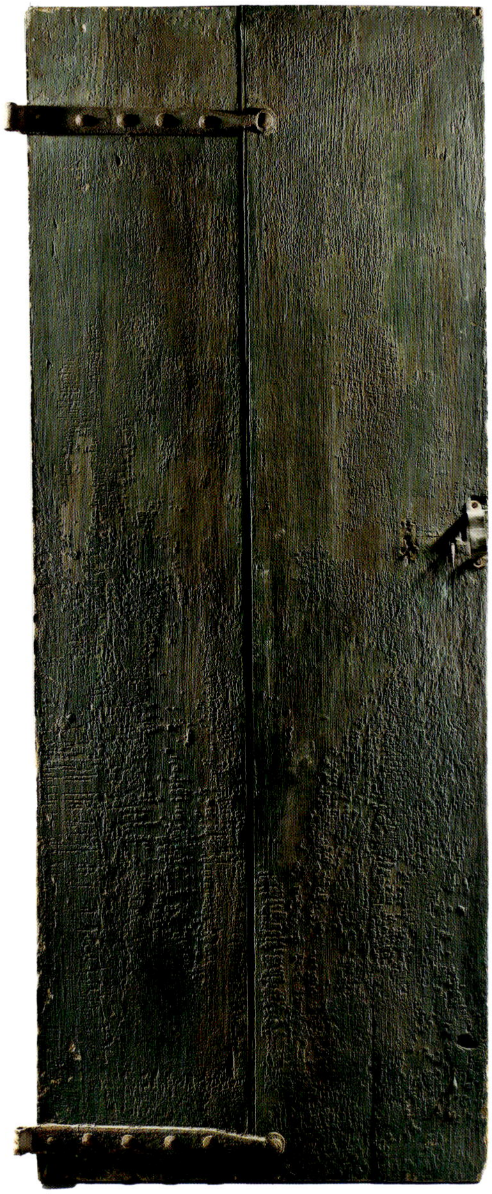

Figure 3.5
Exterior of the front door of the Reuben Hersey House, ca. 1776. H 72 ¾, W 29 ¼. Hersey Farm Collection

Figure 3.6
Interior of the front door of the Reuben Hersey House, ca. 1776. H 72 ¾, W 29 ¼. Hersey Farm Collection

Hersey family trade, producing a wide array of buckets, tubs (fig. 3.7), and pails. He was probably the first craftsman in Hingham to manufacture the lightweight storage containers commonly known as "wafer" boxes, which he delivered to Boston by horse and wagon.[14] The best record of his coopering activities comes from a historical vignette published in the *Hingham Journal* in 1876. While tinged with centennial boosterism, this account still conveys the energy and breadth of Reuben's enterprise:

> *One hundred years ago Reuben Hersey, of Hingham, commenced manufacturing small boxes by hand, cutting out the tops and bottoms with a pocket knife, and shaving out the sides and hoops with a drawing-knife. When he had accumulated a sufficient number, he brought them to Boston on his own wagon and sold them to families, receiving for his pay as much Indian corn as the box would hold. He commenced making wafer boxes, and increased the capacity to what was called the "seven pound box," for sugar, or meal, or rice or any other dry material. From these he went on to making buckets with handles, and pails, and tubs, and churns, dumb-betties for washing clothes, and finally toy pails and other*

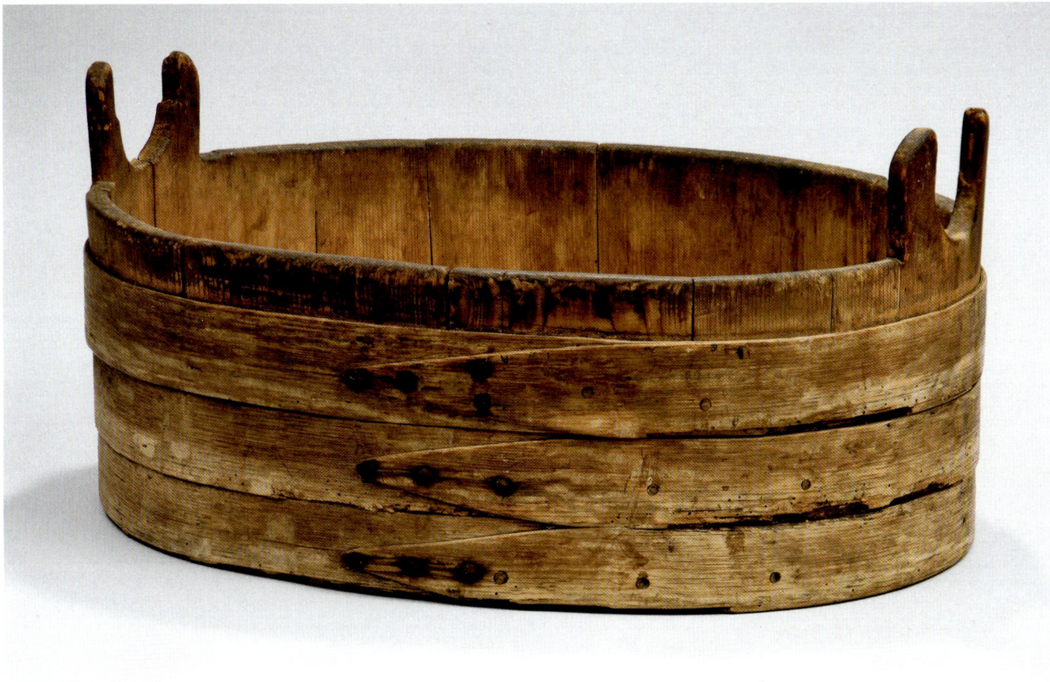

Figure 3.7
Oval tub, by Reuben Hersey, 1780–1830. Incised "RH" on bottom. White pine, oak, iron nails; H 9 ¾, W 20 ¾, D 15 ¾. Hersey Farm Collection

> *toys for children for which Hingham has been, and still is, famous. As business increased, a small vessel, known as the "Hingham Station Packet," was kept moored off the end of the Long Wharf, and was used for the storage and sale of Hingham productions.*[15]

An innovative craftsman and Hingham's first professional toymaker, Reuben laid the foundation for the town's thriving toy trade and the Hersey family's prominence within it. His four sons all built homes and workshops on either side of their father's property, establishing a veritable compound of family industry. Cotton (1792–1881), his third eldest, was locally esteemed for his miniature buckets and tubs and other toy woodenware; Reuben's second son, Samuel French Hersey (1783–1876), operated a similar business, though he later turned to experimental farming, primarily of fruits, and built a cider mill on the property.[16] The multitalented Reuben Jr.—surveyor, mapmaker, cooper, and blacksmith—sold tens of thousands of his iron nails to his relatives for clinching hoops.[17] He was also one of Hingham's first historians. A stone office on his property housed many maps and manuscripts, some contemporary and others copies of seventeenth- and early eighteenth-century documents related to the town's settlement.[18]

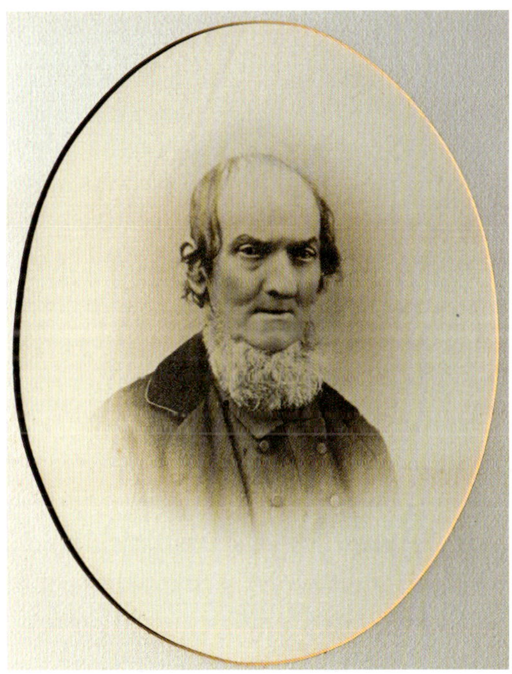

Figure 3.8
Isaiah Hersey, ca. 1865. Hersey Farm Collection

Reuben Hersey's youngest son, Isaiah (1803–68) (fig. 3.8), lived at home for nearly thirty years, working as an apprentice and journeyman cooper in his father's shop.[19] This long, close relationship seems to have instilled in him his father's exceptional craft and business acumen; by age twenty-one

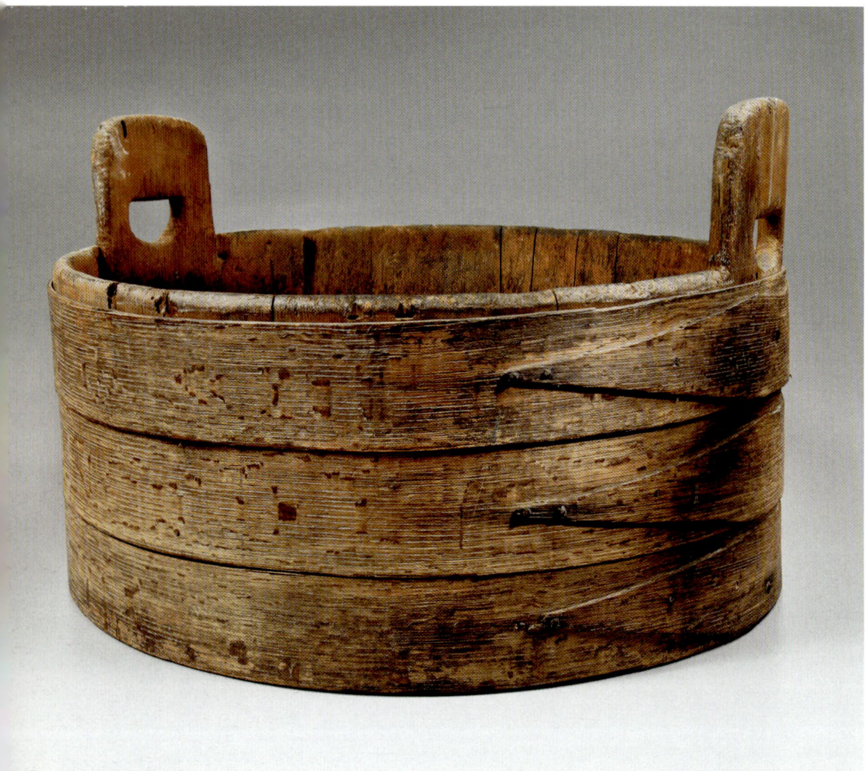

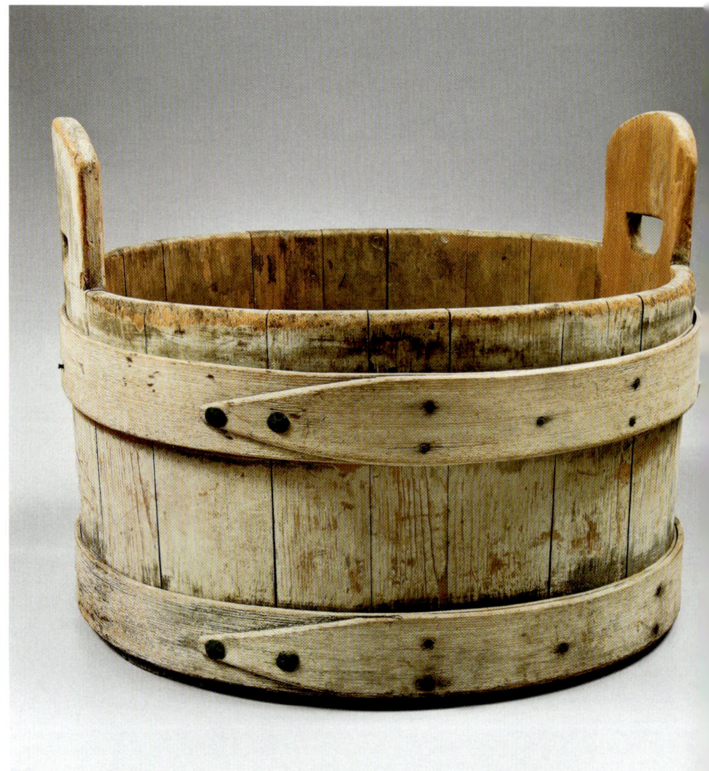

Figure 3.9
Large tub, by Isaiah Hersey, 1820–40. Stamped "IH" on bottom. White pine, oak, iron nails; H 15 ¾, Diam 24 ⅞. Hersey Farm Collection

Figure 3.10
Large tub, by Isaiah Hersey, 1820–40. Stamped "I. Hersey" on bottom. White pine, ash, iron nails; H 15, Diam 19 ¼. Hersey Farm Collection

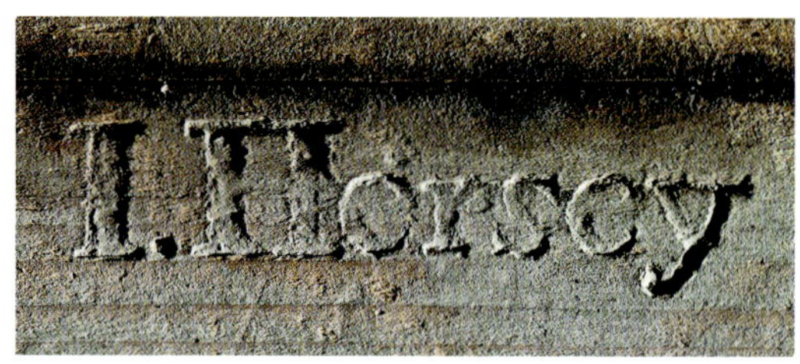

Figure. 3.11
Detail of stamp on tub in fig. 3.10

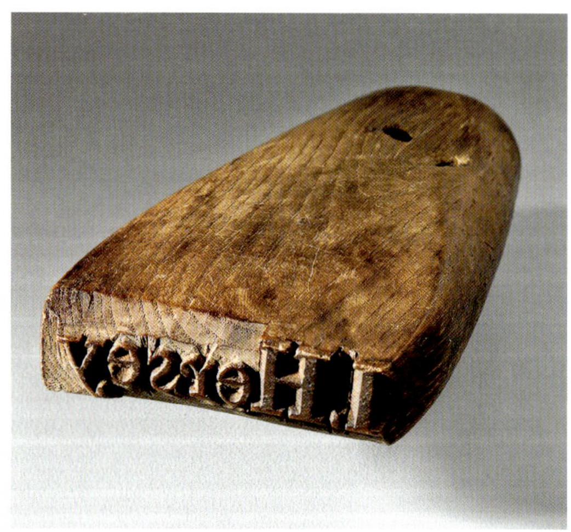

Figure 3.12
Carved stamp used by Isaiah Hersey to mark woodenware, 1820–30. Hickory; L 4 ¼, W 2 ⅛. Hersey Farm Collection

Isaiah had begun to broker his own contracts with Hingham's leading traders and merchant ship captains—men named Whiton, Burr, and Easterbrook. They purchased thousands of pieces of woodenware from the young cooper, often paying with spices, textiles, and other imported household goods. A surviving account book, documenting his activities between 1824 and 1842, indicates that Isaiah produced only buckets and tubs (figs. 3.9–3.12), but in spectacular volume; in 1829 alone he crafted 1,010 of these containers.[20] Such specialization may reflect the demands of the wholesale market or, perhaps, a division of labor within the shop—Reuben was, of course, a noted boxmaker, and he lived to a great old age; he may have still been active in the trade at this time.

Isaiah's accounts also offer a rare glimpse of life on Hersey Farm in the first half of the nineteenth century. In addition to detailing his craft business, entries reveal a complex web of exchange and interdependence among Hersey Street's coopers, all of whom were Hersey relatives by either blood or marriage. They often turned to one another for help in completing orders, supplying parts and materials, and carting "ware" to the wharf—a task Isaiah performed many times for his brethren. These same communal relationships spilled over into the family's agricultural work. Isaiah planted and harvested corn, rye, and potatoes in his father's gardens, mowed and raked the fields, gathered apples, cut and corded wood, slaughtered hogs, transported dung and rockweed to fertilize crops, and carted salt hay to feed cattle. Gathering peat was another typical activity on the farm. This dense swamp compost was cut into bricks, dried, and used to heat buildings; the slow burn was said to have carried with it quite a stench. In addition to agricultural work, the warmer months lent themselves to property maintenance. The Hersey men devoted much time to repairing fences, shingling, and painting each other's buildings.[21]

The Hersey shop (fig. 3.13) was among the buildings that underwent major maintenance, renovations, and even relocation over the years. Built around 1835, the small wooden structure began life at another site, possibly on Reuben's land.[22] By midcentury, however, the family had moved the shop to a hill behind Isaiah's house, where it stands today.[23] A cedar post—later replaced with a granite post—was installed to support its rear wall on the slope. The original weather-beaten door, carved with Isaiah's initials, was boarded up and covered with shingles, and a new entryway was let into the narrow end of the building. The most significant change, however, was the new vantage point. Now situated high up on the property, the shop's windows offered grand, sweeping views of the farm. Isaiah and later generations of Hersey craftsmen could now produce woodenware and wooden toys while keeping a close eye on their crops and cattle. Dozens of *Farmer's Almanacs* found strewn about its interior,

Figure 3.13
Workshop of Isaiah Hersey and Reuben Hersey III, built ca. 1835

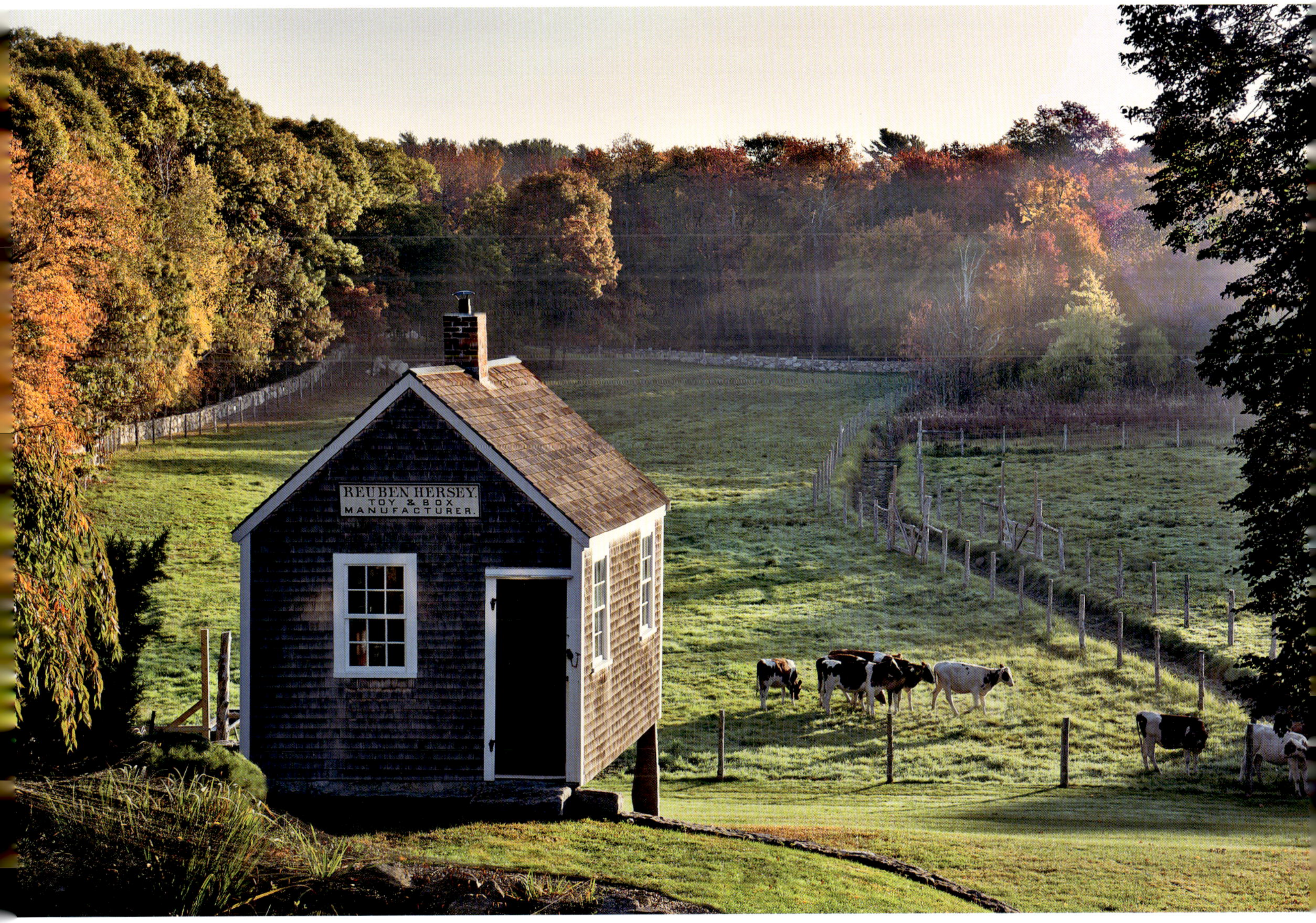

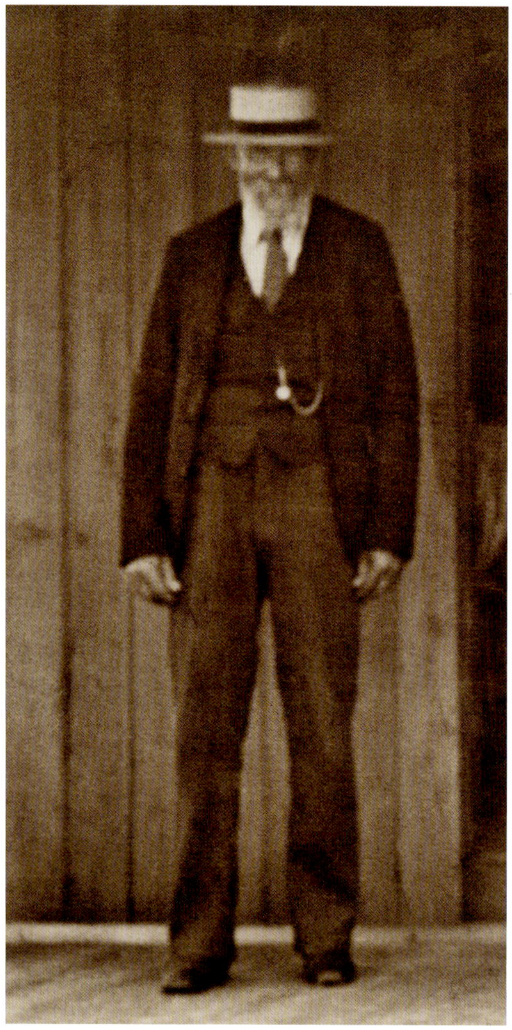

Figure 3.14
Reuben Hersey III, ca. 1895.
Hersey Farm Collection

Opposite:
Figure 3.15
Interior of the Hersey workshop

dating from 1817 to 1899, suggest that the shop functioned as a kind of control tower for farming activities. Those earliest almanacs, inscribed with Reuben Sr.'s name, also hint that the old cooper himself may have occupied the shop for a time.

Reuben's later descendants further populated Hersey Street, establishing fresh iterations of the Hersey creativity and drive. His daughter Polly had two sons who became toymakers; Cotton's son Bela (1817–79) was a fine boxmaker.[24] Samuel French Hersey's son, also named Samuel (1820–1909), became one of the region's leading toymakers in the 1850s and 1860s. Like his father, Samuel had a knack for horticulture. He cultivated the Hersey strawberry, a new breed that became well-known throughout Massachusetts.[25] Samuel French Hersey's second son, Edmund (1823–1909), operated a thriving mechanized box factory across the street. During the 1870s it was the largest maker of boxes in the country, producing 15,000 units a day; not surprisingly, strawberry and other fruit boxes were among its chief goods. Edmund sold the business in 1871 to concentrate on agricultural studies. He served as the first instructor of farming and superintendent of Bussey Farm for Harvard's Bussey Institute of Agriculture and Horticulture.[26]

Isaiah's only child, Reuben III (1837–1914) (fig. 3.14), succeeded his father as owner and operator of Hersey Farm and shop.[27] The shop's contents, still largely intact from his tenure, offer an unprecedented look at the world of a nineteenth-century toymaker (fig. 3.15). Reuben III probably apprenticed with his father, but by age seventeen he had already begun to craft a line of miniatures, including toy mirrors and fireplace bellows painted in bright reds and yellows (figs. 3.16–3.21).[28] He often stamped these goods with his initials. The same markings are found on countless other items in the shop—tools, patterns, and molds—even a painter's box with old, paint-caked brushes (figs. 3.22–3.25). The drawer beneath the built-in workbench is also struck with the initials of both Reuben III and his father, Isaiah. The initials of his cousin Samuel and his uncle Cotton appear on small tool chests and hand tools in the shop, raising interesting questions about how the space was used. Did the men work collaboratively from time to time, or share each other's shops and tools? Or, as the last craftsman on Hersey Street and owner of the old homestead, was Reuben III the obvious inheritor of his relatives' coopering and toymaking tools? These questions will probably never be answered, but they certainly suggest some level of interdependence, or at least cooperation, among the Hersey kinsmen.

Although small and seemingly primitive, the Hersey shop offered reasonable comforts and little wasted space. Wood for the cast-iron stove was stacked under the building and

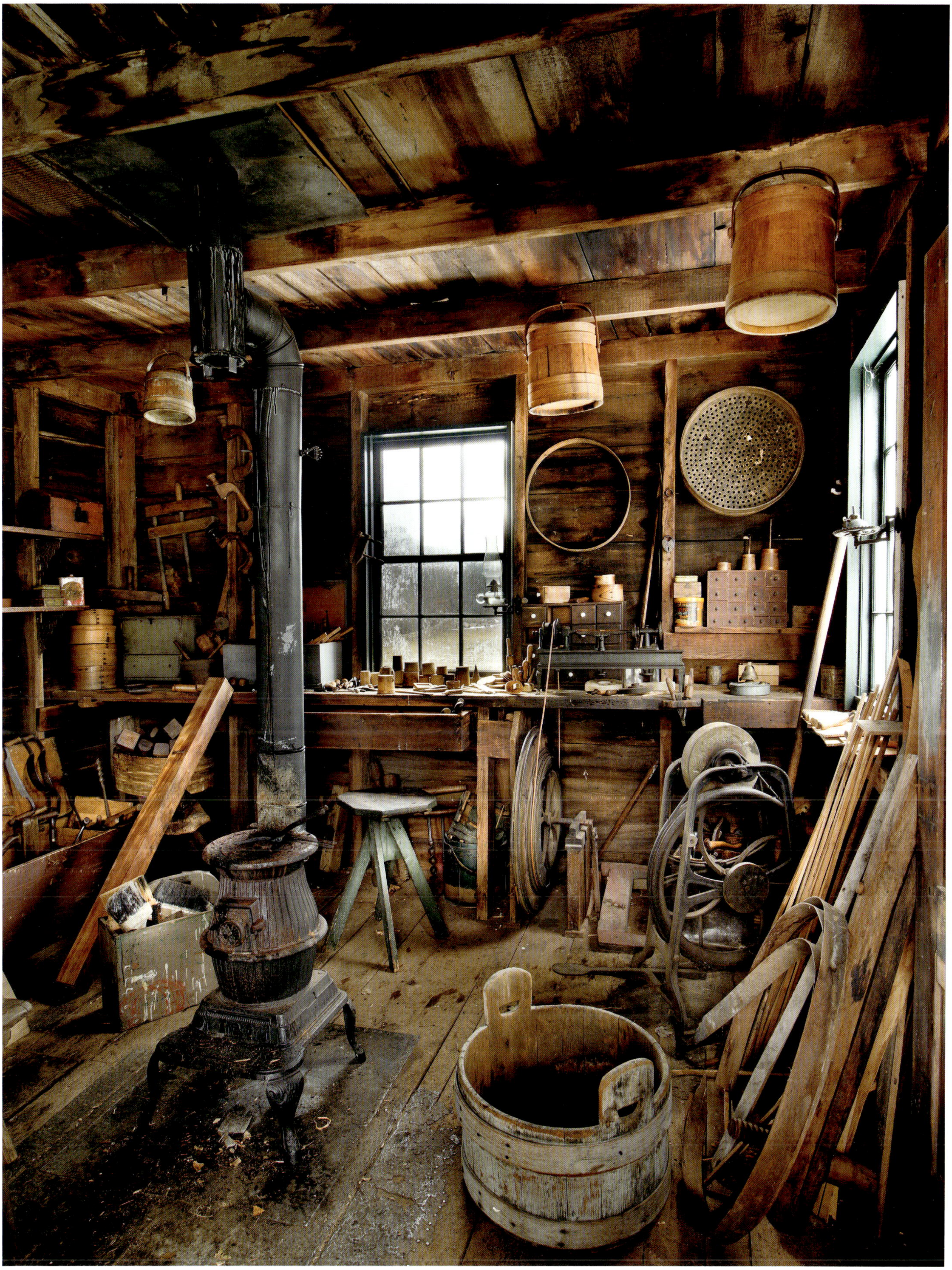

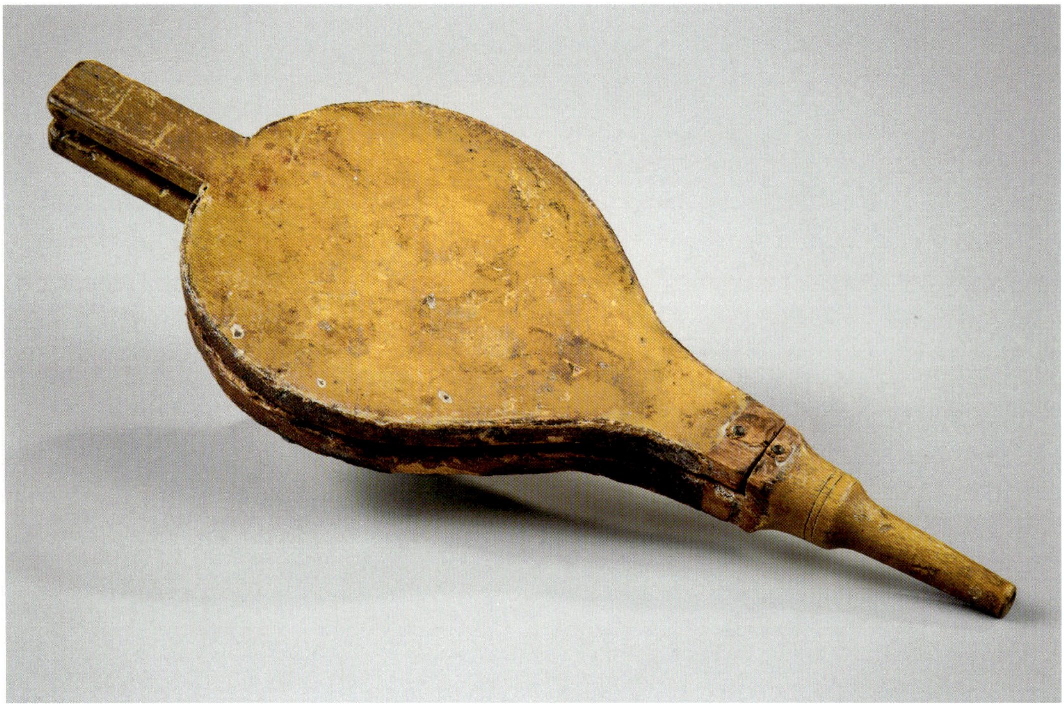

Figure 3.16
Miniature fireplace bellows, by Reuben Hersey III, 1850–70. Stamped "RH" on interior. White pine, leather; L 7 ⅝, W 2 ⅝. Hersey Farm Collection

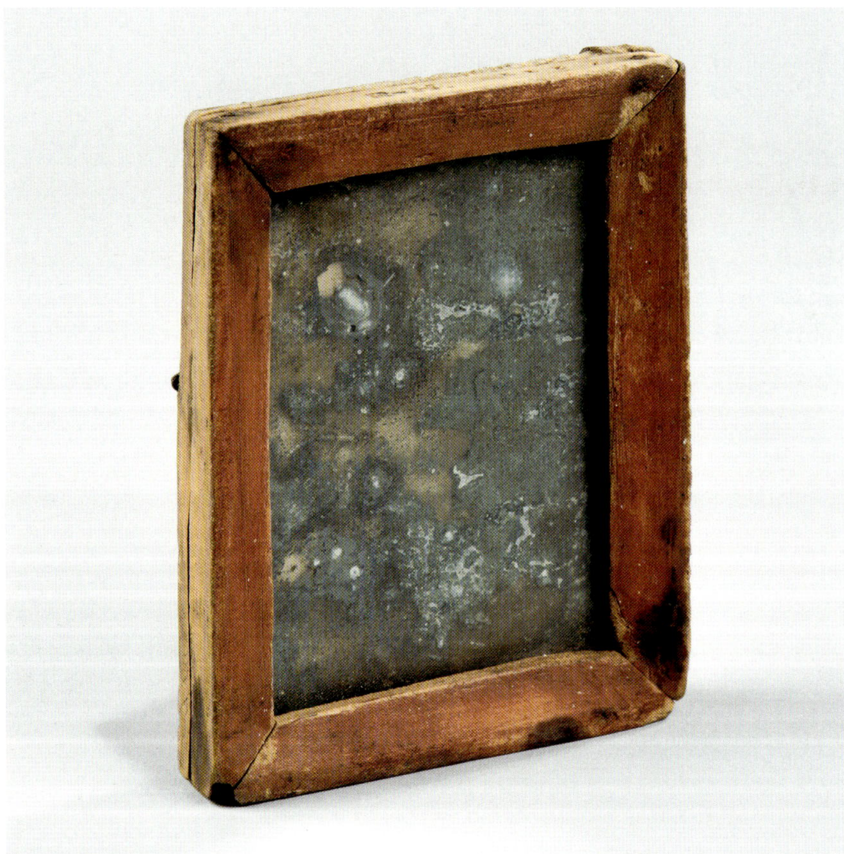

Figure 3.17
Miniature mirror, by Reuben Hersey III, 1854. Stamped "RH" and "1854" on side. White pine, mirrored glass, iron nails; H 4 ¼, W 3 ¼. Hersey Farm Collection

Figure 3.18
Detail of stamps on mirror in fig. 3.17

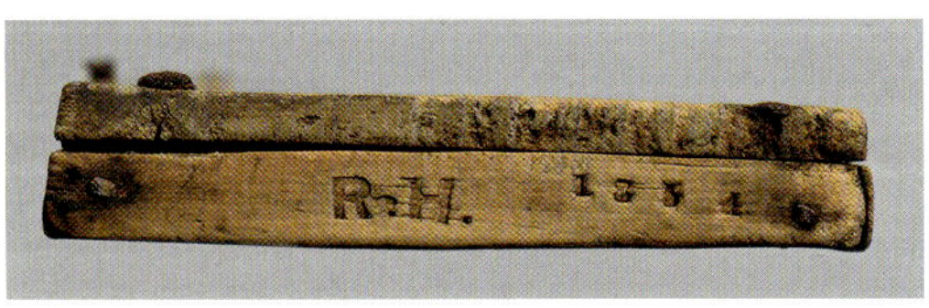

could be brought up through a trapdoor in the floor; additional materials were stored in a small peaked garret, accessed by a ladder built into the wall.[29] This, in fact, is where Reuben III's original painted sign—announcing him as a "Toy & Box Manufacturer" to all who passed by—was discovered nearly a century later (fig. 3.26). Even the back of the closed-in original door was utilized. Reuben III repurposed the paneled boards as a surface on which to brush sample splotches of green and blue paints, the very colors commonly found

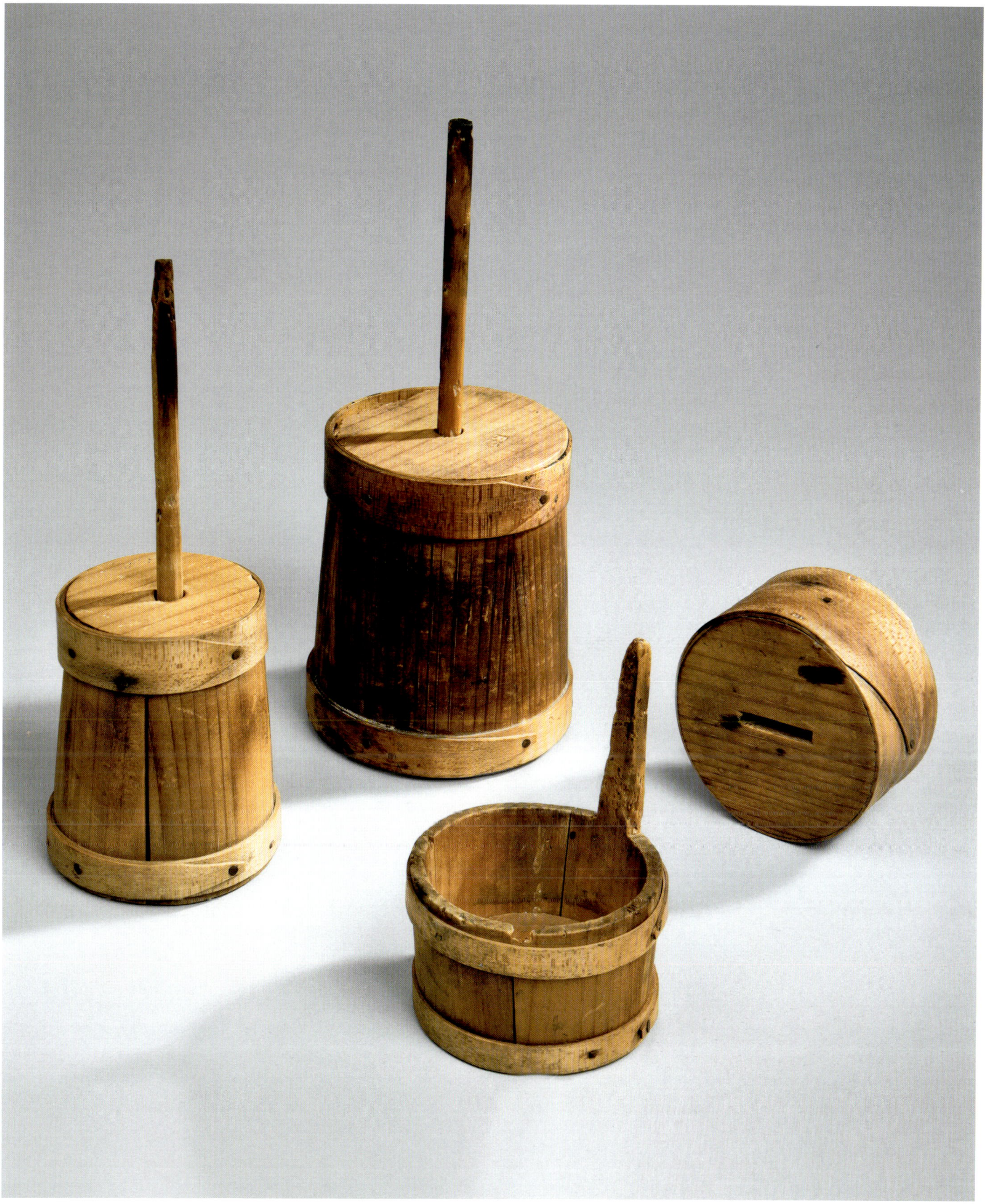

Figure 3.19

Miniature churns, piggin, and coin bank, by Reuben Hersey III, 1850–70. Stamped "RH" on bottoms. White pine, maple, copper tacks; large churn: H 3 ½, Diam 3; small churn: H 2 ⅞, Diam 2 ⅜; piggin: H 3 ⅜, Diam 2 ⅛; bank: H 1 ½, Diam 2. Hersey Farm Collection

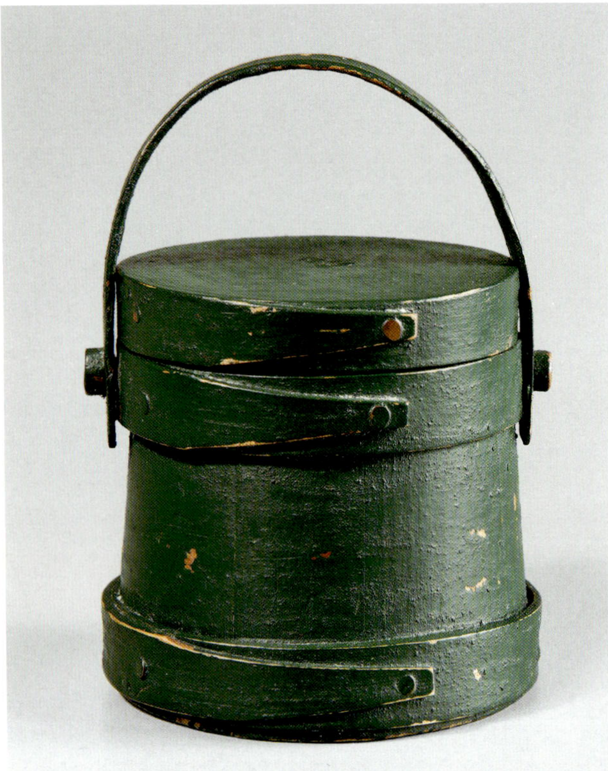

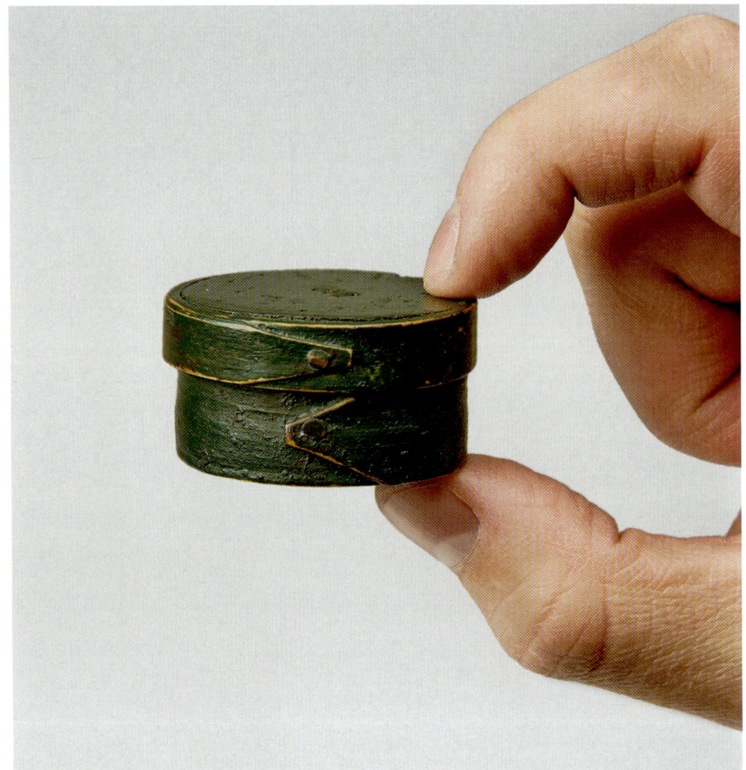

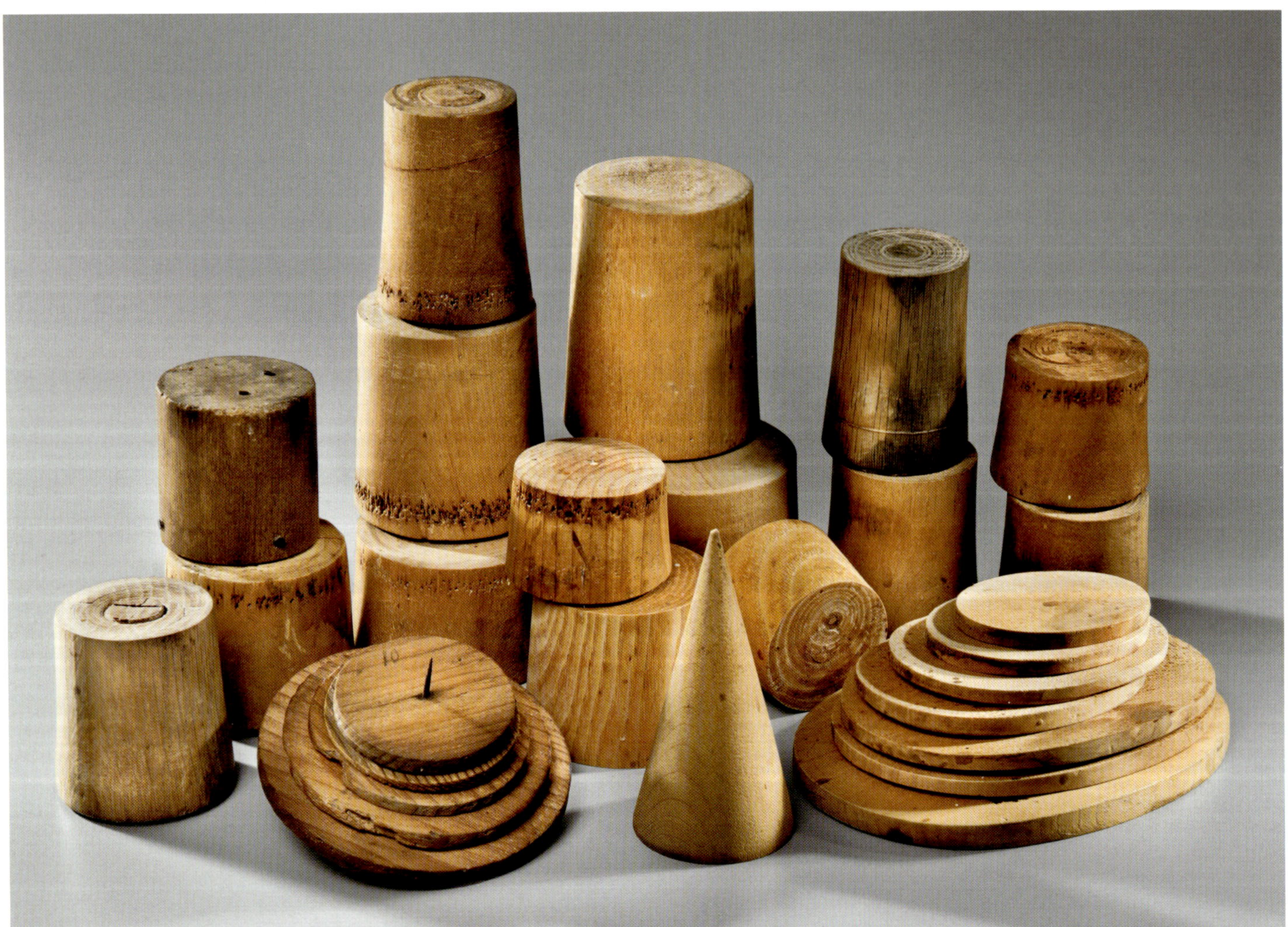

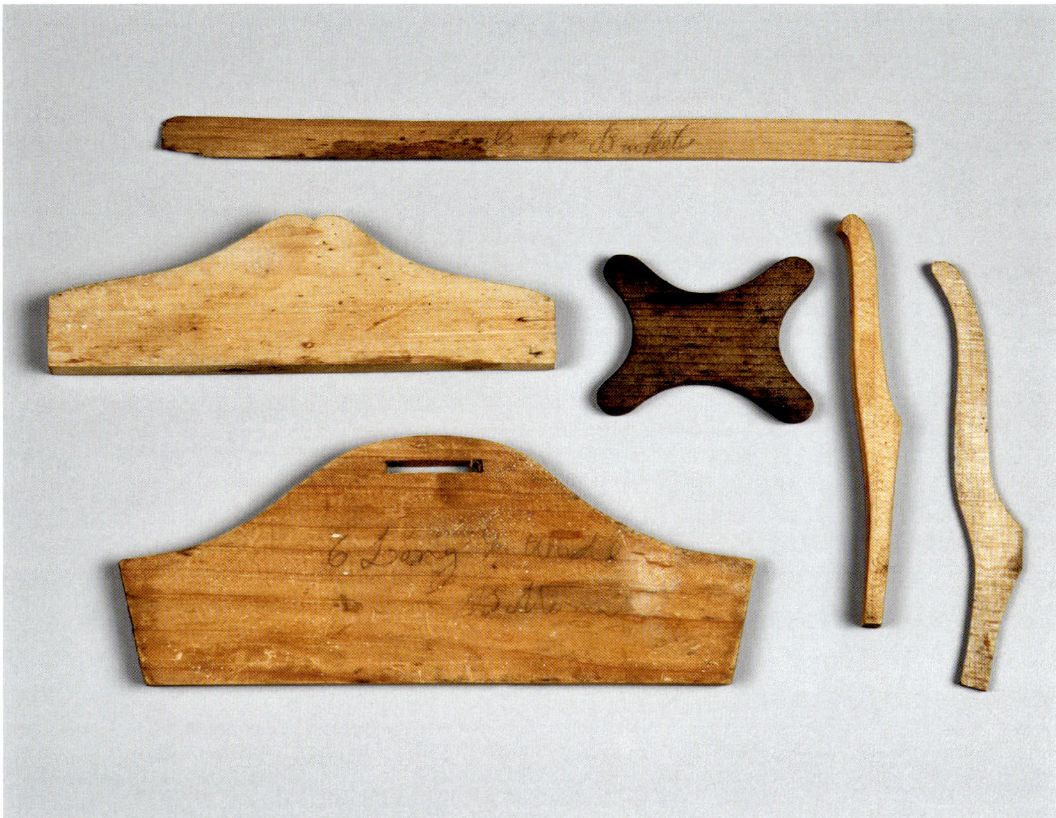

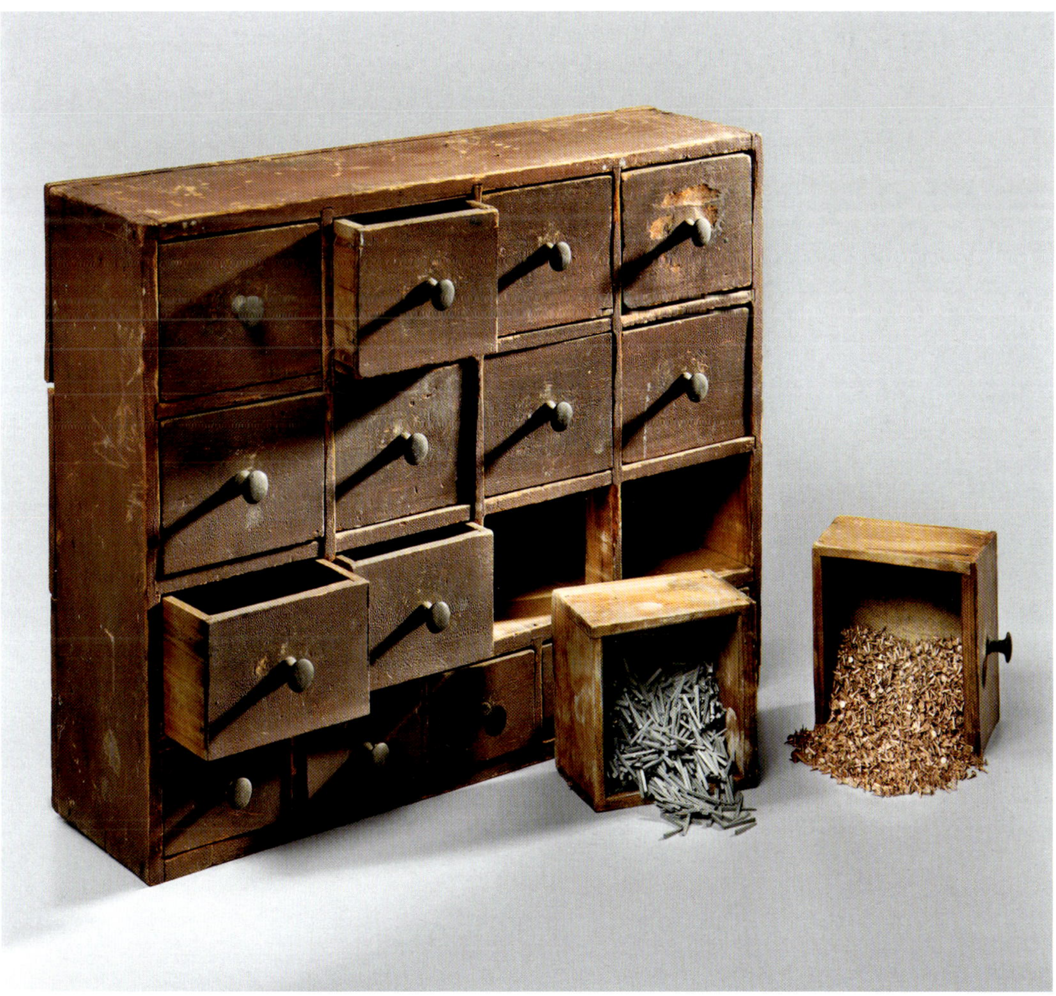

Opposite:

Figure 3.20

Miniature bucket, by Reuben Hersey III, 1859. Stamped "RH" and "1859" on lid. White pine, maple, copper tacks. H 2 3⁄8, Diam 2. Courtesy of the Hersey Family

Figure 3.21

Miniature box, by Reuben Hersey III, 1850–70. Stamped "RH" on lid. White pine, maple, copper tacks; H 13⁄16, Diam 1 5⁄8. Courtesy of the Hersey family

Figure 3.22

Patterns and molds to make boxes and toy woodenware, 1850–70. Stamped "RH" on tops and bottoms of patterns. White pine, maple. Hersey Farm Collection

This page:

Figure 3.23

Patterns for miniature chair legs, knife box handles, dasher bottom, and a bucket handle, 1850–70. "Balls for Buckets" inscribed on pattern for bucket handle. White pine, maple. Hersey Farm Collection

Figure 3.24

Hardware chest with unused iron nails and copper tacks to make boxes and toy woodenware, 1820–70. White pine; H 9 1⁄8, W 12 3⁄4, D 3 1⁄4. Hersey Farm Collection

on Hingham woodenware and toys. Cast-iron brackets attached to the wall studs held glass oil lamps that allowed him to work after dark at his bench or on the foot-powered treadle lathe, turning small dashers for miniature butter churns.

The Hersey shop, however, was far more than a mere place of industry, commerce, and agricultural oversight. Besides nighttime work, the shop's oil lamps also invited long evenings, quiet or otherwise, away from the women and children—tinkering, smoking, drinking, and enjoying the society of men. Evidence from the shop suggests that it was also a powerful locus of male privacy and fraternity. Liquor bottles were found there in numbers, as were several of the mildly pornographic *cartes de visite* of scantily clad French women popular in the 1860s. A photograph of President Abraham Lincoln and his family was prominently displayed in the shop, a focus for political discussion as well as Hersey family and community pride: the Hingham Lincolns were, in fact, related to the president, and Herseys and Lincolns had intermarried over the generations.[30]

The shop's fittings and distinctively masculine "decor" support the notion of sacrosanct male space; family history reinforces it, though more obliquely. For the last ten years of his life, Reuben III's uncle Samuel, by then blind, apparently spent long hours in his own shop, smoking his pipe and dozing.[31] By all accounts, the Hersey men loved their little shops, and family members respected the boundaries of these male sanctums. Later

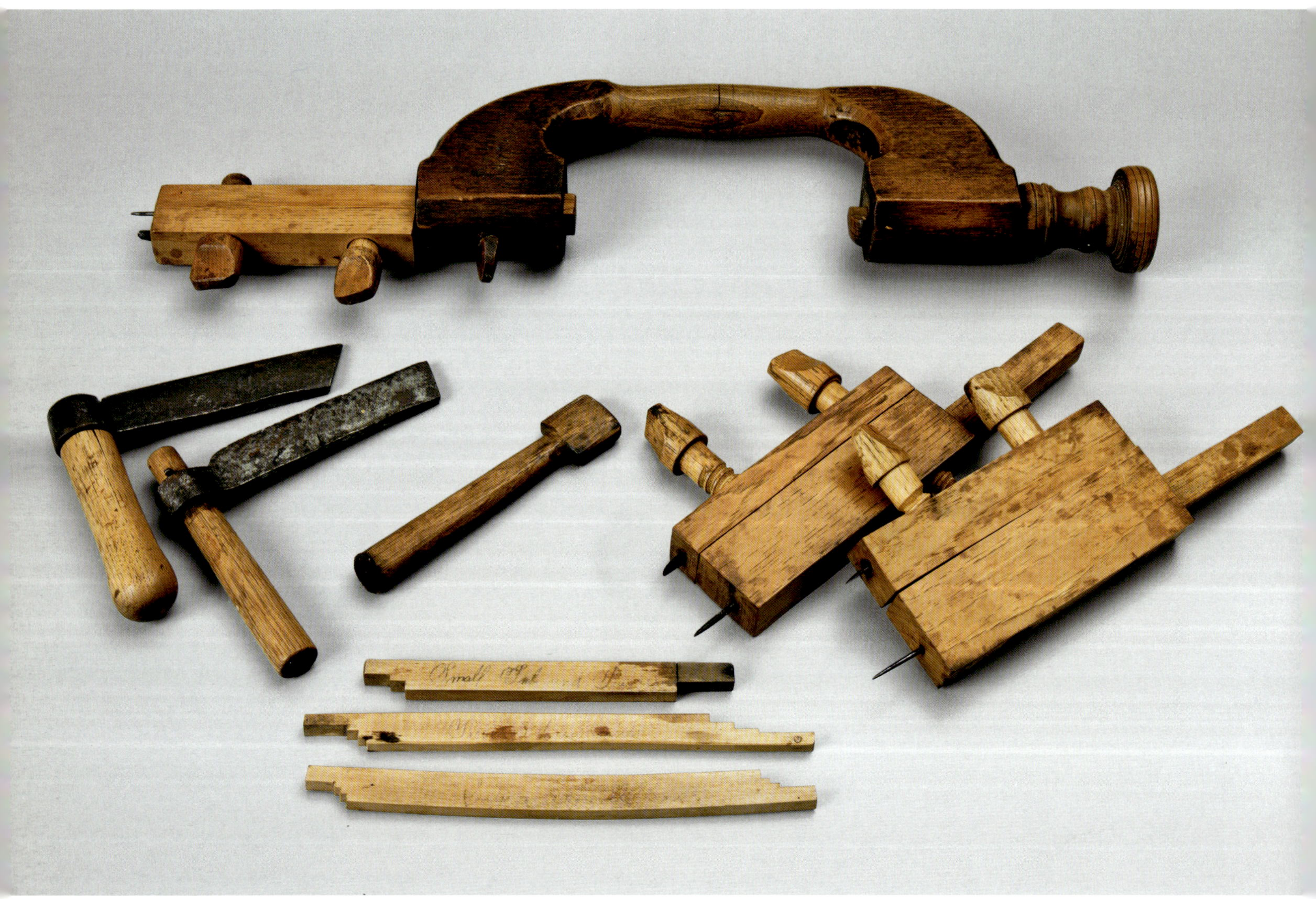

Figure 3.25
Small coopering and toymaking tools, including froes for preparing small staves, a driver for forcing small hoops into place, measuring sticks, and a brace with custom circular cutters to shape the thin tops and bottoms of round boxes and toy woodenware, 1820–70. Carved "RH" and "1826" on one side of the brace, "Small Tub and Pail" and "Sieves" inscribed on sides of measuring sticks. Hersey Farm Collection

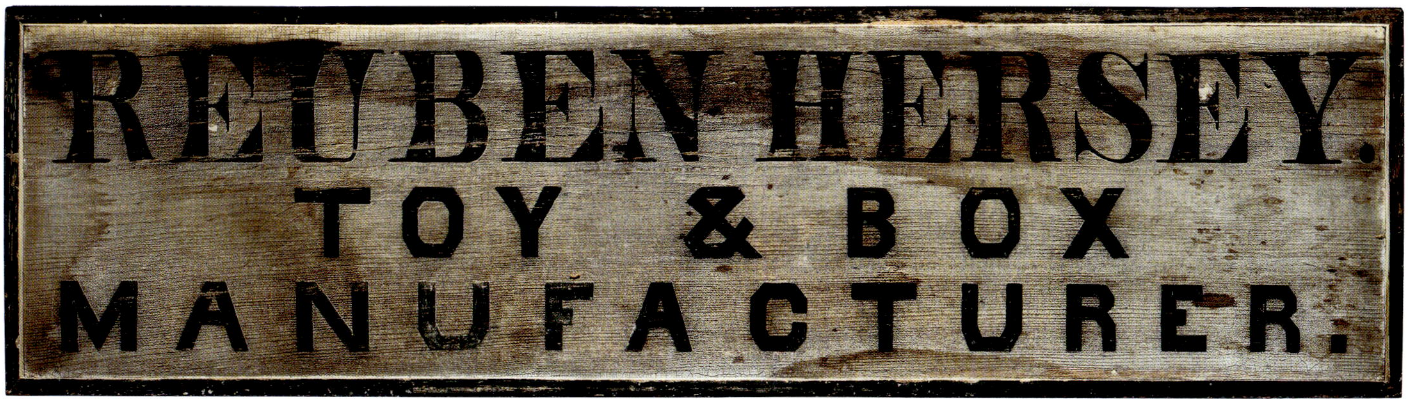

Figure 3.26
Reuben Hersey III shop sign, 1855–75. H 14 ¼, W 60 ¼.
Hersey Farm Collection

generations recall, as children, being warned away from the padlocked, overgrown shed out back; it appears that they continued to honor the Hersey shop's mysteries, even when the reason for such awe had long since been lost.

After 1900 Reuben III no longer listed himself as toy and box manufacturer, an indication that at age sixty-three he had probably retired from the trade.[32] Crafting toys and running a small farm undoubtedly provided him with a modest living, but while still a young man he had begun lending cash to neighbors, holding notes on their homes and charging modest interest rates—what we would recognize today as a home mortgage. Over the years he held mortgage notes on more than two hundred local homes, owned multiple properties in Hingham and surrounding towns, invested in mining and manufacturing stocks, and had tens of thousands of dollars socked away in a dozen local and regional banks. At the time of his death, Reuben Hersey III, the last of the old-time toymakers on Hersey Street, had amassed a personal fortune of over $130,000.[33] It was Reuben's hard work and shrewd investments that kept the Hersey Farm—and many other properties on the street—in the family into the twenty-first century.

NOTES

1. Hersey Farm was established around 1776 by Reuben Hersey (1757–1845); however, circumstantial evidence indicates that at least part of Reuben's land came from his father and grandfather, who settled on Hersey Street, then Austin's Lane, in the 1730s and 1750s. In 1833 Reuben sold a small piece of land to his youngest son, Isaiah (1803–68), who built a house next door. He sold Isaiah another 1.5 acres in 1844. Following Reuben's death in 1845, the remainder of Hersey Farm was divided among his four children and five of his grandchildren—the children of his deceased daughter, Polly. Reuben's sons Cotton and Samuel French each received about 3.75 acres; his grandchildren received a total of about 3.75 acres; his daughter Eliza Lane Hersey received a half acre and Reuben's house; and his son Isaiah received about 3.75 acres with the barn and corn barn; Plymouth Land Deeds, 211:213, 212:131, 225:143. Following the death of Eliza Lane Hersey, Reuben's half acre and house thereon were given to Isaiah; Eliza Lane Hersey, will, recorded July 1, 1869, Plymouth County Probate Records, 140:262. Following his death in 1868, Isaiah Hersey's homestead went to his wife and only son, Reuben III (1837–1914); Isaiah Hersey, will, recorded October 11, 1868, Plymouth County Probate Records, 106:220. The homestead with both houses descended to his oldest son, Reuben Leavitt Hersey (1885–1938); to his son Reuben Woodrow Hersey (1915–1995); to his sons Peter Woodrow Hersey (b. 1948) and Paul Edmund Hersey (b. 1949); to Peter W. Hersey.

2. George Lincoln et al., *History of the Town of Hingham, Massachusetts, Genealogical* (Hingham: By the town, 1893), 2:309. For a partial record of Reuben Hersey's military service, see Office of the Secretary of State, *Massachusetts Soldiers and Sailors of the Revolutionary War,* vol. 7 (Boston: Wright & Potter, 1900), 875.

3. For more information on the 1830 map of Hingham, see Joseph G. Garver, *Surveying the Shore: Historic Maps of Coastal Massachusetts, 1600–1930* (Beverly, Mass.: Commonwealth Editions, 2006), 72–73.

4. According to the only surviving child of Reuben L. Hersey, in the 1920s and 1930s the shop was opened only on rare occasions to fetch tools. The garret of the shop appears to have survived completely undisturbed. Stanley Lincoln Hersey (b. 1925), interview by the author held at Hersey Farm, June 22, 2011.

5. Cushing Homestead on East Street in Hingham is another remarkable survival among local historic sites. It has remained in the Cushing family since 1678 and retains its original barn, corn crib, and forge, but there is no example of the small workshops that were once common throughout Hingham.

6. Lincoln et al., *History of Hingham, Genealogical,* 2:301.

7. Ibid., 3:289. Daniel Hersey's grandfather William Hersey settled in Hingham in 1635; for a complete genealogy, see Stephen E. Hersey, *The Hersey Family: Tracing Descendants of William Hersey of Hingham, Massachusetts, 1635–1998,* 7th ed. (Westford, Mass.: Stephen E. Hersey, 1998). Several members of the Whiton family of coopers settled near the Herseys on parts of South Street, Main Street, and Austin's Lane. Matthew's brothers, James (1651–1725) and Thomas (1662–1708), were also coopers but lived farther away, at Liberty Plain. Curiously, Thomas Whiton's father-in-law, Samuel May of Roxbury, was the grandfather of Daniel Hersey's wife. Daniel's son, Jonathan (1713–60), also a cooper, married a Whiton.

8. In 1866 the Hingham historian George Lincoln wrote, "Daniel Hersey built the first house on Hersey Street in 1730 which is still standing. Daniel's son, Isaiah, built his house on the opposite side of Hersey Street beside the cedar swamp. Nearly opposite is Isaiah Hersey's farm and cooper shop. The house was built by Daniel's grandson, Reuben Hersey, in 1776. The house stood partly boarded while Reuben was away in the war with England. Further along is Cotton Hersey. The lower part of Hersey Street from South Street to near its junction with Elm Street was, in the memory of several now living, referred to as Austin's Lane because it was over the lot granted to Jonas Austin. The upper part of Hersey Street south of Elm Street was referred to as The Way to Hobart's or The Way to Hobart's Corner over cedar swamp bridge"; see John P. Richardson research notes, which contain Lincoln's original handwritten notes, Hersey Family Collection. Lincoln's original notes on Hingham houses can be found at the Hingham Public Library and the Hingham Historical Society.

9. For a history of the settlement of Hersey Street, see John P. Richardson research notes, Hersey Family Collection. For additional information and early photographs of houses on Hersey Street, see the Julian Loring's notes on Hingham houses, Hingham Historical Society, Hingham.

10. For a list of Hersey coopers and toymakers see appendices C and D. The Corthell family of coopers, who intermarried with the Herseys, also lived and worked on Hersey Street.

11. Lincoln et al., *History of Hingham, Genealogical,* 2:309. For information on Reuben Hersey's house, see note 8. Reuben's brother, Zadock (1753–1850), may have played a role in the settlement of Hersey Farm. Tax records for 1779 list Zadock as the owner of a house on Hersey Street, whereas Reuben is not listed at all. According to the 1788 tax records—the next year of records to survive—Zadock and Reuben each owned half a house and shop on Hersey Street; Zadock owned no land and Reuben owned three acres; see Hingham Valuations, John P. Richardson Manuscript Collection, Hingham Historical Society, Hingham. Zadock moved to Dennysville, Maine, with his family in 1788; see Benjamin Richard Jones, "Memoir of Zadock Hersey of Dennysville, Maine," *Bangor Historical Magazine* 6, no. 8 (February 1889): 160.

12. John P. Richardson research notes, Hersey Family Collection. See also Julian Loring's notes on Hingham houses, Hingham Historical Society, Hingham.

13. George Lincoln et al., *History of the Town of Hingham, Massachusetts, Historical* (Hingham: By the town, 1893), 1.1:291–92.

14. State Board of Agriculture, *Twenty-sixth Annual Report of the Secretary of the Massachusetts Board of Agriculture for 1878* (Boston: Rand, Avery, & Co., 1879), 11.

15. "The Box and Toy Business," *Hingham Journal,* August 11, 1876, 1.

16. Lincoln et al., *History of Hingham, Genealogical,* 2:316. For additional information on Cotton Hersey, see chap. 2 and cat. 11. Samuel French Hersey obituary, *Hingham Journal,* April 13, 1876, 2.

17. Reuben Hersey Jr. accounts, December 1824 to January 1825, in Isaiah Hersey, account book, 1824–42, 12–13, Hersey Farm Collection.

18. Reuben Hersey's will requested that his brothers and nephew "keep the books, manuscripts, and maps in said office so that the citizens of Hingham may have access to the same"; Reuben Hersey Jr., will, recorded March 19, 1837, Plymouth County Probate Records, 86:164. Hersey's transcriptions of Hingham vital records, genealogies from the diary of David Hersey, and the diary of Daniel Cushing (1619–1700) are now at the New England Historic Genealogical Society, Boston. Hersey's copy of a 1672 map of Hingham is at the Cohasset Historical Society.

19. Lincoln et al., *History of Hingham, Genealogical,* 2:303.

20. Isaiah Hersey, account book, 1824–42, Hersey Farm Collection.

21. Ibid. See also Reuben Hersey Jr., journal, 1841–43, Hersey Farm Collection.

22. The shop exhibits an unusual and early combination of timber framing and balloon framing.

23. Isaiah Hersey built a modest Cape next to his father's house on Hersey Street. Construction on the building probably began in 1833, when Reuben sold him a small piece of land carved out from the homestead; Plymouth County Deeds, 211:213. It is likely that Isaiah built the house in preparation for his marriage to Polly Leavitt on January 12, 1834. In the 1930s during the depths of the Depression, Reuben Leavitt Hersey remodeled the home into a two-story Colonial Revival.

24. Polly's sons Joseph Jr. and Edward Lewis Hersey were toymakers; see appendix D. For additional information on Edward Lewis Hersey, see cat. 13. For Bela Hersey, see appendix C.

25. The Hersey Strawberry took the premium at the 1890 fair of the Hingham Agricultural and Horticultural Society. Three years later it was awarded first prize in the Strawberry Exhibition of the Massachusetts Horticultural Society; see Vaughn's 1894 *Gardening Illustrated* (Chicago: Vaughn's Seed Store, 1894): xv. For an 1898 advertisement for the Hersey strawberry, see John P. Richardson research notes, Hersey Farm Collection. For additional information on Samuel Hersey, see chap. 2 and cat. 12.

26. Elroy S. Thompson, *History of Plymouth, Norfolk, and Barnstable Counties, Massachusetts,* vol. 1 (New York: Lewis Historical Publishing Co., 1928), 418; "The Box and Toy Business." Seymour & Cain, account book, 1871–72, Hingham Bicentennial Microfilm Collection, Hingham Public Library. Bussey Farm was part of what is now the Arnold Arboretum in Jamaica Plain, Mass.

27. In this book Reuben Hersey (1837–1914) has been designated Reuben Hersey III to avoid confusing him with his grandfather and uncle of the same name. Hersey, *The Hersey Family,* 64.

28. A miniature toy mirror discovered in the shop is stamped "RH" and "1854."

29. The space underneath the shop was still used to store and stack cords of wood through the mid-twentieth century; see Peter Woodrow Hersey (b. 1948), interview by the author, Hersey Farm, August 15, 2012. The stove currently in the shop is one of several that were discovered inside the building. It was refurbished by Tuck Wadleigh and reinstalled in 2008.

30. In 1848 the Hingham historian Solomon Lincoln corresponded with Abraham Lincoln, then a congressman, to gather genealogical information about his family. This was how Lincoln learned of his ancestral roots in Hingham; see Roy P. Basler, comp., *The Collected Works of Abraham Lincoln,* vol. 1 (New Brunswick, N.J.: Rutgers University Press, 1953), 456, 460. Reuben Hersey Sr.'s second wife was Celia Lincoln; see Hersey, *The Hersey Family,* 20.

31. Samuel French Hersey obituary, *Hingham Journal,* April 13, 1876, 2.

32. Reuben Hersey III consistently listed himself as a toy and box manufacturer in town and regional directories from 1867 to 1880. Afterward he lists himself as a farmer, though the 1900 federal census identifies him, once more, as a toymaker.

33. For Reuben Hersey III's extensive property holdings, see Plymouth County Registry of Deeds. See also Reuben Hersey, inventory, filed January 19, 1915, Plymouth County Probate Records, 254:198, 160:289.

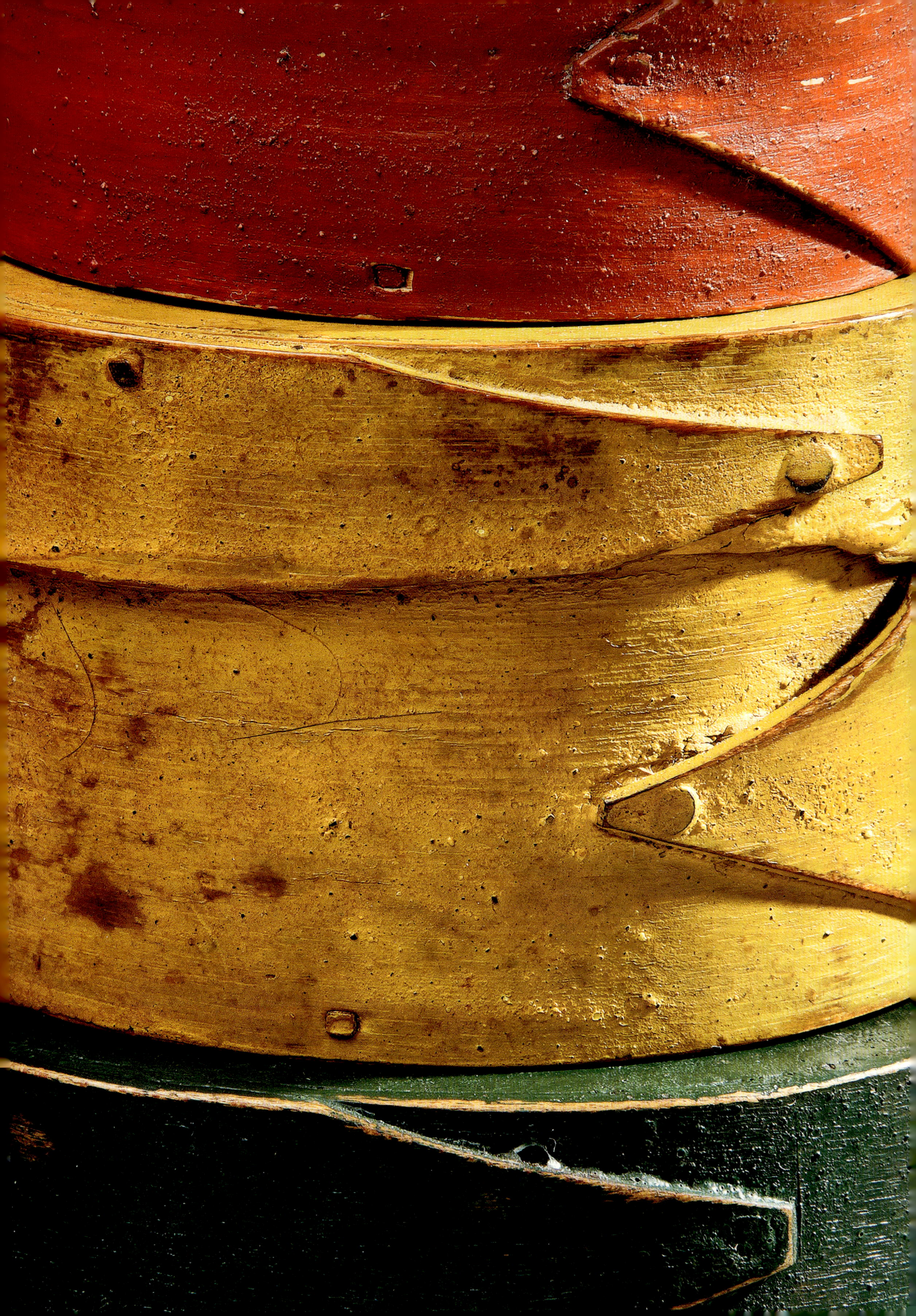

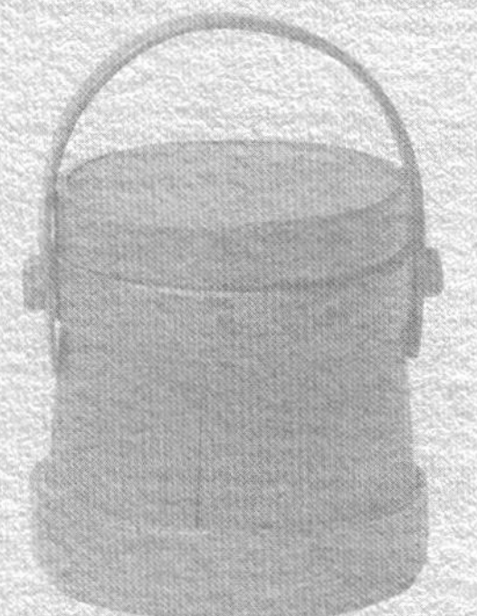

CATALOGUE

THIS CATALOGUE BRINGS TOGETHER DOCUMENTED EXAMPLES OF WOODENware and wooden toys made in Hingham, Massachusetts, from the 1780s through the 1940s. Although a helpful reference tool, it is not intended as a guide to all known works; that is beyond the scope of this study. Rather, it features objects that best demonstrate the breadth of goods manufactured in Hingham. Entries are organized by craftsman and arranged for the most part chronologically. They include contextual histories of use and ownership, construction techniques, materials, identifying marks or labels, and overall dimensions; note, however, that buckets are measured with handles in the down position and churns without dashers. Whenever possible objects are described using the standardized language of Hingham coopers and toymakers; for example, bucket is used in place of firkin, a collector's term that became popular in the twentieth century. Items produced before the Revolution either no longer survive or lack sufficient documentation, which accounts for their absence here. Similarly, several forms mentioned in written and printed sources, such as fish kits, rattlers, and whirligigs, have not yet come to light. Additional information for many of the craftsmen can be found in the preceding essays. For a list of known Hingham coopers and toymakers, see appendices C and D.

1
PETER HERSEY (1757–1844)

IN 1784 MADAM SARAH DERBY FOUNDED THE PRESTIGIOUS DERBY SCHOOL IN HINGham, one of the first coeducational institutions in the country.[1] Its young ladies received a fine ornamental education to both grace a drawing room and administer a large household; its young men were destined for skilled trades and professions or groomed to attend Harvard College. Decades later, when recounting the school's early development, one historian noted the peculiar manner in which the endowment was delivered: "When the money for the . . . academy was brought from Salem by Nathaniel Lincoln, nephew of the donor, it filled a 'Hingham Bucket' standing on the floor of the chaise between the feet of Lincoln and his wife."[2] Whether this anecdote is based in fact or fantasy, the scenario is certainly believable. By the mid-eighteenth century buckets made in Hingham were used throughout the colonies; their production so dominated local commerce and culture that Hingham became known as Bucket Town.[3]

Several families stood at the center of this vigorous industry—notably the Whitons, Burrs, Spragues, and Towers—but none was as large or influential as the Herseys.[4] In 1730 the cooper Daniel Hersey (1682–1766) built a house on Austin's Lane, later known as Hersey Street.[5] The property offered a modest parcel of farmland and access to nearby swamps teeming with white cedar, a durable wood coveted by coopers.[6] Within a few generations the homes and workshops of Daniel's progeny lined both sides of the mile-long road. Between 1730 and 1850 more than forty members of the Hersey family manufactured buckets and other utilitarian woodenware in Hingham.[7]

Peter Hersey, Daniel's nephew, also resided on Hersey Street. He was among the first coopers in Hingham to brand his work, a practice that took hold following the Revolution—probably to distinguish the products of dozens of shops working in the same tradition. Some time in the late eighteenth or early nineteenth century Peter crafted this exceptionally large bucket (pl. 1.1), a form commonly described in period records as a "nailed pail." Hersey prominently stamped his full name on the lid and painted the exterior bright blue—a popular color for woodenware that has since faded to gray. The bucket bears all the hallmarks of hand-coopering. It is made of white pine staves pegged invisibly at the tops, butt-joined, and set within oak hoops. Each hoop has a distinctive triangular tip—a tail—held in place with rose-head copper nails. Unlike iron hardware, copper fasteners withstood rust and rot and were better suited for the storage of liquids or use in damp environments. The top of the bucket, or head, is made of white pine and secured in place with a series of tiny wooden pegs visible from the exterior. The bottom of the bucket, also called a head, is set into the croze, a groove cut into the staves. Larger white pine cotter pins, or pail buttons, attach the oak swing handle to the body of the bucket.

Although buckets were Hingham's principal manufactured export, local coopers used the same basic construction techniques to produce a variety of oval and circular containers—from dippers and boxes to washtubs and butter churns (pl. 1.2). Perhaps the most unusual item made by Peter Hersey is a pair of fire buckets (pl. 1.3) crafted, probably as a special order, for the Mordecai Lincoln House on South Main Street in Cohasset, formerly Hingham.[8] The underside of each is impressed with his name and the year 1825. Members of local fire societies were required to keep a pair of fire buckets within easy reach, as fires were an imminent danger, and dousing them often called for the efforts of an entire community.[9] Though examples are typically made of stitched leather, Hersey's coopered versions—painted green with red interiors—must have been just as serviceable. They are the only wooden fire buckets known to have been made in Hingham.

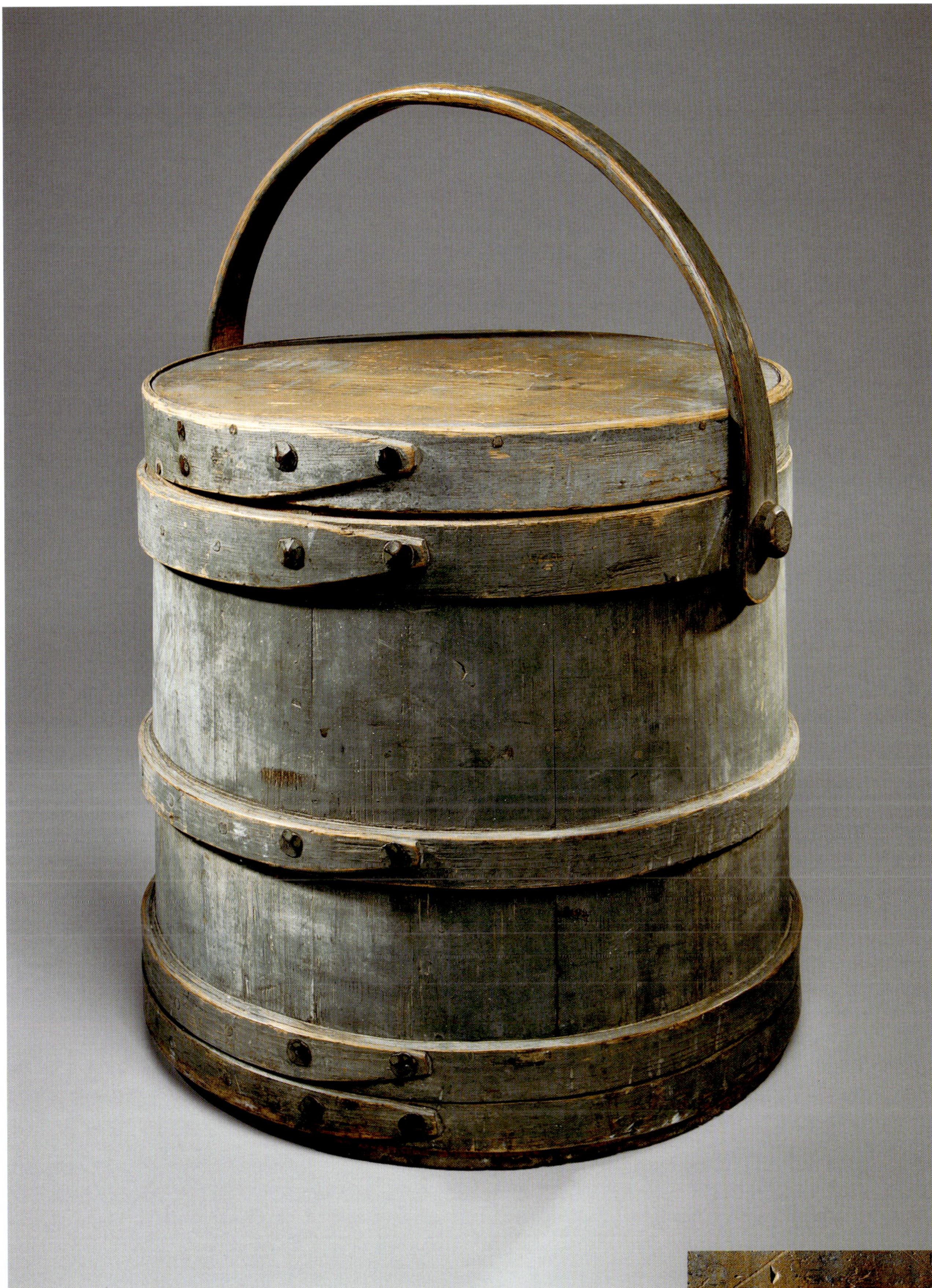

Plate 1.1

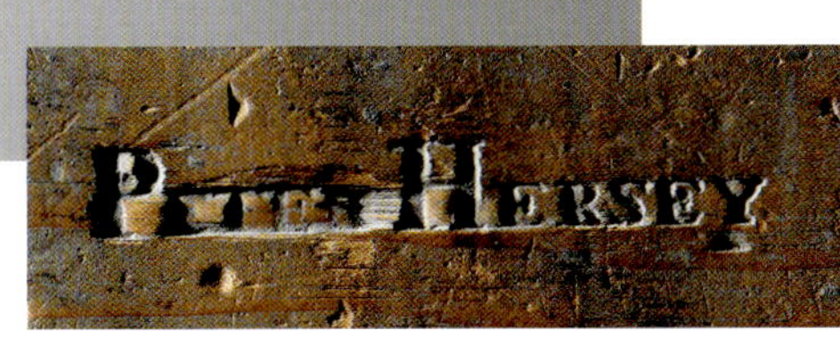

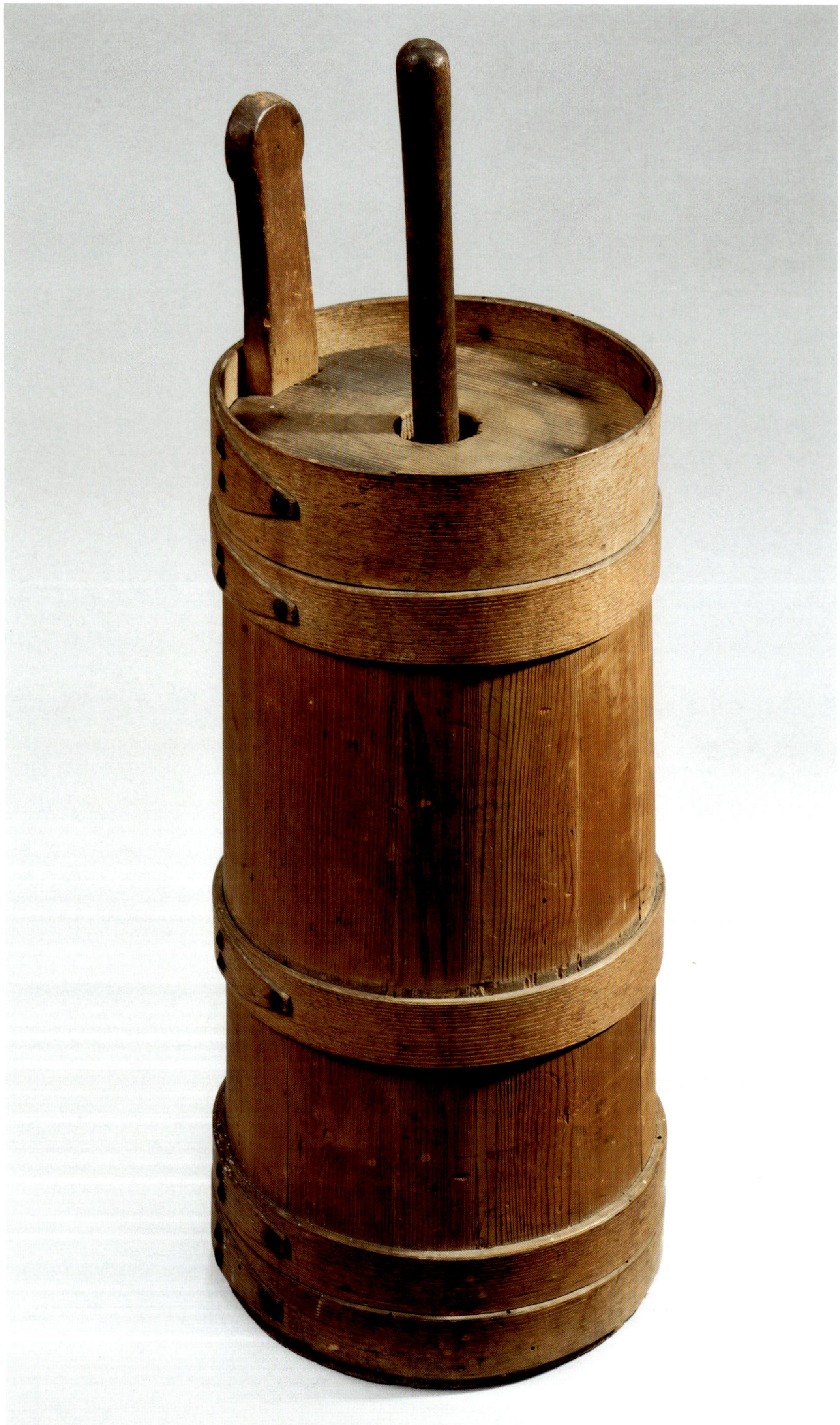

Plate 1.2

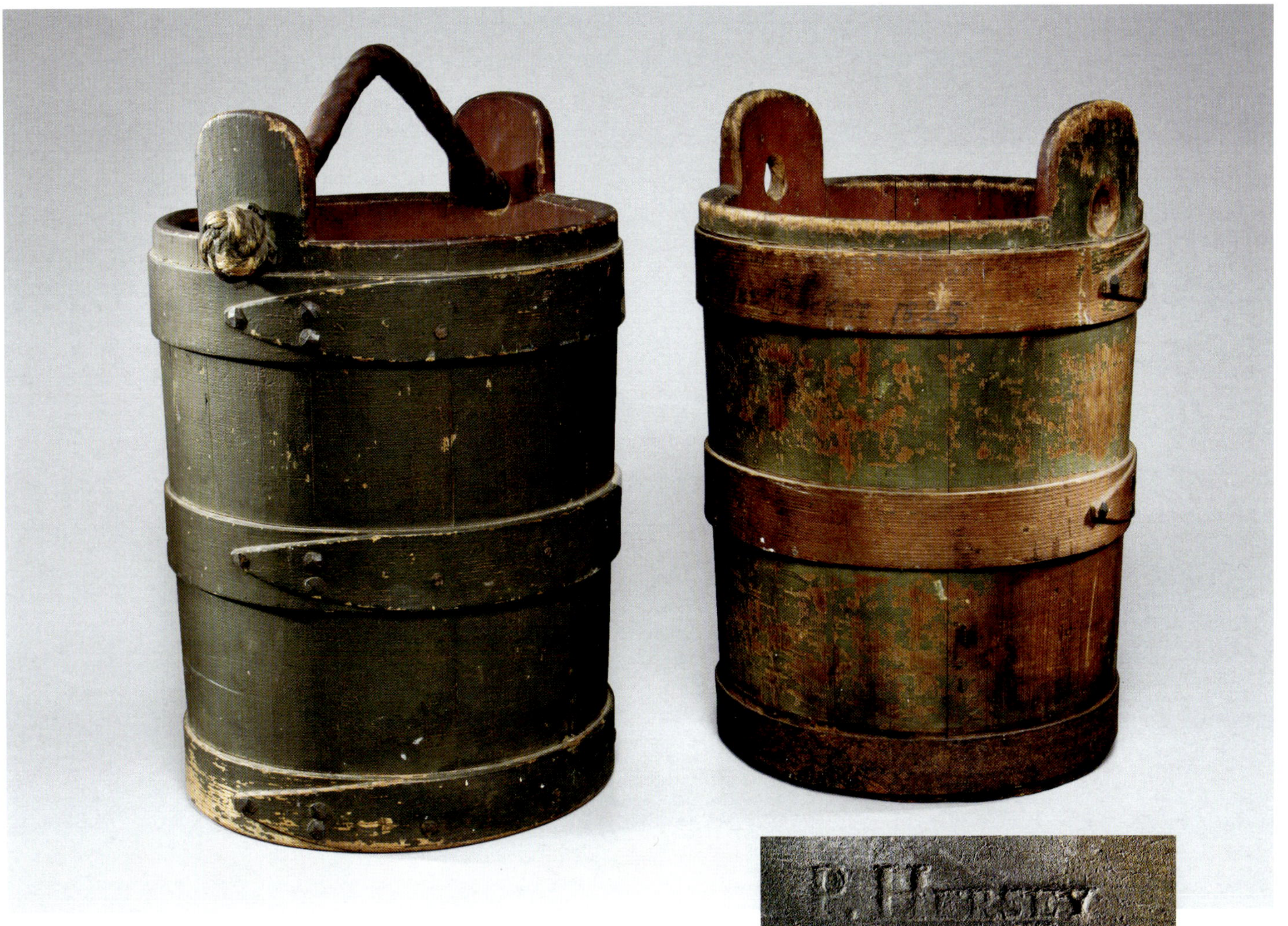

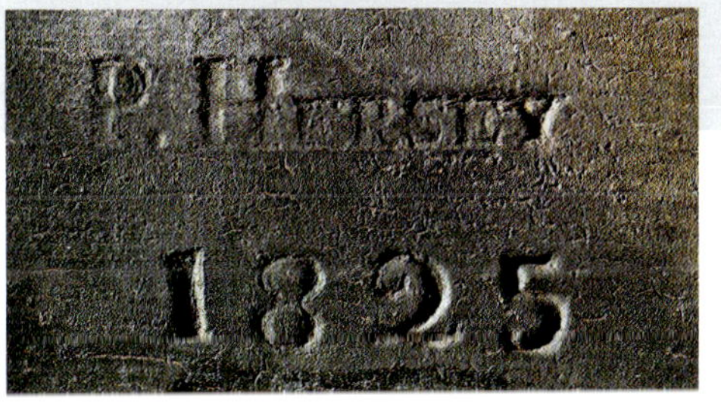

Plate 1.3

Plate 1.1
Bucket, Peter Hersey, 1780–1830
Materials: White pine staves, heads, and buttons; oak hoops and handle; copper nails
Dimensions: H 14 ¼, Diam 14 $^{9}/_{16}$
Inscription: "PETER HERSEY" stamped on lid
Provenance: Courtesy of the Hersey family

Plate 1.2
Churn, Peter Hersey, 1780–1830
Materials: White pine staves, lid, bottom, and dasher bottom; ash hoops; maple handle; iron nails
Dimensions: H 23 ⅝, Diam 9
Inscription: "P. HERSEY" stamped on bottom
Provenance: Courtesy of Hingham Historical Society

Plate 1.3
Fire buckets, Peter Hersey, 1825
Materials: White pine staves and bottoms; ash hoops; copper nails; rope and leather handles
Dimensions: H 15, Diam 9 ¾
Inscriptions: "P. HERSEY" and "1825" stamped on bottoms
Provenance: Given to the Cohasset Historical Society in 1934 by Henry N. Lincoln. Courtesy of Cohasset Historical Society

2

JARED LANE (1752–1839)

LIKE MANY HINGHAM CRAFTSMEN DURING THE AMERICAN REVOLUTION, JARED LANE divided his time between the workbench and the battlefield. On April 19, 1775, he was among the volunteers who marched to Concord and Lexington for the first military engagements of the war.[1] It was back to business a few months later, when he sold dozens of tubs and boxes to Hawkes Fearing, a prominent local trader.[2] The transaction was typical for both men. In fact, Fearing maintained similar relationships with over twenty coopers, carting loads of purchased woodenware—especially pails and canteens—to the Continental Army's encampment at Roxbury.[3] Lane may have recognized some of his own goods among the army's supplies when he returned to Boston in January 1776 to help defend Dorchester Heights.[4]

Lane's small Hingham farm included a cooper's shop and three acres of maple, pine, and cedar of "uncommonly large growth."[5] He harvested the trees for many of the household containers he stamped with his full name, including this well-used water pail (pl. 2.1). The form wholesaled for around two dollars a dozen, a considerable price for the average laborer, who earned about a dollar a day.[6] It is crafted by hand with white pine staves butt-joined and set within oak hoops. Each hoop is secured with small wooden pegs as well as flat-head and rose-head iron nails to hold in place the triangular ends, or tails. Lane applied a coat of blue paint to the exterior, a signature color of Hingham woodenware.[7]

Some time in the late eighteenth or early nineteenth century Jared Lane also crafted this large oval box (pl. 2.2), possibly as a special commission. It measures nearly seventeen inches in length and twelve inches in width, nearly triple the size of a standard example. The decoratively painted surface, intended to simulate wood grain, is the work of an early owner or ornamental painter.[8] The box is constructed of a single ash side bent into the shape of an oval and lap-joined with a row of iron rose-head nails. A white pine bottom and top, also known as heads, are held in place with wooden pegs. The rim, also a narrow strip of ash, features traditional triangular-shaped ends held in place with nails and wooden pegs. The underside of the box is boldly stamped with Lane's name.

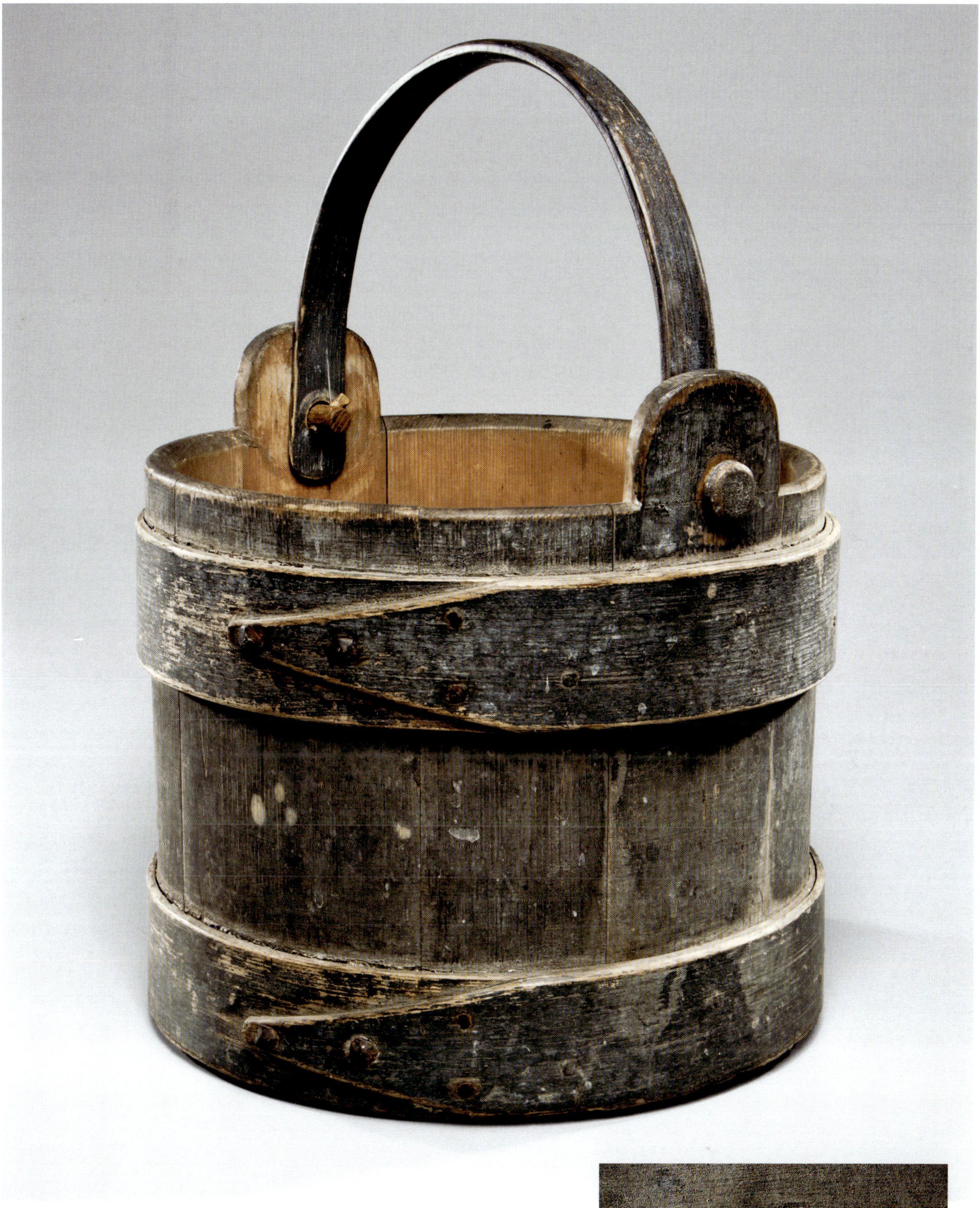

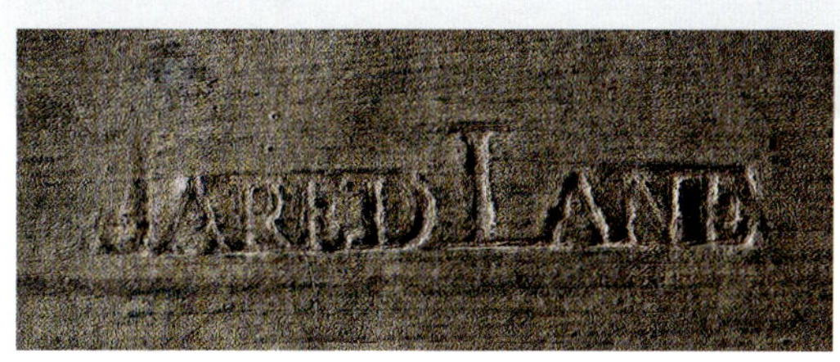

Plate 2.1

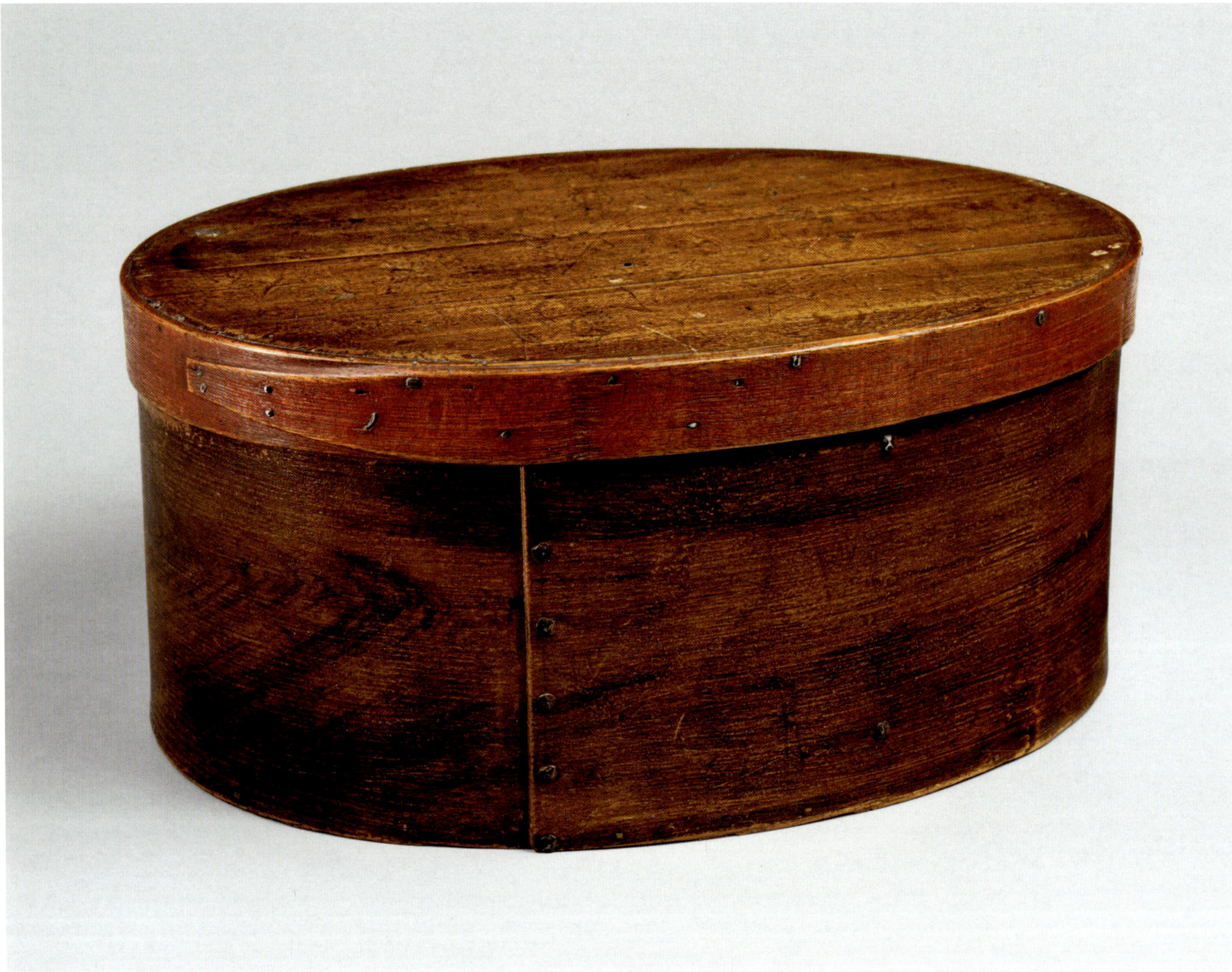

Plate 2.2

Plate 2.1

Pail, Jared Lane, 1780–1830

Materials: White pine staves and bottom; oak hoops; iron nails

Condition: Replaced handle

Dimensions: H 10 ⅞, Diam 11 ½

Inscription: "JARED LANE" stamped on bottom

Provenance: Courtesy of Hingham Historical Society

Plate 2.2

Oval box, Jared Lane, 1780–1830

Materials: White pine heads; ash sides; iron nails

Dimensions: H 7 3⁄16, W 16 7⁄16, D 12 ⅛

Inscription: "JARED LANE" stamped on bottom

Provenance: Courtesy of Rick Young

3

ELIJAH LEWIS JR. (1762–1819)

IN THE 1930s THE HINGHAM HISTORIAN MASON FOLEY DISCOVERED AN OLD SET OF coopering tools and woodenware, languishing in a derelict building at Hingham Harbor.[1] That structure was razed, but not before Foley collected the items and documented their history. They remain the only known artifacts connected to Elijah Lewis Jr., a cooper who also worked as packet man in the early nineteenth century. In addition to running his shop, Lewis owned a store at the head of Union Wharf, just across the street from where his tools were found. At the time of his death in 1819 he was also part owner of the sloops *Traveler* and *Russell,* which probably freighted his wares—buckets, barrel covers, and other coopered goods—to wholesale markets in Boston and other coastal communities.[2]

Among the items salvaged by Foley is a tall, worn pail that, remarkably, survives with its original locked hoops (pl. 3.1). Throughout the eighteenth century and well into the 1840s, Hingham's famous buckets and pails were of two types: nailed and locked. The circular hoops of nailed pails were secured by wooden pegs and iron and copper nails; the hoops of locked pails, in contrast, were notched on one end and inserted into a triangular hole in the other, locking them into place without the use of wooden or metal fasteners.[3]

Although both techniques were widely used in Hingham, few examples of locked buckets and pails are known today. An 1840 price list for woodenware indicates that they were more expensive—$2.00 for a dozen "common nail pails" and $2.25 for a dozen "common lock pails"—probably owing to increased labor and material costs. It is unclear why consumers paid more for containers that were seemingly less durable—or at least had a shorter lifespan—than their nailed counterparts. Perhaps locked buckets and pails were intended for heavy-duty use, wore more quickly, and were discarded. Unfortunately, period records tell us little about how these commonplace items were used.

Lewis's handcrafted pail shows other evidence of a long life on the wharf. Its swing handle has long since disappeared, probably around the same time its ear was broken. Rather than repair the old container—a common practice—it was tossed into the corner of a building with other cast-off items. The initials "EL," presumably for Elijah Lewis, are roughly inscribed on the bottom along with "W. DAVIS" in a neat impression the owner's mark of William Davis (1805–85), a local maker of masts, spars, pumps, and blocks (pl. 3.2).[4] In the early 1830s Davis purchased a house and shop on the wharf and, apparently, acquired items that had belonged to Lewis. Davis's mark is also found on woodenware and wooden items recovered by the local antiquarian John P. Richardson from the loft of another Summer Street commercial building.

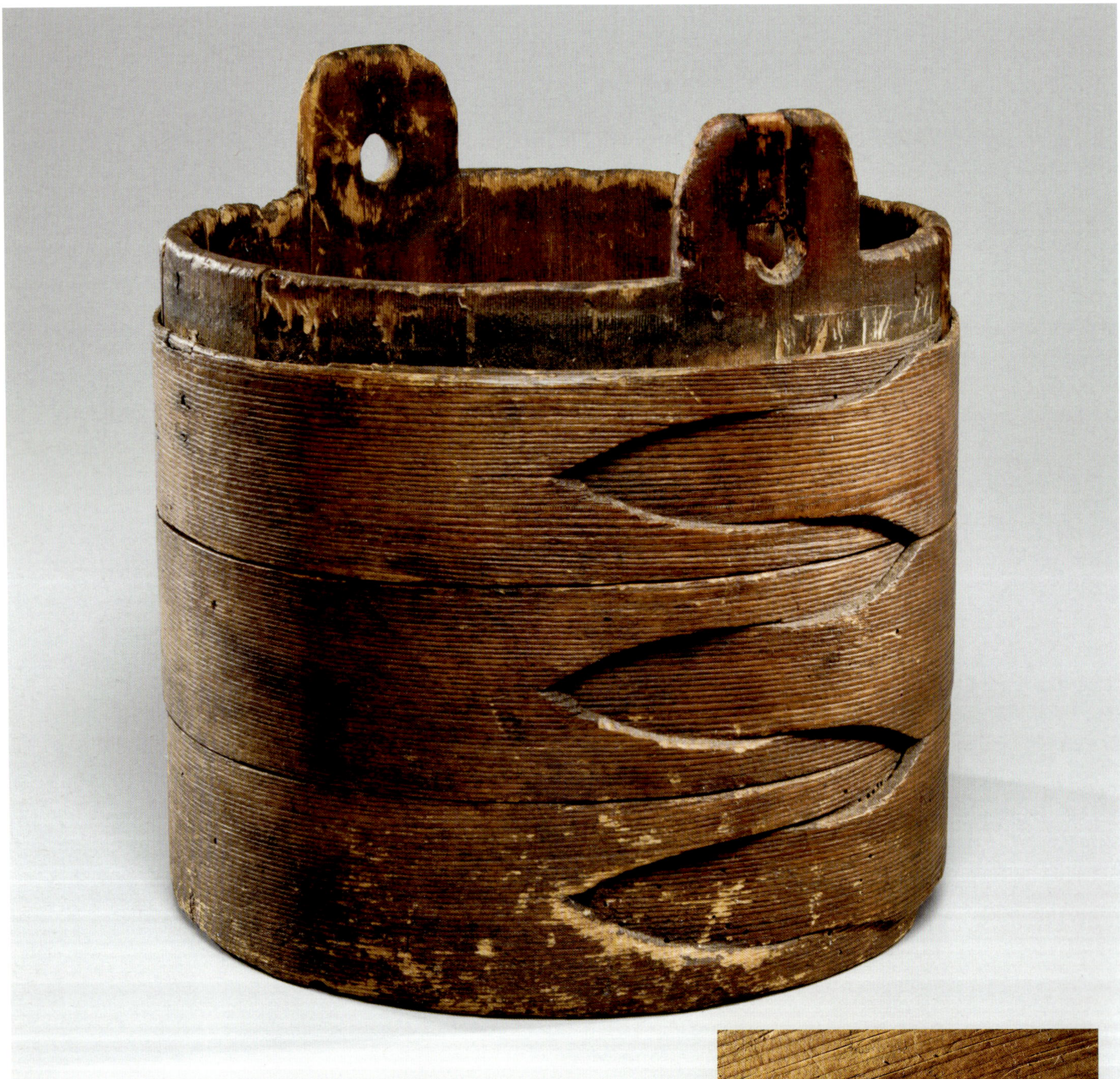

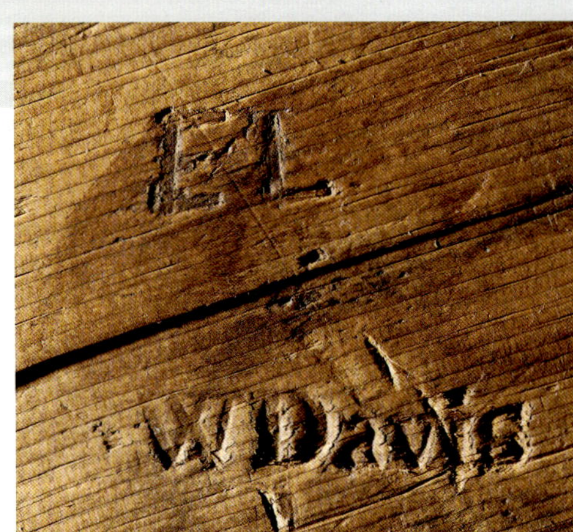

Plate 3.1

Plate 3.1

Pail, Elijah Lewis, 1780–1819

Materials: White pine staves and bottom; ash hoops

Condition: Missing handle

Dimensions: H 10 ¾, Diam 10 ⅞

Inscriptions: "EL" inscribed on bottom; "W. DAVIS" stamped on bottom

Provenance: Mason A. Foley, Hingham, Massachusetts; to his widow, Frances Foley; acquired in 1968 by John P. Richardson, Hingham; to current owner.

Courtesy of the Hersey family

4

R. BATES (ACT. 1780–1830)

There was one labor-saving machine, the product of their skill, which deserves notice. This was a washing machine, constructed wholly of wood and known by the name of the "Dumb Betty." It had a local habitation, and was hardly known outside of the town [of Hingham] and its immediate vicinity. But among the townspeople it was held to be essential to the proper performance of the week's washing; and a household would be considered deficient in its equipment if it were wanting in one of these implements.[1]

WHEN THIS ANECDOTE WAS PUBLISHED IN 1891, HINGHAM DUMB BETTYS WERE ALREADY regarded as relics of the preindustrial age. Today only a few are known, including this impressive example branded on the bottom "R. BATES" by a still-unidentified craftsman working in the late eighteenth or early nineteenth century (pl. 4.1).[2] Like dozens of other local coopers, Bates offered this large domestic tool as part of a general line of woodenware.[3] Its considerable size and heft—it weighs nearly thirty pounds—leave little doubt that it was the most expensive item produced in his shop.

It consists of two main parts: a large washtub and a robust washing mechanism, both of which retain their original blue paint. The mechanism is secured to a large cover that rests on the rim of the tub. A frame constructed of vertical and horizontal boards allows a plunger, or dolly stick, to move side to side and up and down.[3] Attached to the bottom of the plunger is a block with six legs, or a dolly, to beat and stir laundry (pl. 4.2). This is manipulated by a handhold at the top of the plunger and a ladder-shaped handle tenoned through its sides. Not surprisingly, the handgrips show wear from what must have been years of vigorous and tiring use.

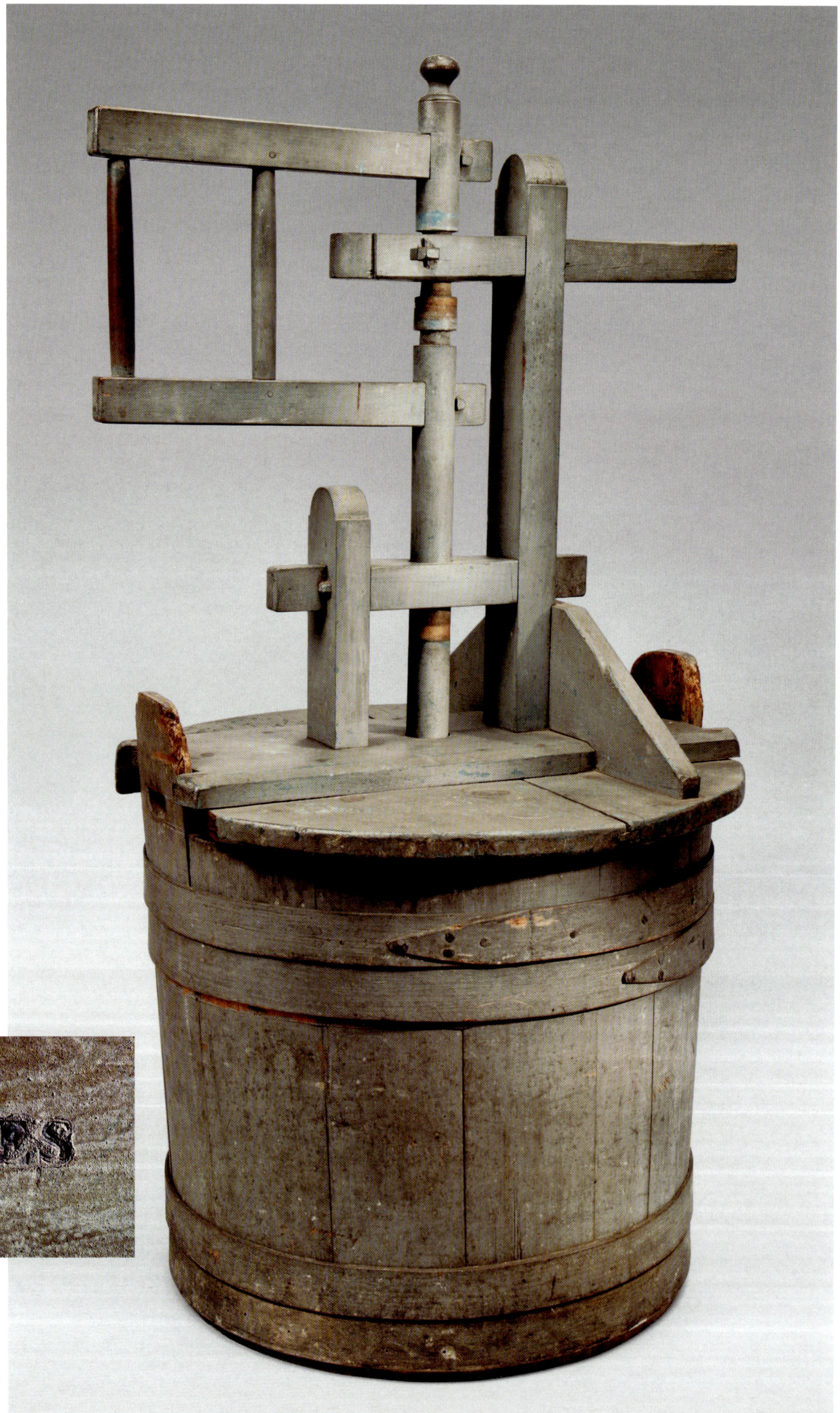

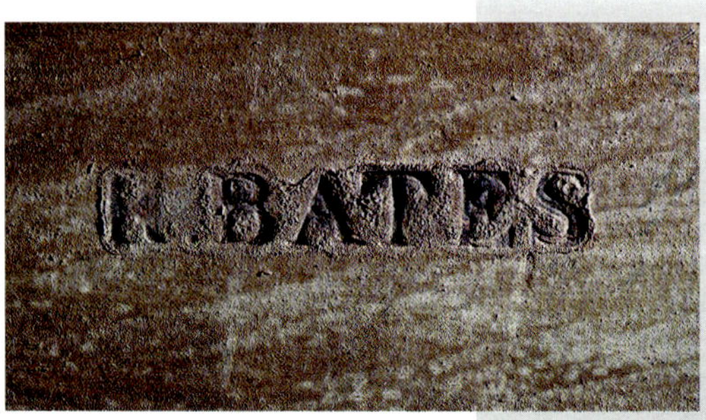

Plate 4.1

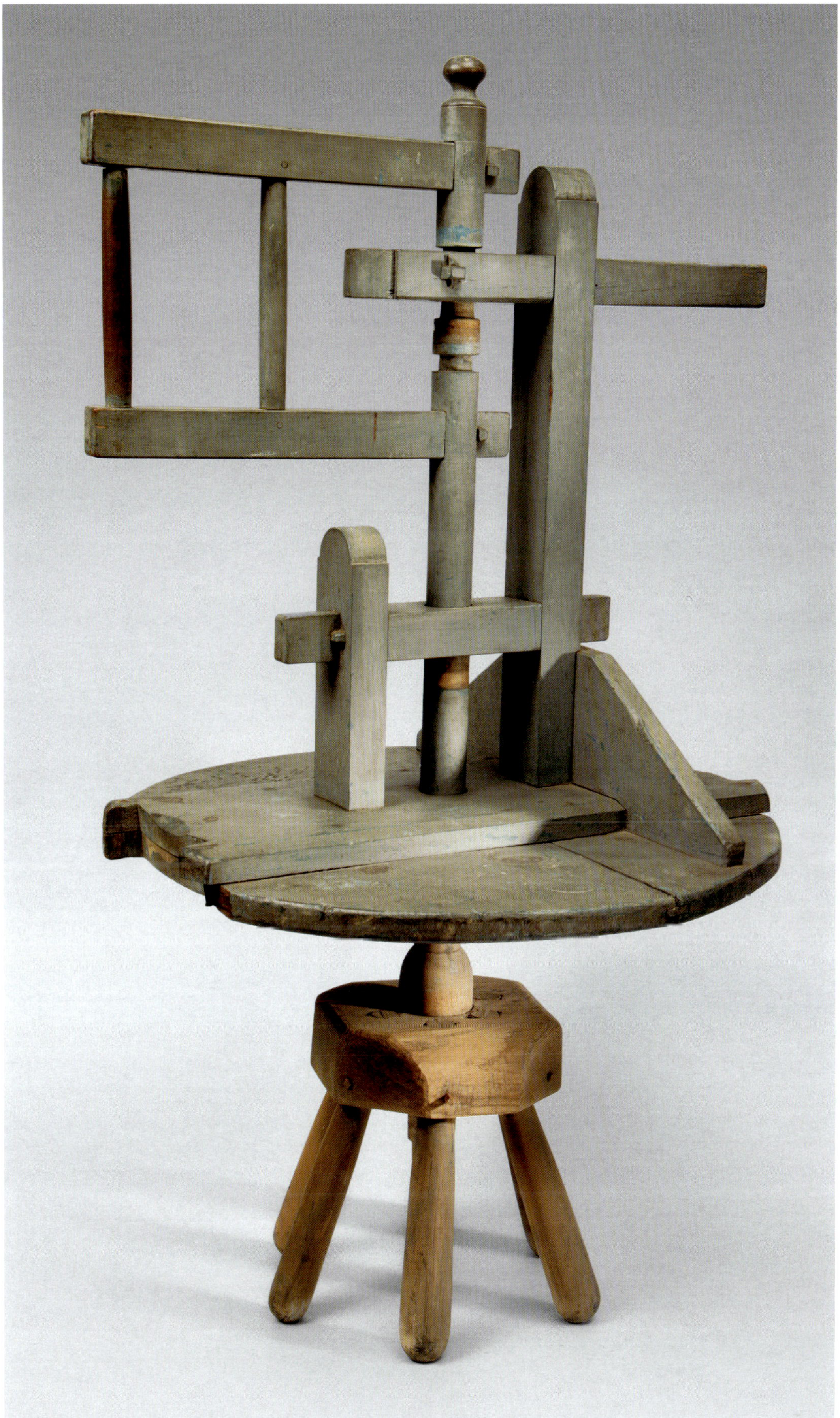

Plate 4.2

Plate 4.1

Dumb betty, R. Bates, 1780–1830

Materials: White pine staves, bottom, and cover; oak hoops; white pine, oak, and maple washing mechanism; copper nails

Dimensions: H 43 ½ (overall); Diam 23 ¼

Inscription: "R. BATES" branded on bottom

Provenance: Courtesy of Hingham Historical Society

Plate 4.2

Detail of washing mechanism in dumb betty in pl. 4.1

5

ELISHA BURR (1776–1863)

ON JUNE 19, 1819, THE HINGHAM RIFLE COMPANY VOTED TO PURCHASE NEW CANTEENS "painted green and ornamented like the knapsacks."[1] For the citizen-soldiers of Bucket Town, acquiring these vital and fashionable accoutrements must have been an easy task. Military canteens were a staple of local coopers, many of whom had produced them for decades. This was especially true of brothers John, Thomas, and Samuel Burr and their nephew David. During the War for American Independence, they crafted thousands of canteens and other woodenware for troops encamped at nearby Roxbury.[2]

In the early nineteenth century Elisha Burr emerged as one of Hingham's most prolific manufacturers of wooden canteens. He was a fourth-generation boxmaker, a specialty among some coopers, and like his father, David, operated a small handcraft shop on Leavitt Street in Hingham Centre.[3] He maintained close business relationships with traders in Hingham and Boston and, on at least one occasion, sold one thousand canteens at fifteen cents apiece to the firm of Matthew & Fearing Burr, run by his cousins.[4] Today his maker's mark—"E. BURR"—is found on dozens of examples, including a decorated canteen (pl. 5.1) that once belonged to David Sprague (1754–1832), a local farmer and veteran of the Revolutionary War.[5]

The brightly painted face of Sprague's canteen is a reminder of the dashing image struck by New England's elite volunteer militiamen, many of whom were featured attractions at balls, parades, and other events where people admired their flamboyant uniforms and accessories (pls. 5.2, 5.3).[6] Sprague's initials, "DS," appear in calligraphic lettering within a red sunburst border. Such embellishments were probably the work of a local ornamental painting firm like that of R. & C. Lane, run by two brothers, Rufus (1788–1863) and Charles (1790–1854).[7] For generations the Lanes were the town's leading decorators of furniture, fire buckets, and shop signs;[8] they even performed work for the Hingham Rifle Company, of which Charles served as captain.[9]

Smaller, oval-shaped canteens were not as popular, probably because they held less refreshment—be it water or the soldier's all-important rum ration. An example of the form bearing Burr's mark (pl. 5.4) is one of only a few known to exist.[10] It is finely painted with the initials "IL" or "LI," possibly an abbreviation for Light Infantry. Its basic construction is similar to that of the oval boxes made by Burr and dozens of other Hingham coopers: the side, a thin piece of oak, is lap-joined with a row of four rose-head iron nails. The white pine back and front, or heads, are secured to the hoop with a series of tiny wooden pegs. Once filled with liquid, the wood expanded, making a watertight seal. Three strips of leather nailed to the bottom and to either side of the canteen's mouth formed loops for a narrow carrying strap, probably made of leather.

At least five other craftsmen made wooden canteens in Hingham between 1800 and 1840.[11] Because all these men worked in the same tradition with the same materials, their canteens can be distinguished only by their stamped names and initials; remarkably, one of Elisha Burr's original iron brands survives (pl. 5.5).[12] Since the tops and bottoms of boxes and canteens are typically made of soft eastern white pine, it took little force to make a clear impression. In the spirit of Yankee thrift, the tool was repurposed by his nephew, also named Elisha Burr (1839–1909), who used it to mark the duck and shorebird decoys he carved by hand a generation later (pl. 5.6).[13]

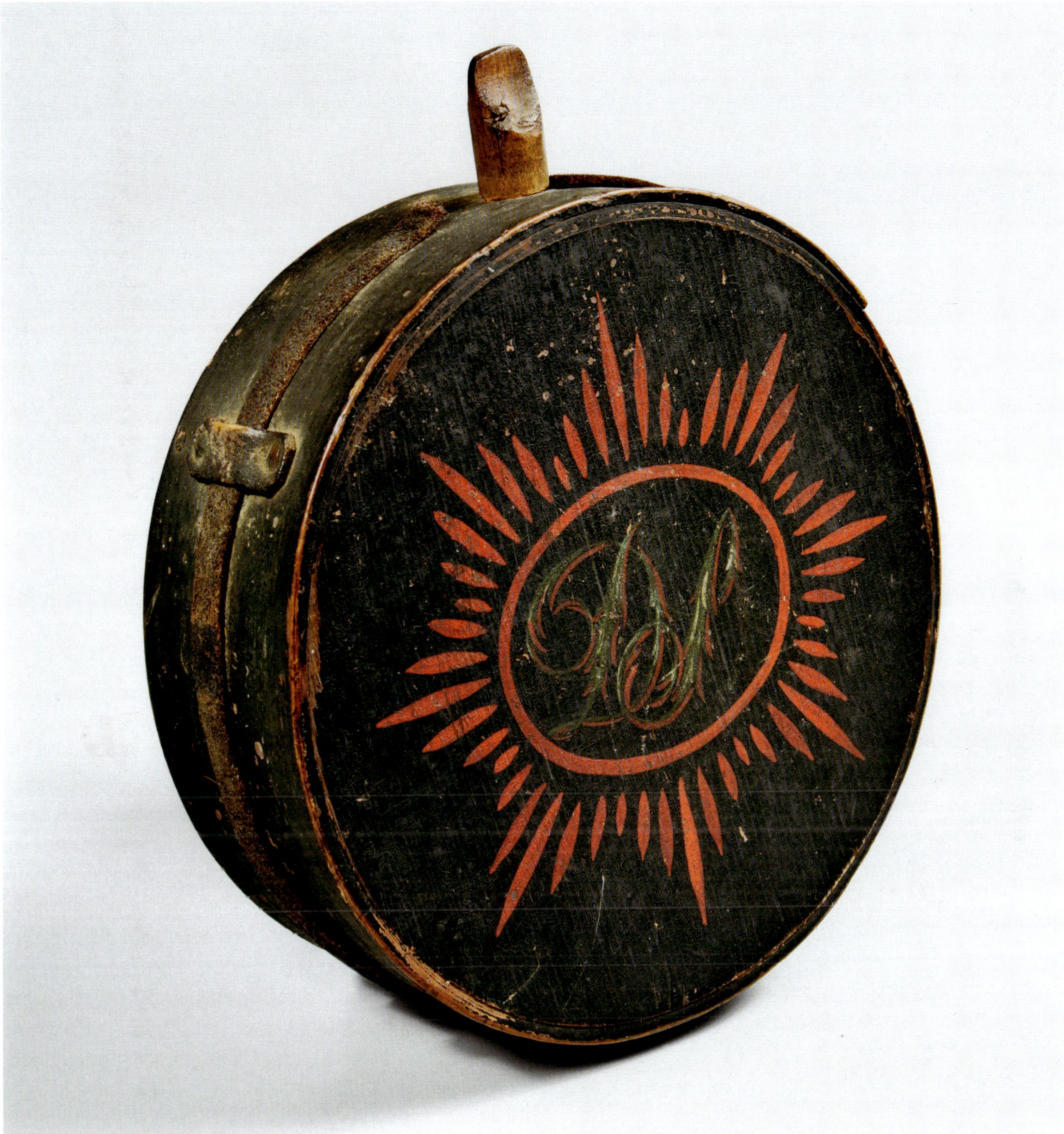

Plate 5.1

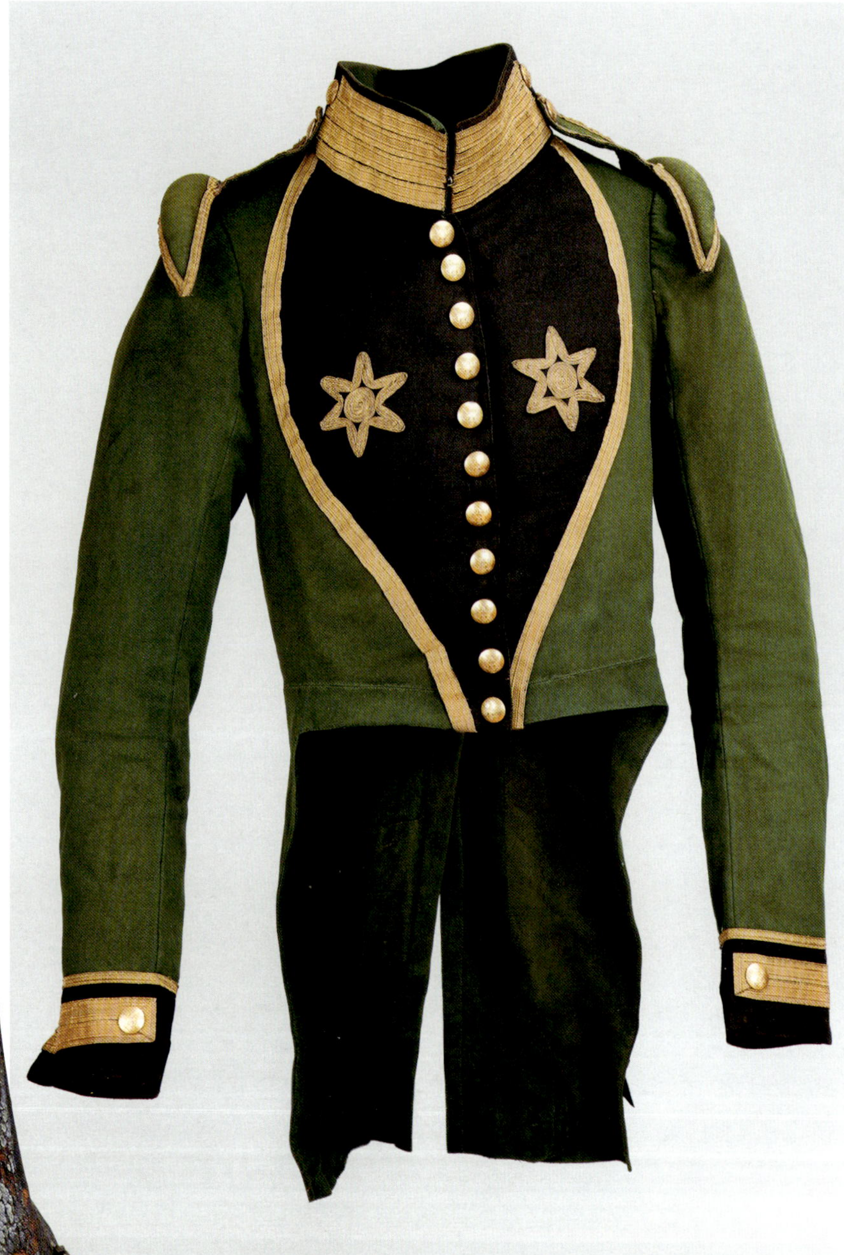

Plate 5.2

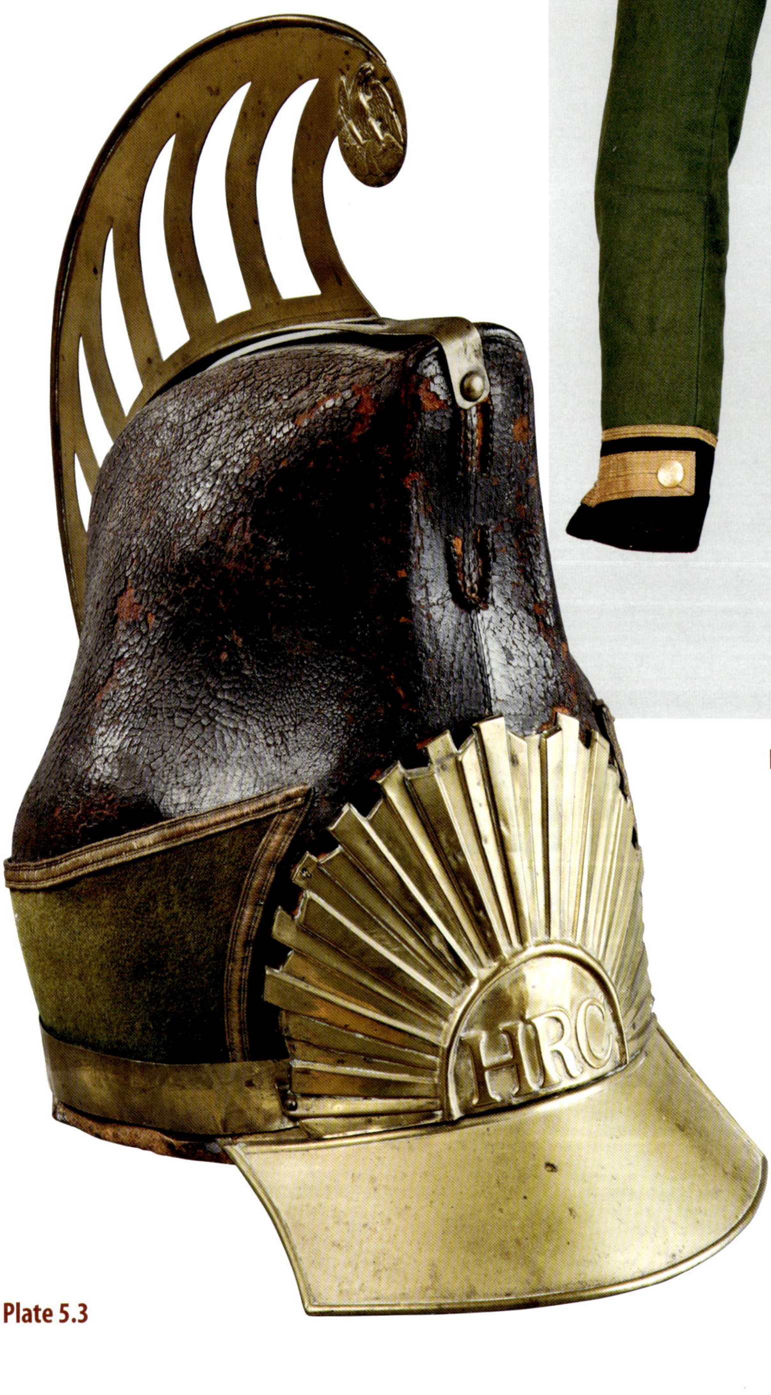

Plate 5.3

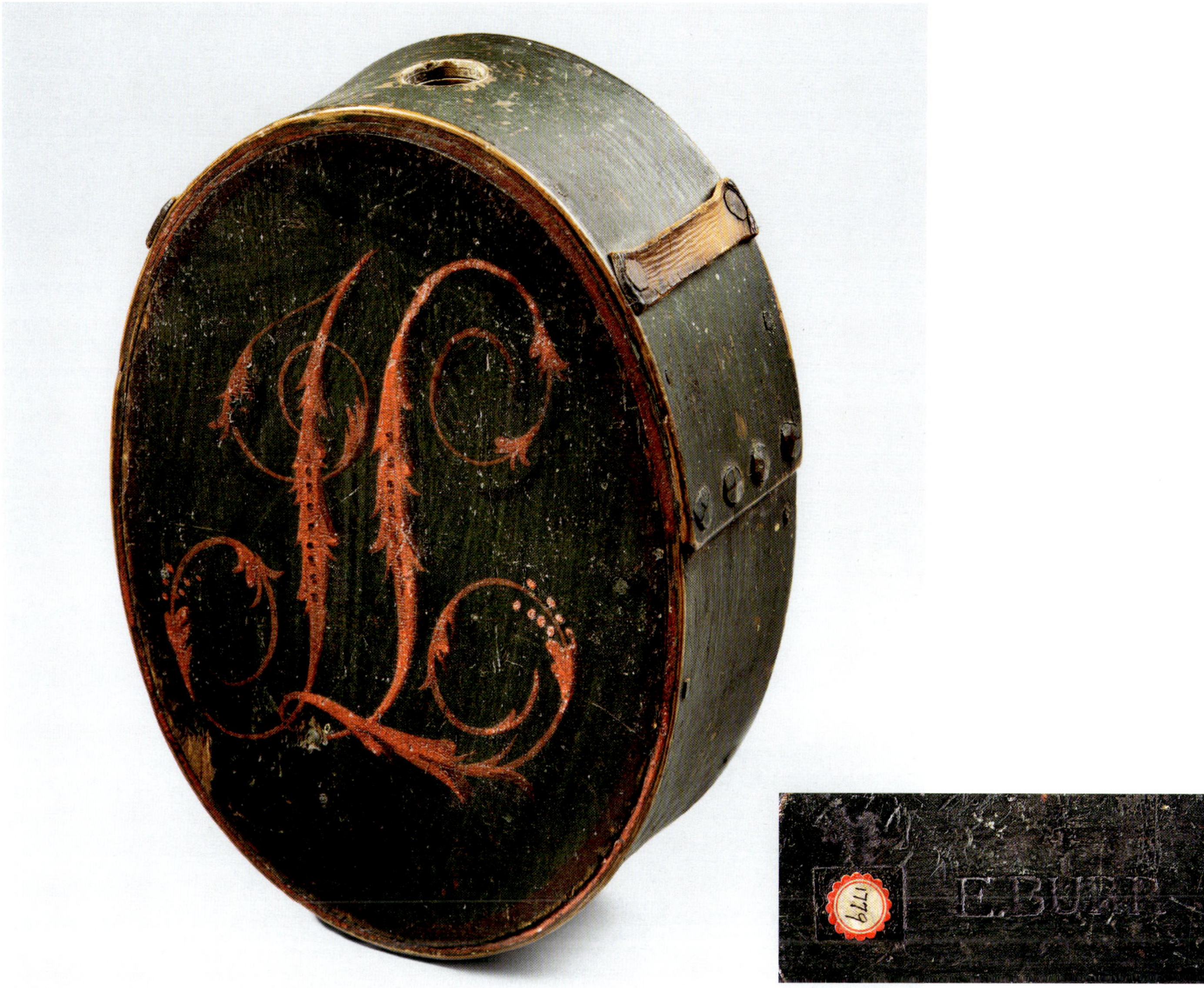

Plate 5.4

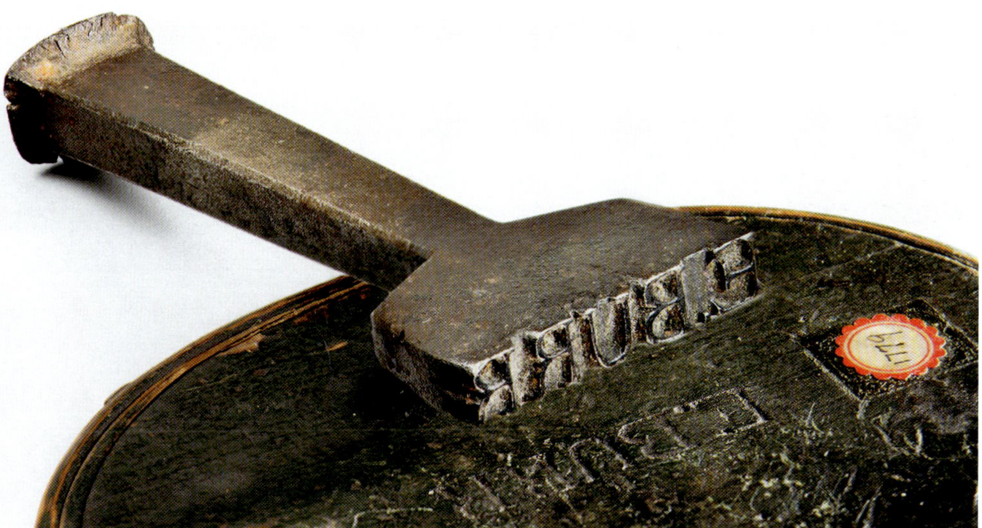

Plate 5.5

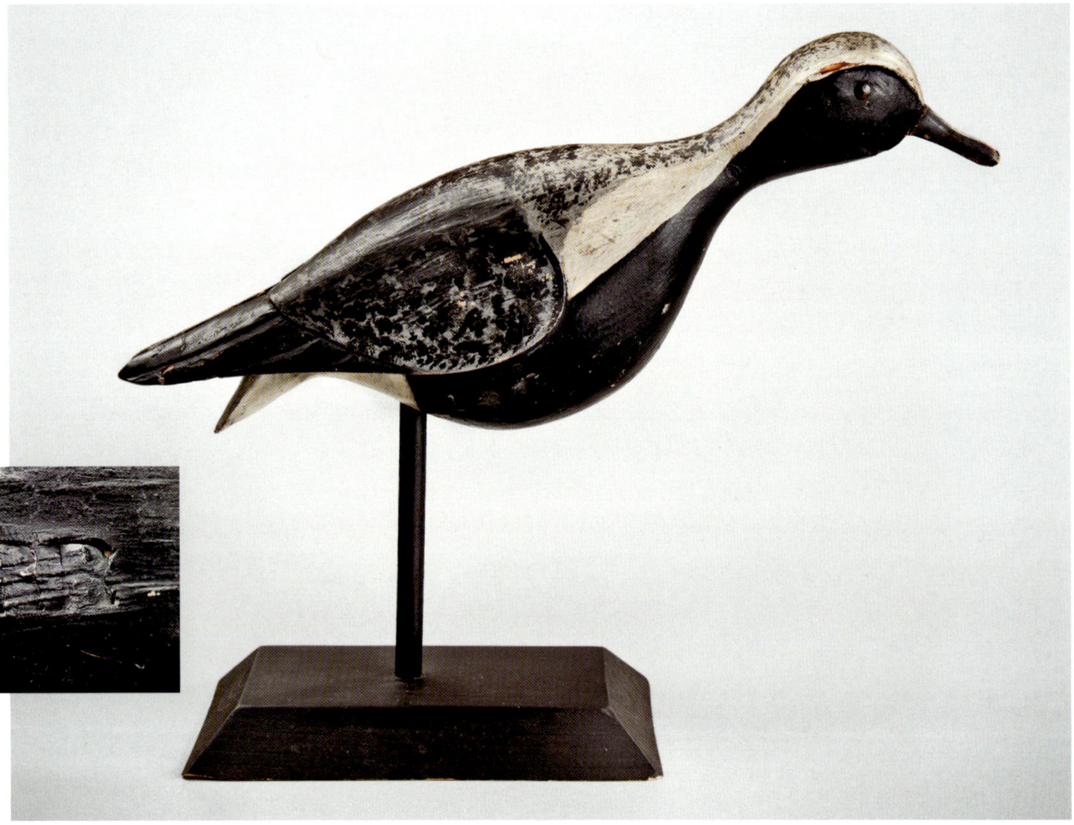

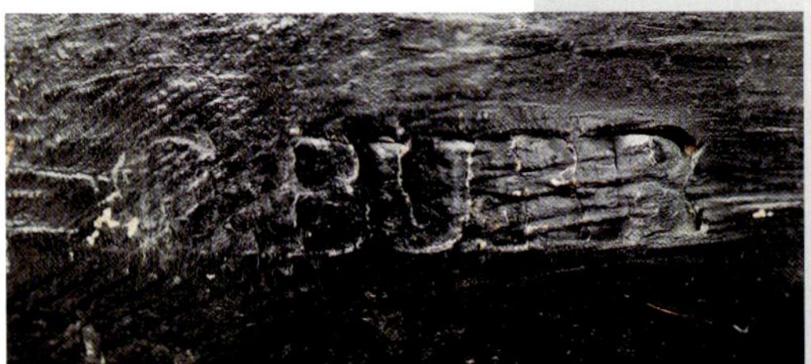

Plate 5.6

Plate 5.1
Round canteen, Elisha Burr, 1800–30
Materials: White pine heads; oak hoop; iron nails; leather strap and strap loops
Dimensions: Diam 7 ⅛, D 2 ½
Inscriptions: "DS" painted on front; "E. BURR" stamped on back
Provenance: Daniel Sprague; given to the Hingham Historical Society by Mrs. William R. Burr, November 15, 1917. Courtesy of Hingham Historical Society

Plate 5.2
Militia coatee, Hingham Rifle Company, 1820–40
Provenance: Given to the Hingham Historical Society in 1932–33 by Isaac Sprague of Boston. Courtesy of Hingham Historical Society

Plate 5.3
Militia helmet, Hingham Rifle Company, 1820–40
Provenance: Sold in the 1960s or 1970s at the Bouvé family estate auction in Hingham; purchased shortly thereafter by Frank W. Klay, Marshfield, Massachusetts; to Derin Bray, Portsmouth, New Hampshire; to current owner. Courtesy of the Hersey family

Plate 5.4
Oval canteen, Elisha Burr, 1800–30
Materials: White pine heads; oak side; iron nails; leather strap loops
Dimensions: H 5 ⅜, W 4, D 1 ⅝
Inscriptions: "IL" painted on front; "E. BURR" stamped on back; "1779" in ink on paper label glued to back; "1779" in white paint on side
Provenance: Crosby Milliman, Eaton Center, New Hampshire; to William H. Guthman, Westport, Connecticut; purchased at Sotheby's, January 19, 2003, lot 159, by David Good, Camden, Maine; to Derin Bray, Portsmouth, New Hampshire; to current owner. Courtesy of the Hersey family

Plate 5.5
Branding iron, Elisha Burr, 1800–30
Materials: Iron
Dimensions: L 4 9⁄16
Provenance: Elisha Burr; to his nephew Elisha Burr; to his son Carl Burl (1884–1961). Courtesy of Bob and Peggy Mosher

Plate 5.6
Black-bellied plover, Elisha Burr, 1880–1909
Materials: White cedar
Dimensions: H 4 ¼, L 12 ¼
Inscription: "E. BURR" stamped on bottom
Provenance: Collection of Bob and Peggy Mosher

6

MOSES SPRAGUE (1776–1850)

TRACES OF AN OLD BLUE PAINT ARE STILL VISIBLE ON THE DECORATED BOX (pl. 6.1) of Merrick Ryder (1789–1858), a prominent ship captain from Plymouth, Massachusetts.[1] During the nineteenth century Ryder was among an elite class of mariners who sailed merchant vessels to Boston and other coastal markets where fresh goods such as fish, farming produce, and whale oil were bought and sold daily. In 1829 he captained the two-masted brig *Elder Brewster,* named for one of Plymouth's most revered pilgrim settlers.[2] It is probable that aboard this ship, or one like it, Ryder carved his initials and the date 1831 into the lid of a Hingham box, further embellished with a large, compass-scribed pinwheel and geometric border (pl. 6.2). Sailors commonly produced such whimsies during long and often boring stretches of inactivity at sea.[3]

The underside of Ryder's box is boldly impressed with another set of initials—"M.S"—probably those of the cooper Moses Sprague.[4] Sprague lived in Hingham Centre, where he and several other members of the Sprague family worked primarily at boxmaking—a fact confirmed by the variety of round and oval boxes and canteens that still bear their marks (pl. 6.3).[5] This example is constructed of simple ash hoops, lap-joined and secured in place with a column of five iron nails for the body and two for the lid. The corners of the joints are chamfered at forty-five-degree angles, a decorative feature that also reduces the chance of the hoop ends splitting. Although this method of construction was favored by a number of Hingham coopers, by itself it is not a reliable diagnostic tool for identifying Hingham woodenware; similarly made boxes were produced throughout New England and probably elsewhere.[6] In typical fashion, the pine heads are held in place with small wooden pegs.

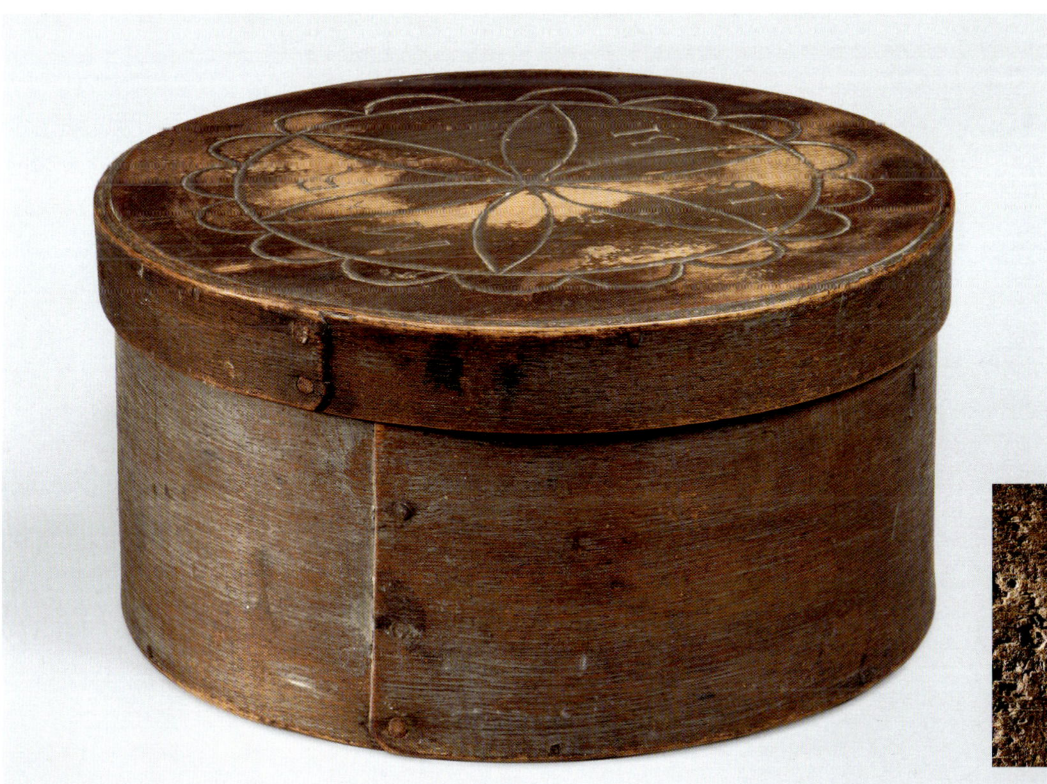

Plate 6.1

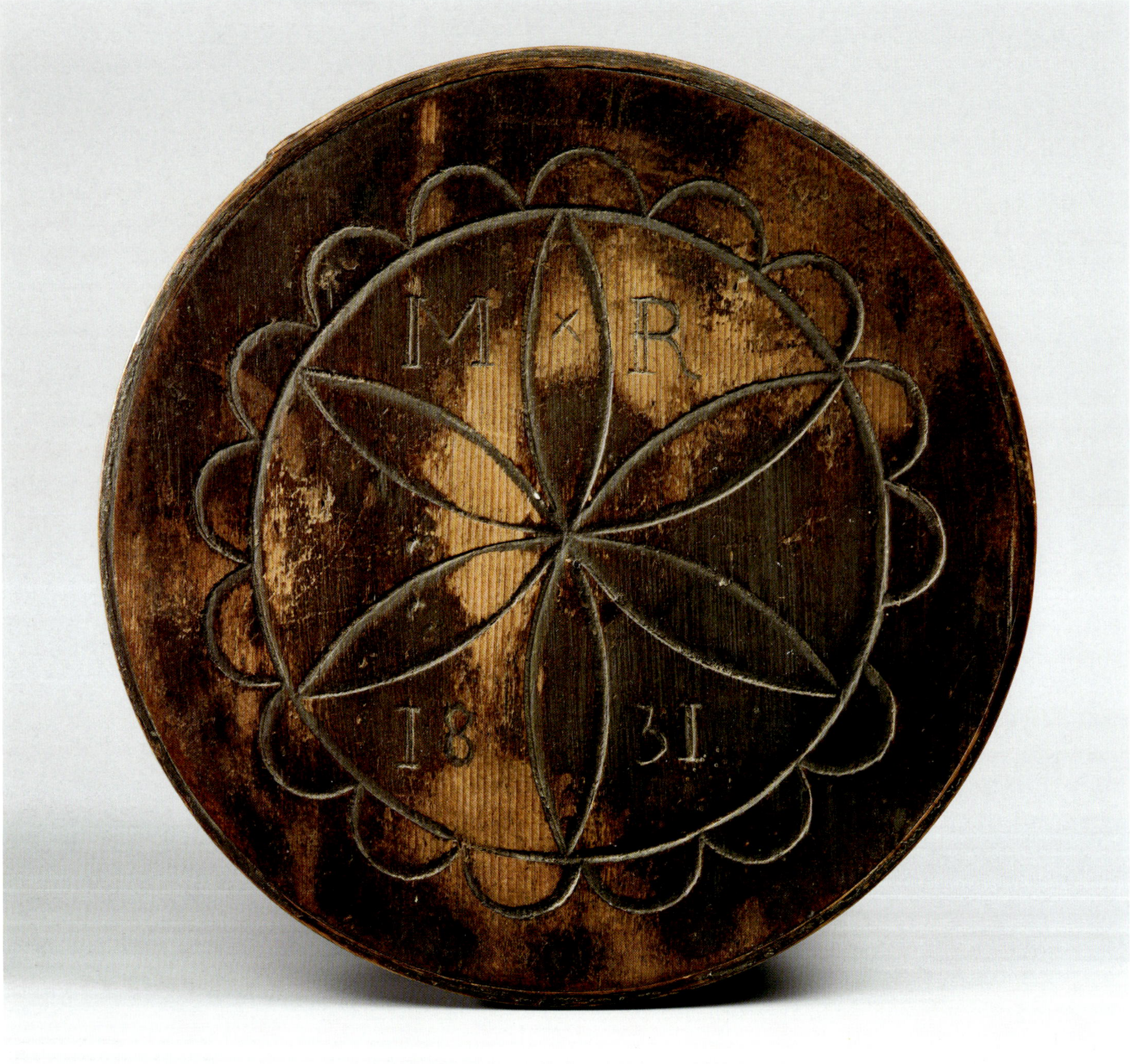

Plate 6.2

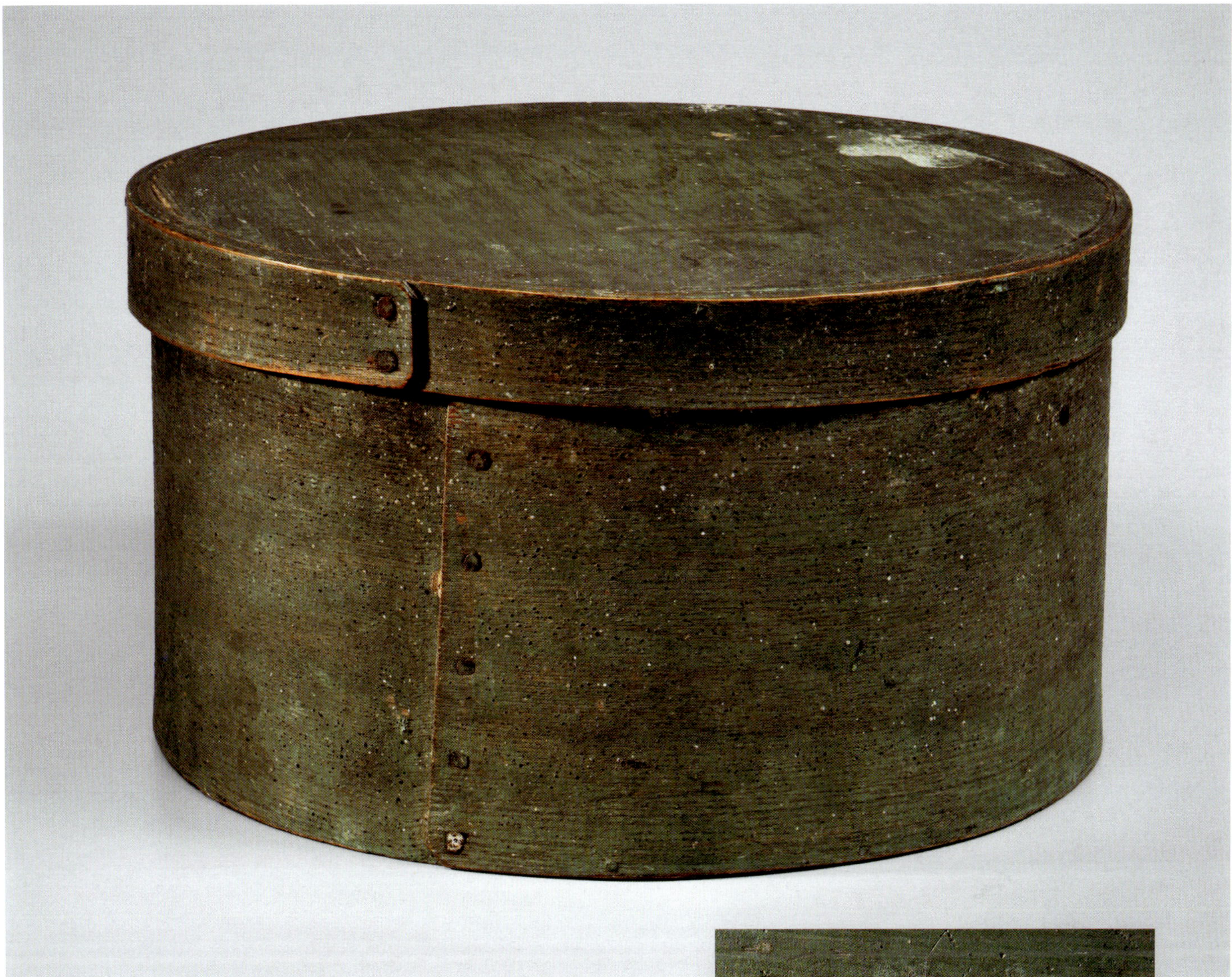

Plate 6.3

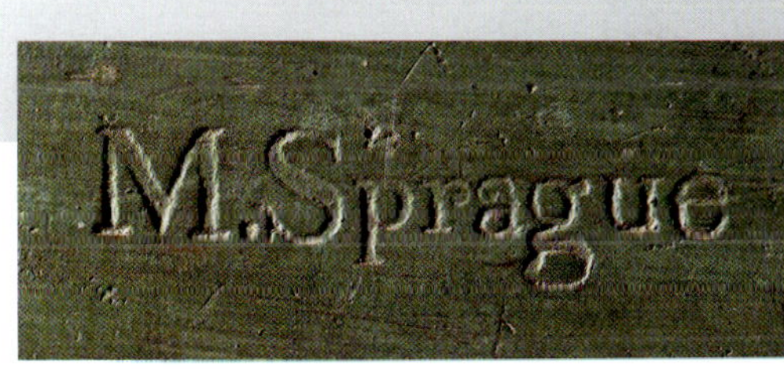

Plate 6.1

Round box, Moses Sprague, ca. 1831

Materials: White pine heads; ash hoops; iron nails

Dimensions: H 4, Diam 8 ⅜

Inscriptions: "M.S" stamped on bottom; "MR" and "1831" carved in lid

Provenance: Merrick Ryder; given to the Plymouth Antiquarian Society in 1945 by Miss E. H. Faunce. Courtesy of Plymouth Antiquarian Society

Plate 6.2

Detail of lid for box in pl. 6.1

Plate 6.3

Round box, Moses Sprague, 1800–40

Materials: White pine heads; ash hoops; iron nails

Dimensions: H 5 ½, Diam 10 ⅜

Inscription: "M. Sprague" stamped on bottom

Provenance: Private collection

7
BELA SPRAGUE (1804–1878)

BOXES RIVALED BUCKETS AS HINGHAM'S CHIEF EXPORT IN THE EIGHTEENTH AND nineteenth centuries. Local coopers, sometimes identified as boxmakers, produced thousands of these lightweight, multipurpose containers for a large Boston market.[1] They were equally popular with merchants and ship captains, who profited from wholesaling boxes along coastal trade routes from Quebec to South Carolina.[2] In May 1791 a concerned resident of Philadelphia posted a notice in the *Federal Gazette* describing a particularly nefarious use of Hingham boxes: "We hear that a most infamous traffic is now carrying on between apprentice boys, &c. belonging to this city, and the master of a vessel said to be from Hingham, lying at or near Chestnut-Street wharf. This vessel came loaded with . . . wooden sugar boxes; these the master exchanges with apprentices, servants, vagrants, &c., &c. for old iron, which in all probability, is generally stolen from the stores, yards, kitchens, cellars, and vaults of the citizens."[3]

The cargo at the center of this petty thievery probably resembled the work of Bela Sprague, a member of the large and industrious Sprague family of coopers.[4] Some time between 1820 and 1860 he crafted this small oval box (pl. 7.1) at his shop on Union Street, where he worked for nearly five decades.[5] It bears all the hallmarks of classic Hingham construction, including the use of thin ash sides—for the body and the lid—with triangular ends, or tails, that point in opposite directions.[6] Each tail is chamfered at a slight angle and held in place with copper tacks. To distinguish his work from that of other manufacturers, Sprague stamped his name—"B. SPRAGUE"—in the center of the box's lid (pl. 7.2). The green paint and stenciled leaf adorning the exterior are probably the work of an early owner. Hundreds like this example survive with no decoration and often no markings, which makes it difficult to identify the maker.

Like many of his brethren, Bela Sprague offered his clientele a wide range of handcrafted woodenware, including buckets. This example in light blue paint (pl. 7.3) is notable for its small size, measuring only seven inches in height. It is probably the "Small Pail, 3 Hoops" described in the coopers' price agreement of 1840, selling for $1.53 per dozen.[7] Sprague stamped the underside with his initials—"B.S"—a marking system used by at least three other members of the Sprague family.[8]

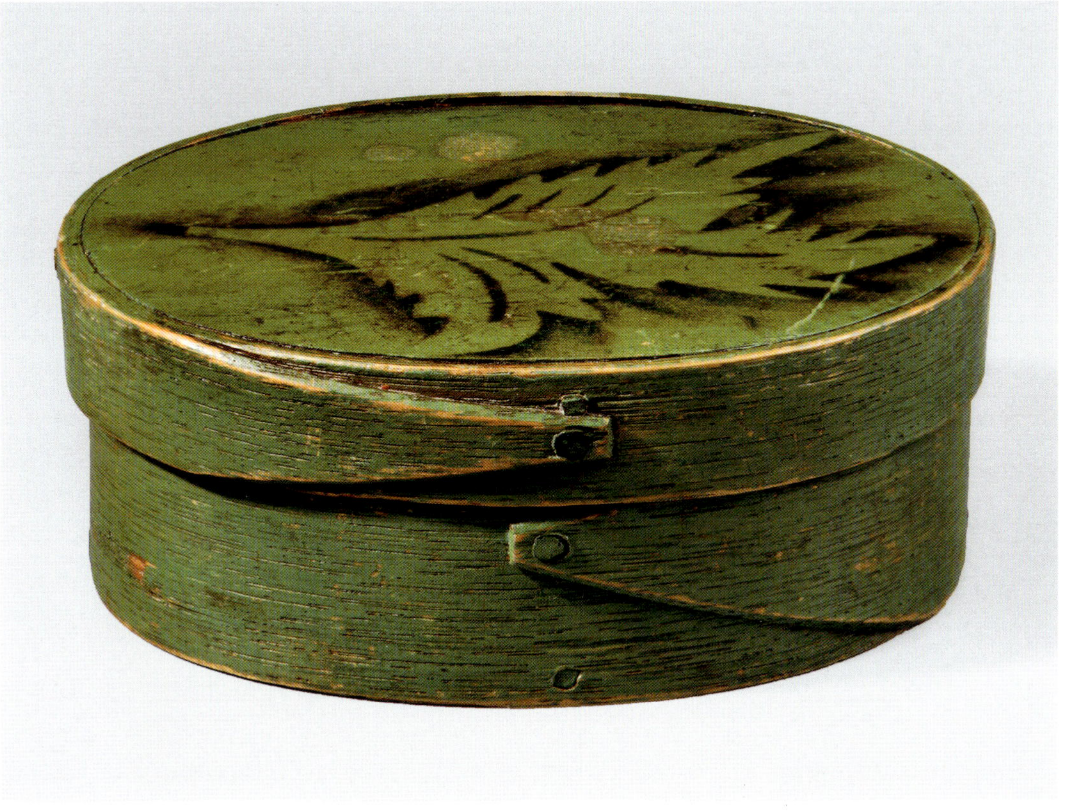

Plate 7.1

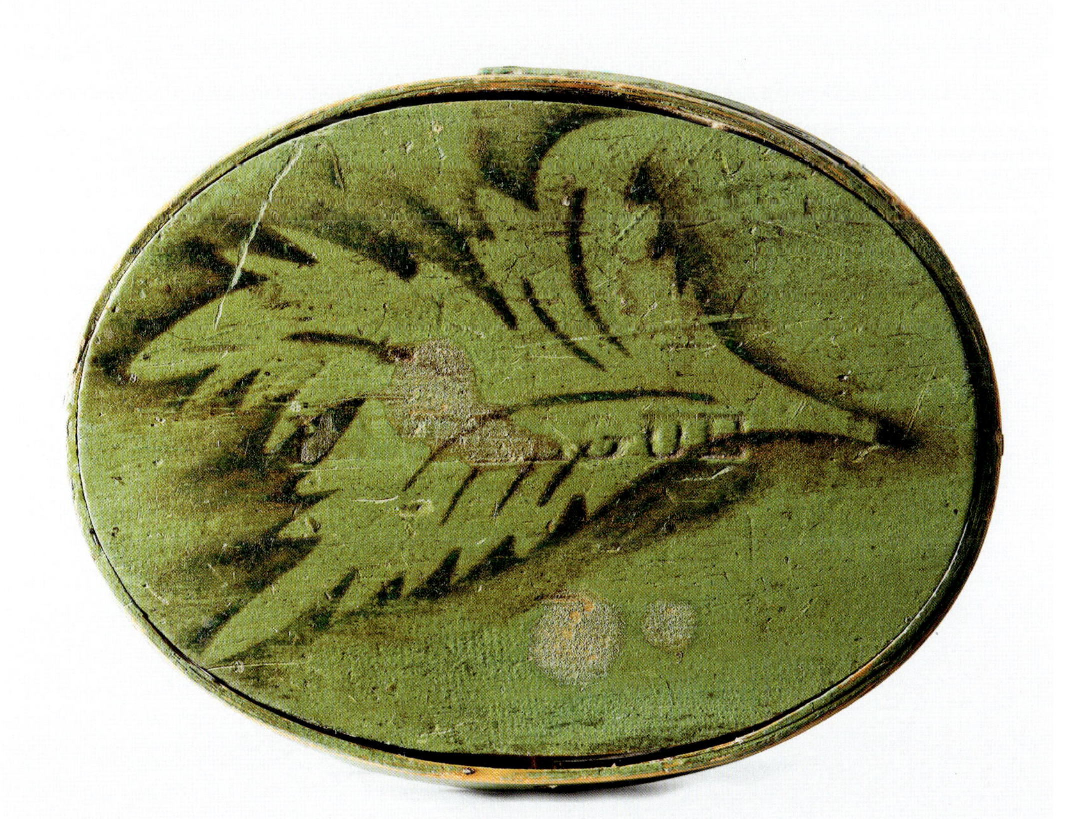

Plate 7.2

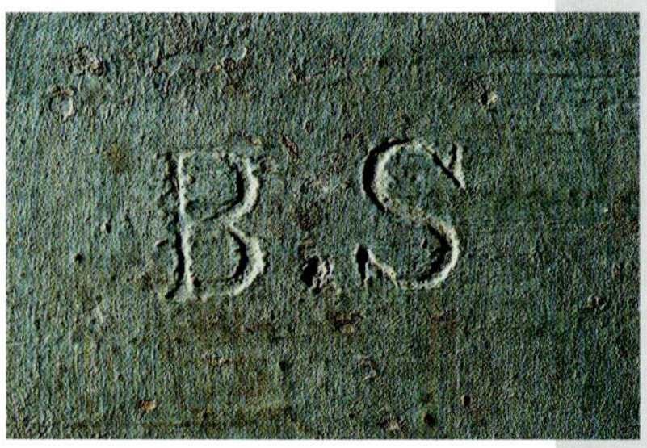

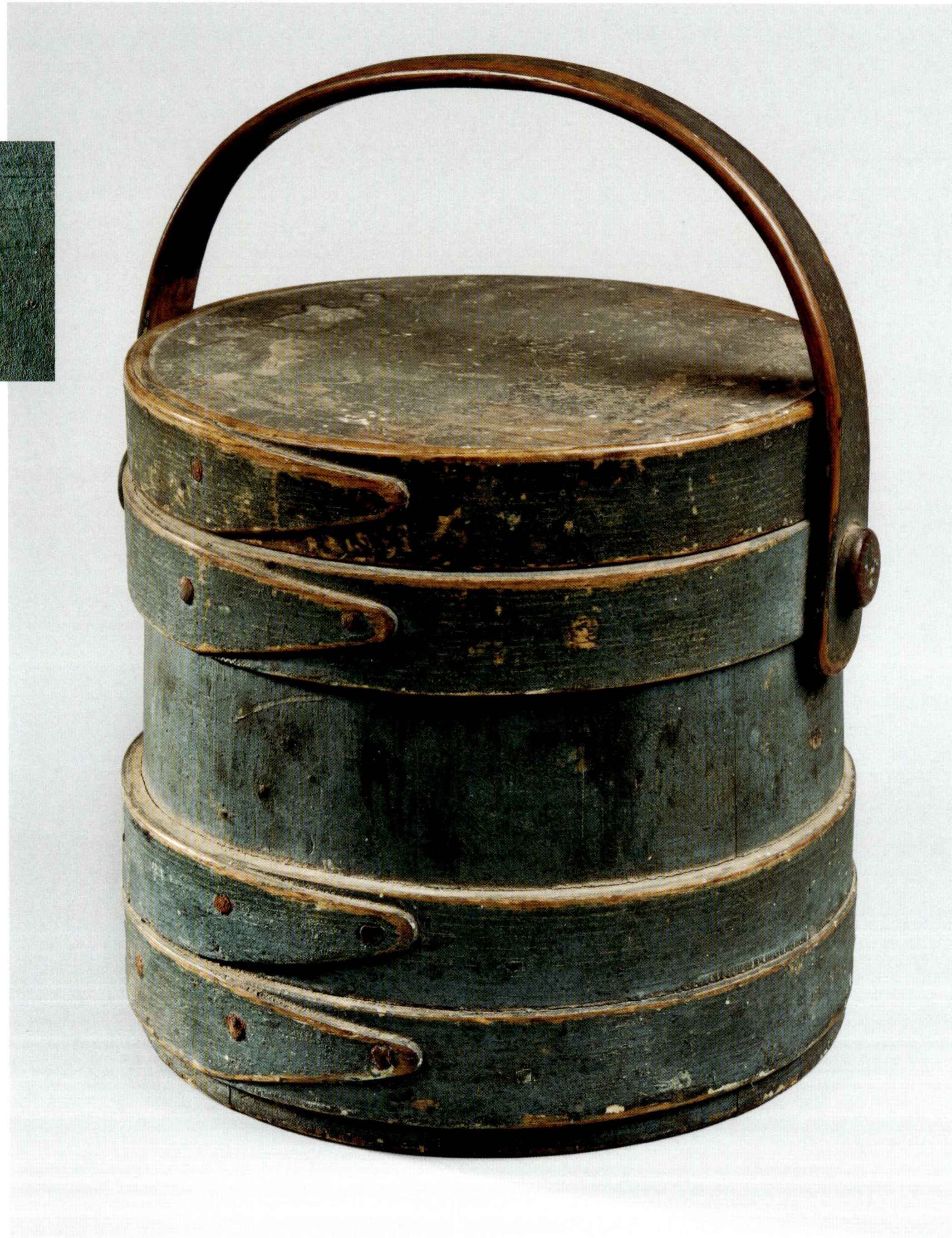

Plate 7.3

Plate 7.1
Oval box, Bela Sprague, 1820–60
Materials: White pine heads; ash sides; copper tacks
Dimensions: H 1 ⅝, W 4 ¼, D 3 ⅛
Inscription: "B. SPRAGUE" stamped on lid
Provenance: Courtesy of the Hersey family

Plate 7.2
Detail of stamp on lid of box in pl. 7.1

Plate 7.3
Bucket, Bela Sprague, 1820–60
Materials: White pine staves, heads, and buttons; oak hoops and handle; iron nails
Dimensions: H 7, Diam 7 ⅝
Inscription: "B.S" stamped on bottom
Provenance: Derin Bray, Portsmouth, New Hampshire; to current owner. Courtesy of the Hersey family

8
WILDER BUCKET FACTORY (1830–1891)

MECHANIZATION REVOLUTIONIZED THE WOODENWARE INDUSTRY IN THE NINETEENTH century. Buckets and tubs once crafted by hand were now produced with stave jointers, cylinder saws, and other specialized equipment powered by water or steam.[1] The first cooper in Hingham to embrace this technology was the precocious and visionary Crocker Wilder (1802–75).[2] In 1830, at age twenty-eight, he opened a factory on Friend Street near Main.[3] It boasted a steam engine by the pioneering engineer Holmes Hinkley, a noted designer of locomotives and coopering machinery.[4] Wilder could now fashion the heads, hoops, and staves of his products with greater speed and precision than his competitors, most of whom continued to work by hand in small shops.[5] Not surprisingly, he became the town's leading manufacturer of these goods.

Wilder was perhaps best known for his buckets. This handsome example (pl. 8.1), which retains its original blue paint, was made between 1830 and 1850, when the firm stamped its goods "C. WILDER & SONS / HINGHAM."[6] Its distinctive machine-cut staves dressed with tongue-and-groove joints (pl. 8.2) locked the sides of the bucket together like a puzzle, forming a stronger and more durable seal than earlier butt-joined staves. Rose-head nails securing the hoop ends confirm its age; cheaper, mass-produced flat-head nails and copper tacks had largely supplanted these hand-hammered fasteners by midcentury.

In 1845 Crocker Wilder & Sons erected a larger factory (see fig. 1.15) at the former site of Cushing's saw- and gristmill at Cushing Pond.[7] The main building was supported by a lumberyard, dry house, and paint shop. Waterpower drove the machinery, though a steam boiler was used to dry lumber and heat the factory during winter.[8] Around 1850, Alden Wilder (1804–89), Crocker's younger brother, became a partner in the firm.[9] Together the men oversaw a workforce of nearly thirty men and boys who produced huge quantities of woodenware, primarily for the Boston market, from which it was shipped all over the world. They also did a brisk business in the Oregon-Washington Territory and in California, where their sturdy buckets did heavy duty in frontier settlements and mining camps.[10] Items made by the Wilders during this period from 1850 to 1870 are stamped "C. & A. WILDER / SO. HINGHAM / MASS." (pl. 8.3).

In 1870 William C. Wilder (1829–91), Crocker's only son, took the helm.[11] He renamed the company and began marking goods "C. WILDER & SON / SO. HINGHAM, MASS."[12] Although buckets dominated the factory's production, tubs, fish kits, pails, and other household articles were also in the general line. Its butter churns, also called dash churns, were especially popular; during the factory's peak years five thousand to six thousand were turned out annually.[13] The underside of this tall example in green paint (pl. 8.4) is stamped with the updated brand. Its most noteworthy feature is the use of riveted iron hoops, rather than traditional wooden hoops, to bind the staves. Iron- and brass-hooped goods were made in the factory as early as 1834, though most known examples date to this later period.[14]

Following the death of William C. Wilder in 1891, the factory, with "all the machinery, tools, belts, and stock in hand," was sold for $9,600 and rebranded under the name A. J. Sprague.[15] A local carpenter, Anthony Jacob Sprague (b. 1855), failed to achieve the same success as his predecessor and shut down the operation after only a few years.[16] The enterprise was then taken over by Alvin Streeter of Streeter Manufacturing Company of Winchendon. He installed up-to-date machinery of his own design and resumed production of buckets. He also erected a new building for the manufacture of coopering machinery, now in demand throughout the country. After a fire destroyed the factory in 1902,[17] Streeter never regained momentum. Within a few years it was permanently closed, marking the end of large-scale manufacture of woodenware in Hingham.[18]

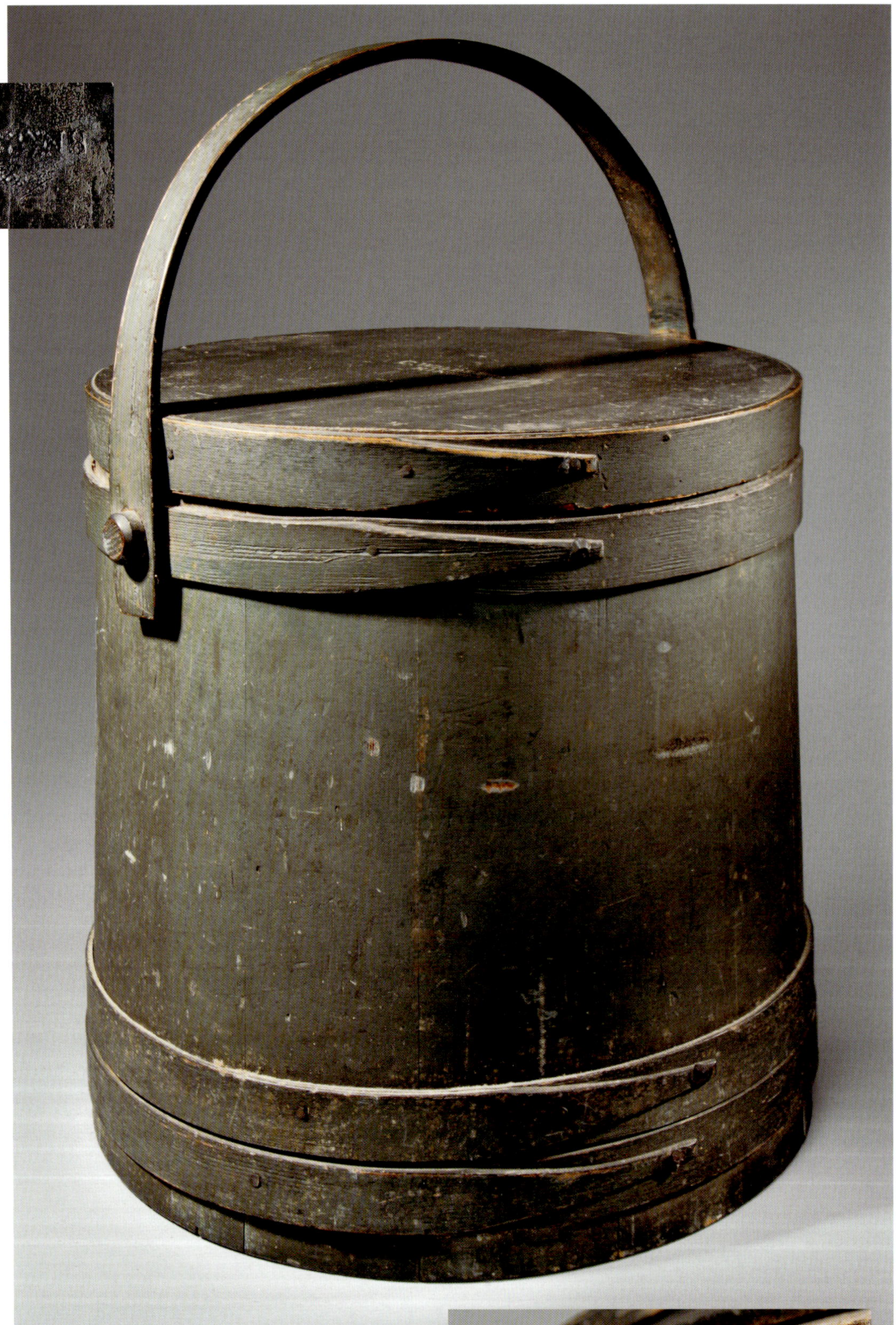

Plate 8.1

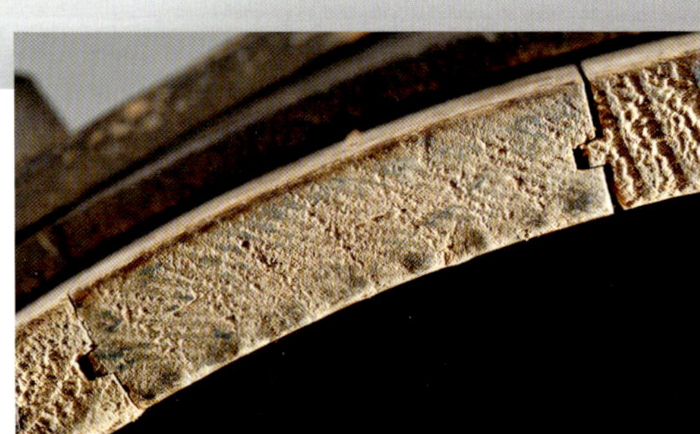

Plate 8.2

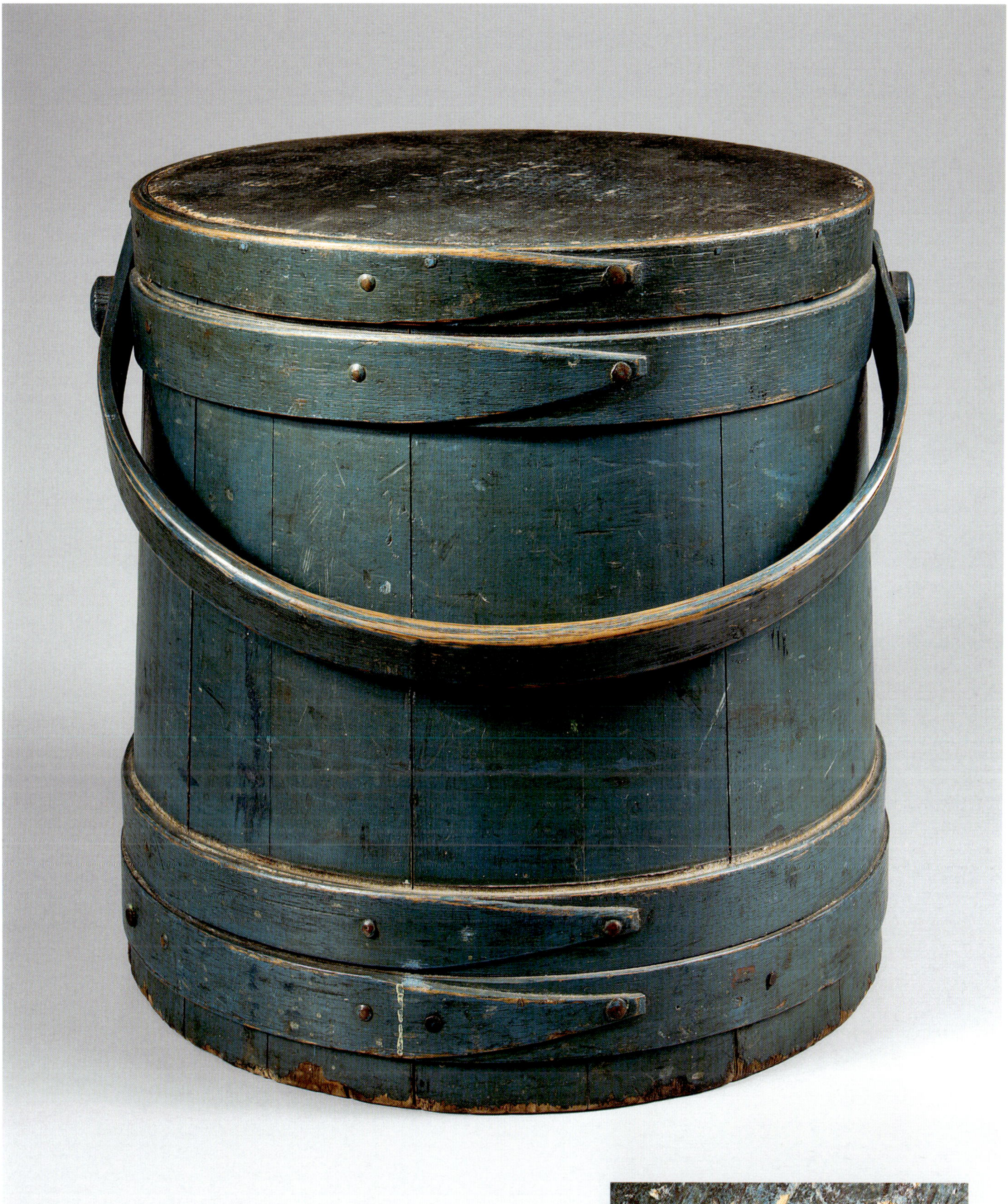

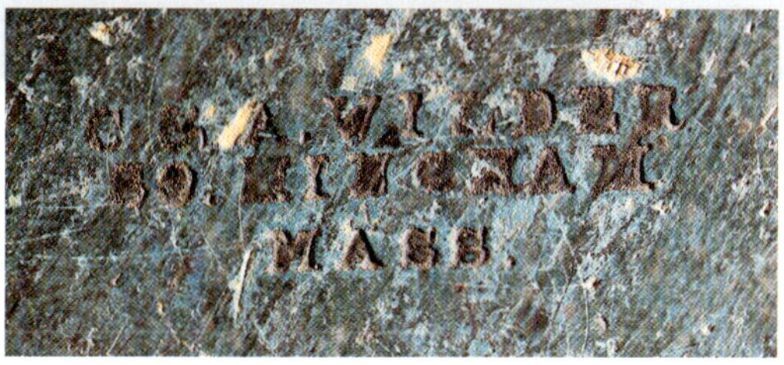

Plate 8.3

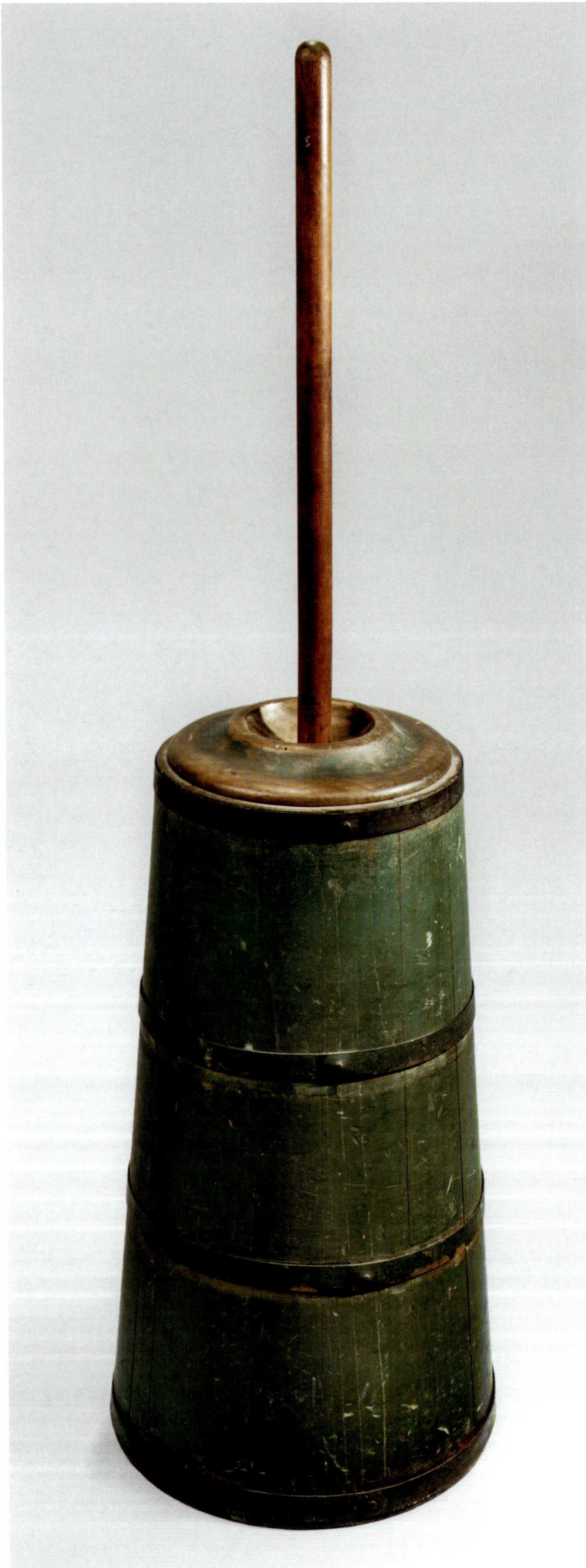

Plate 8.4

Plate 8.1

Bucket, C. Wilder & Sons, 1830–50

Materials: White pine staves, heads, buttons; ash hoops; oak handle; iron nails

Dimensions: H 14 ¼, Diam 14 ¾

Inscription: "C. WILDER & SONS / HINGHAM" stamped on lid

Provenance: Courtesy of Rindy and Gerry Bennett

Plate 8.2

Detail of stave joints on bucket in pl. 8.1

Plate 8.3

Bucket, C. & A. Wilder, 1850–70

Materials: White pine staves, heads, buttons; ash hoops; oak handle; copper nails

Dimensions: H 12 ⅛, Diam 12 ¼

Inscription: "C. & A. WILDER / SO. HINGHAM / MASS." stamped on lid

Provenance: Courtesy of Hingham Historical Society

Plate 8.4

Churn, C. Wilder & Son, 1870–91

Materials: White pine staves, bottom, and lid; chestnut dasher bottom; probably birch handle; iron hoops

Dimensions: H 22 ½, Diam 12 ½

Inscription: "C. WILDER & SON / SO. HINGHAM, MASS." stamped on bottom

Provenance: Courtesy of Peter and Diana Bennett

9
LEONARD GARDNER (1809–1890)

IN 1885 THE *BOSTON GLOBE* COMMENDED LEONARD GARDNER FOR HIS "ASTONISHING achievements in mathematics" and "perseverance in scientific research."[1] One might expect such accolades for a college professor, but Gardner, the son of a small-time Hingham cooper, was a bucketmaker whose formal education did not extend beyond grade school. Remarkably, this modest craftsman had taught himself advanced mathematics in his spare hours—a feat that required him to translate the work of French astronomers, often by candlelight. He gained particular recognition for his calculation of the transit of Venus across the face of the sun; a noted Harvard scientist remarked that "there were not but six men in the United States who could have done it."[2]

By all accounts Leonard Gardner had a modest upbringing. He was raised on a farm in South Hingham where his father, Charles, ran a small cooper's shop.[3] As was customary for the sons of rural craftsmen, he left school at the of age fourteen to work full-time at the bench and plow.[4] His father's account book for the years 1796 to 1801 offers a glimpse of what young Leonard's life might have been like. Winters were spent splitting staves from white wood and assembling large numbers of pails and tubs; the warmer months brought the return of a heavy cycle of planting, tending, and harvesting crops and repairing fences and buildings; coopering was relegated to side work.

In 1835 Gardner married the sister of an upstart industrialist named Crocker Wilder. Just five years earlier Wilder had opened the first mechanized woodenware factory in Hingham, quickly becoming one of the largest and most profitable bucket producers in the region—and a major employer of local coopers.[5] Although it is unknown whether Gardner worked there, the factory must have had some influence on his career; his other brothers-in-law, Elijah Whiton and Joshua Tower, would go on to establish similar enterprises modeled after Wilder's success.[6]

In search of new opportunities, Gardner packed up his family in 1846 and moved to the neighboring town of South Weymouth, where he commenced work at the box mill of Reuben Loud.[7] After only a few years he opened a water powered factory of his own, where his crew of five men turned out 16,000 buckets annually—many with wire bail handles and hoops of brass or iron. Within a decade annual output more than doubled, which established Gardner as one of New England's leading coopers.[8] Many goods made by him during this period, from 1850 to 1866, are impressed with the maker's mark "L. GARDNER / S. WEYMOUTH."[9]

Gardner returned to Hingham around 1866, when he built a new factory on Gardner Street near the old family homestead.[10] There he made classic Hingham buckets—and continued production of the metal-hooped containers he had turned out by the thousands in Weymouth. Among the best-preserved of these is an example in its original bright blue paint (pl. 9.1), constructed of machine-cut tongue-and-groove staves and riveted iron hoops; its lid is iron-hooped as well, with round-headed copper tacks. The wire handle is secured to the bucket's sides with diamond-shaped bail plates. This feature is often mistaken for Shaker construction, but it was also commonly used by coopers in Hingham and probably elsewhere.[11] The underside of the bucket is stamped "L. GARDNER / S. HINGHAM."

Large, round boxes appear to have been the most distinctive items produced in Gardner's factory. This deluxe model in gray paint (pl. 9.2), remarkable for its unusual construction, was probably made as a presentation box—possibly for James Smith, whose name is calligraphically inscribed in ink on the top. The body is formed of vertical staves like a bucket; the two middle hoops, however, are purely decorative. Other Gardner boxes, usually without painted surfaces, are also made with staved sides and typically feature three hoops with ends pointed in opposing directions. The underside of this example is stamped in small letters "L. GARDNER / S. HINGHAM / MASS."

Leonard Gardner (pl. 9.3) remained in Hingham for only a few short years; by 1870 he had left the coopering industry altogether. He began to identify himself variously as a carpenter, painter, and town official—and to devote more of his time to his greatest passion: astronomy (pl. 9.4).[12] At least one account suggests that he performed commissioned work for the Nautical Almanac Office, later the U.S. Naval Observatory in Washington, D.C.[13] Around 1890 his vacant woodenware factory was repurposed by the residents of the Gardner and Whiting streets neighborhood for use as a church and Sunday school; it was known as the United Social Society of South Hingham.[14]

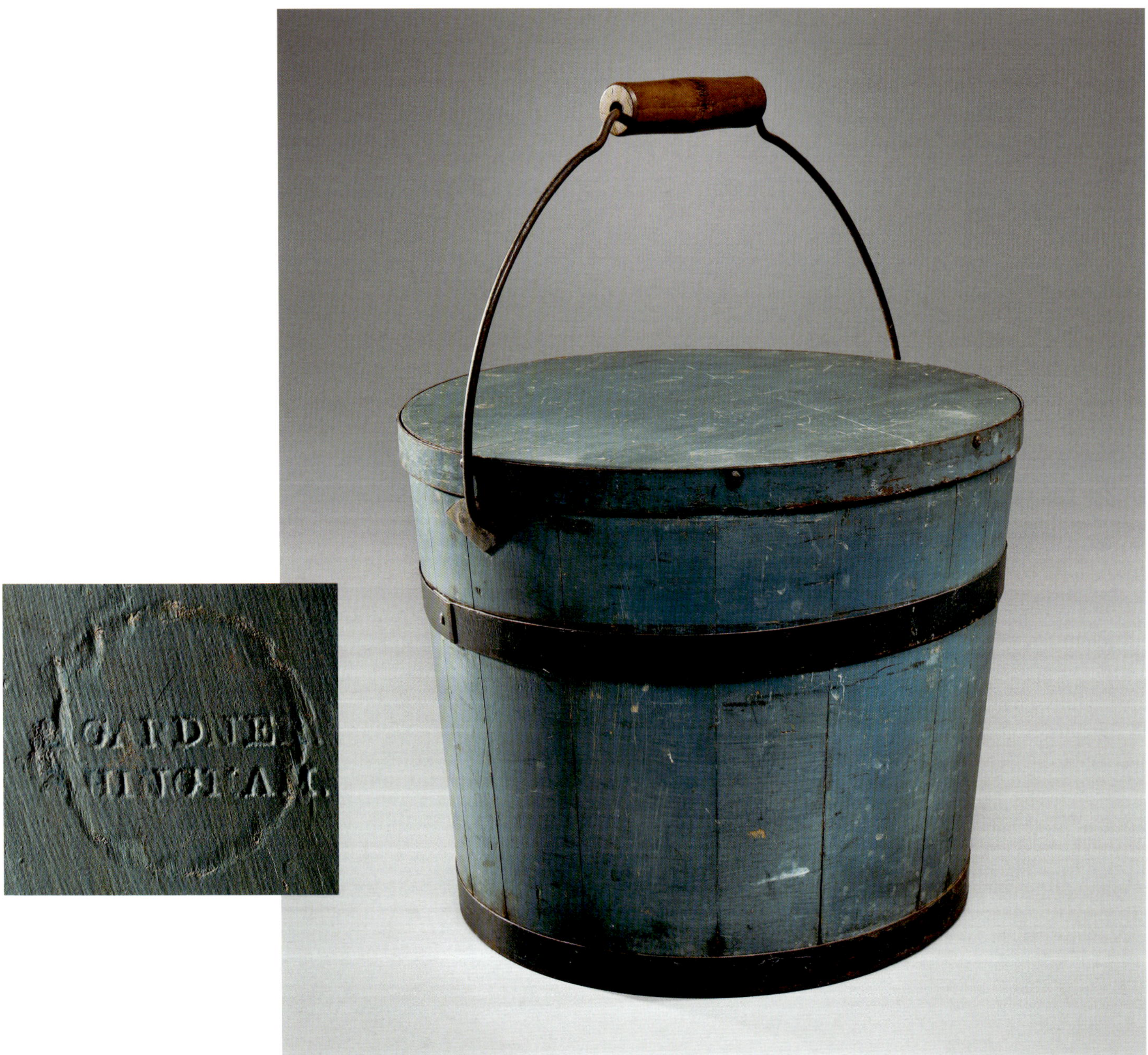

Plate 9.1

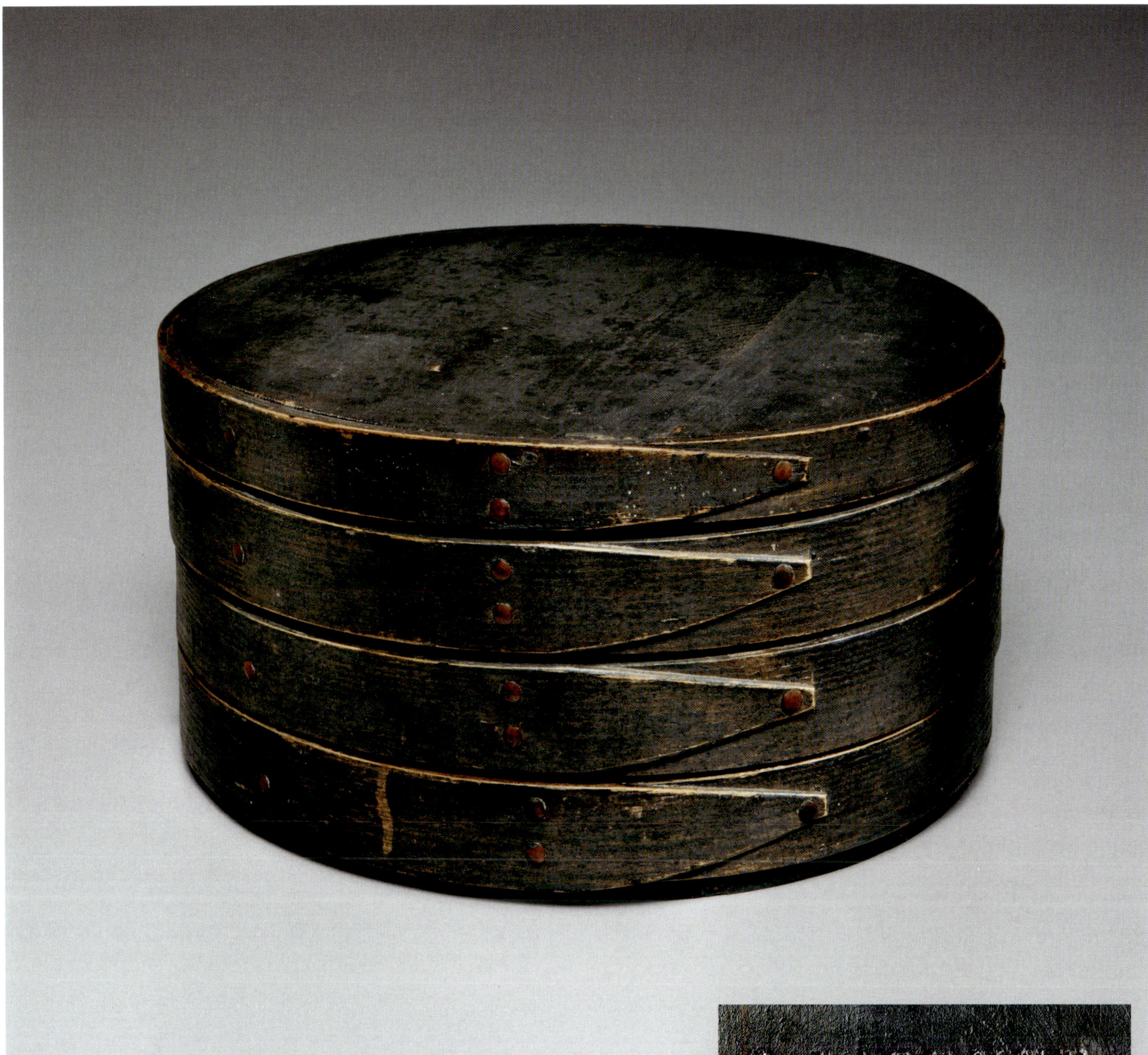

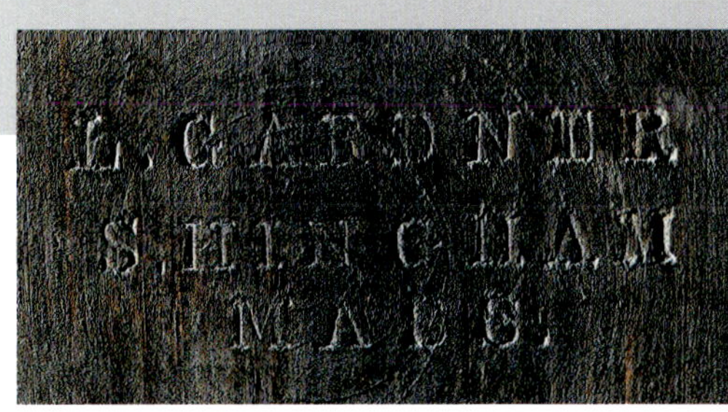

Plate 9.2

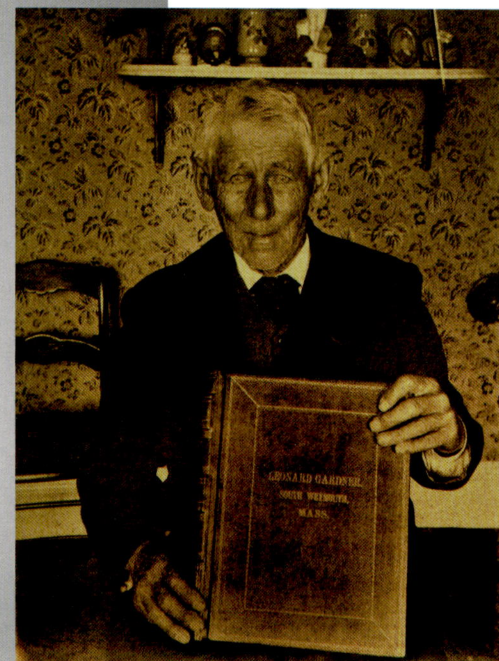

Plate 9.3

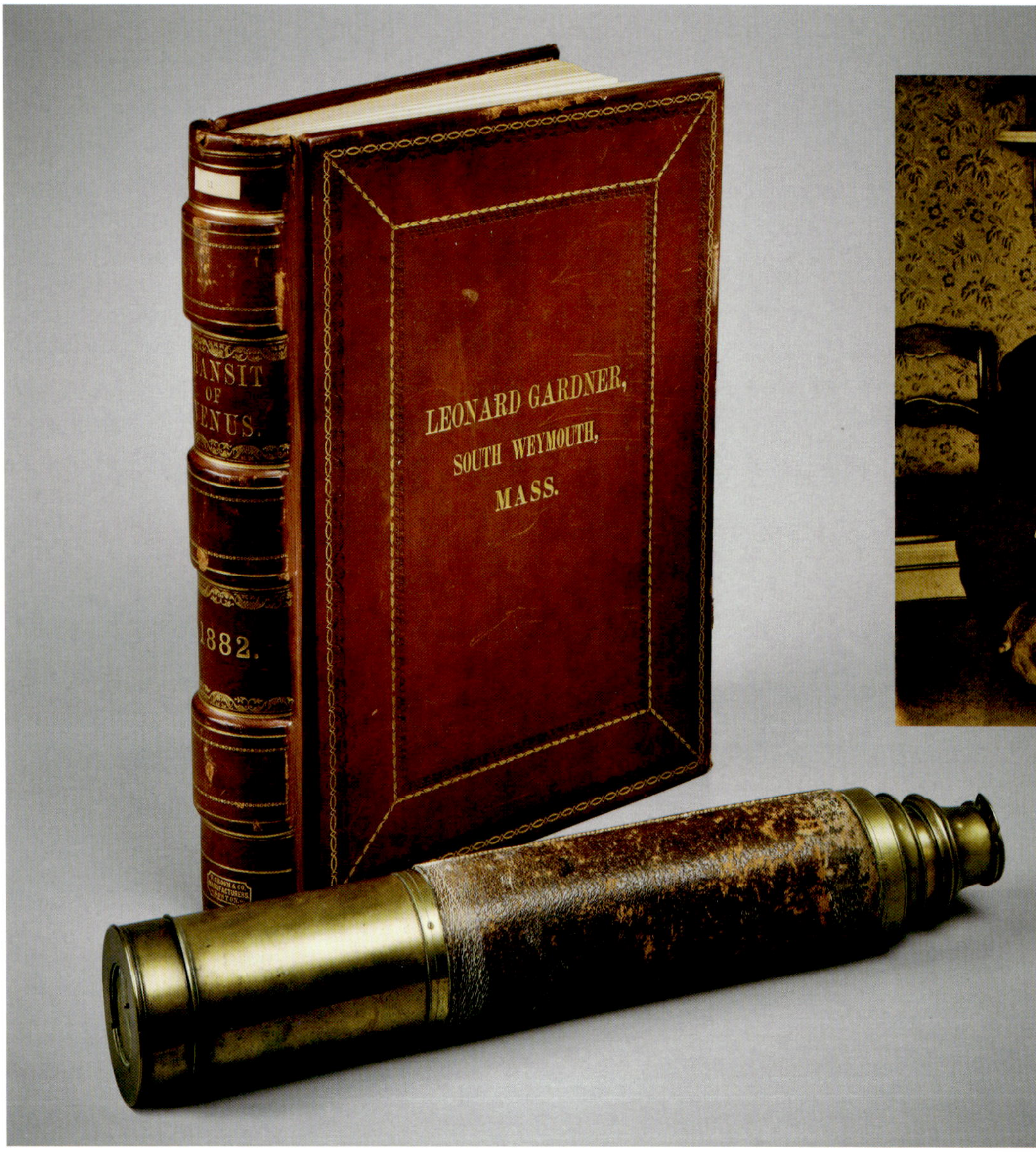

Plate 9.4

Plate 9.1
Bucket, Leonard Gardner, 1866–70
Materials: White pine staves and heads; maple and wire handle; iron hoops, bail plates; copper tacks
Dimensions: H 9 ½, Diam 10 ⅜
Inscription: "L. GARDNER / S. HINGHAM" stamped on bottom
Provenance: Courtesy of Peter and Diana Bennett

Plate 9.2
Round box, Leonard Gardner, 1866–70
Materials: White pine staves and heads; ash hoops; copper nails
Dimensions: H 6 ¾, Diam 13 ¼
Inscriptions: "L. GARDNER / S. HINGHAM / MASS." stamped on bottom; "James Smith" in ink on lid
Provenance: Courtesy of the Hersey family

Plate 9.3
Leonard Gardner with his book of calculations, 1870–90
Provenance: Courtesy of Weymouth Historical Society

Plate 9.4
Leonard Gardner's spyglass and book of calculations, 1865–85
Provenance: Courtesy of Weymouth Historical Society

10
NELSON CORTHELL (1815–1900)

BY 1880 ONLY TEN COOPERS IN HINGHAM STILL MADE BUCKETS AND TUBS BY HAND. Their peers, who once numbered in the hundreds, had either left the trade or gone to work in factories.[1] Nelson Corthell was among the few craftsmen who continued to produce traditional woodenware (pl. 10.1), from his small workshop on Hersey Street.[2] He exhibited several of these goods at the Hingham Agricultural and Horticultural Society fair in 1877; a judge for the show lauded their many fine and preindustrial qualities: "Of the 'home-made wooden ware' for which Hingham is so distinguished, we had two contributions consisting of sixteen pieces. They were samples of the old-fashioned 'pails' and 'piggins,' such as our fathers made in 'ye olden time.' The work showed skill, which did not lose sight of durability—a feature in workmanship hard to find in these 'shoddy' days. Such specimens are hard to find unless made by Hingham's old style artisans who work with frow, brake, and draw-shave."[3]

Corthell survived the industrial age by adopting a new business model. He no longer made woodenware for export to Boston and beyond—markets now firmly the domain of C. Wilder & Son and other large factories. Instead, he tapped into a newly emerging clientele of antiquarians and collectors—people interested in American history. They purchased his buckets not as household necessities, but as novelties that harked back to a simpler time. The throngs of tourists and summer residents who descended on Hingham and nearby Nantasket Beach were a large and steady clientele.[4] Corthell probably had them in mind when, in 1894, he nostalgically identified himself in the town directory as "the last of many small shops."[5]

This water pail and piggin (pl. 10.2) were made during this later period. Both are stamped "N. CORTHELL" on the underside and are masterfully crafted with butt-joined staves and wooden pegs in the hand-tool tradition. The hoops are held in place with round-head copper nails—fasteners that were not used until the mid-nineteenth century. Both containers survive in excellent condition, proof that they led a quiet life on a mantel or shelf, as was intended. Piggins—one of the more distinctive forms made by coopers—originally served a variety of uses, from catching milk drips to carrying food for pigs and chickens.[6]

The Hingham historian and collector John P. Richardson recovered this small bucket (pl. 10.3) from a collapsed barn on Emerald Street, just a short distance from where Nelson Corthell worked. He also salvaged other coopered goods from the building—some stamped by Corthell and others by their original owner, Michael Burns, an Irish immigrant who moved to Hingham in the 1860s.[7] The bucket's underside is marked N. CORTHELL, each letter struck with an individual stamp. Curiously, this example exhibits many of the construction techniques associated with factory production, including machine-cut staves, a wire handle with tin bail plates, and copper staples to secure the hoop-ends—this last feature introduced around the end of the nineteenth century.[8] No clear explanation for the bucket's construction has come to light. Perhaps Corthell was the owner, not the maker; perhaps he retailed the work of other coopers.

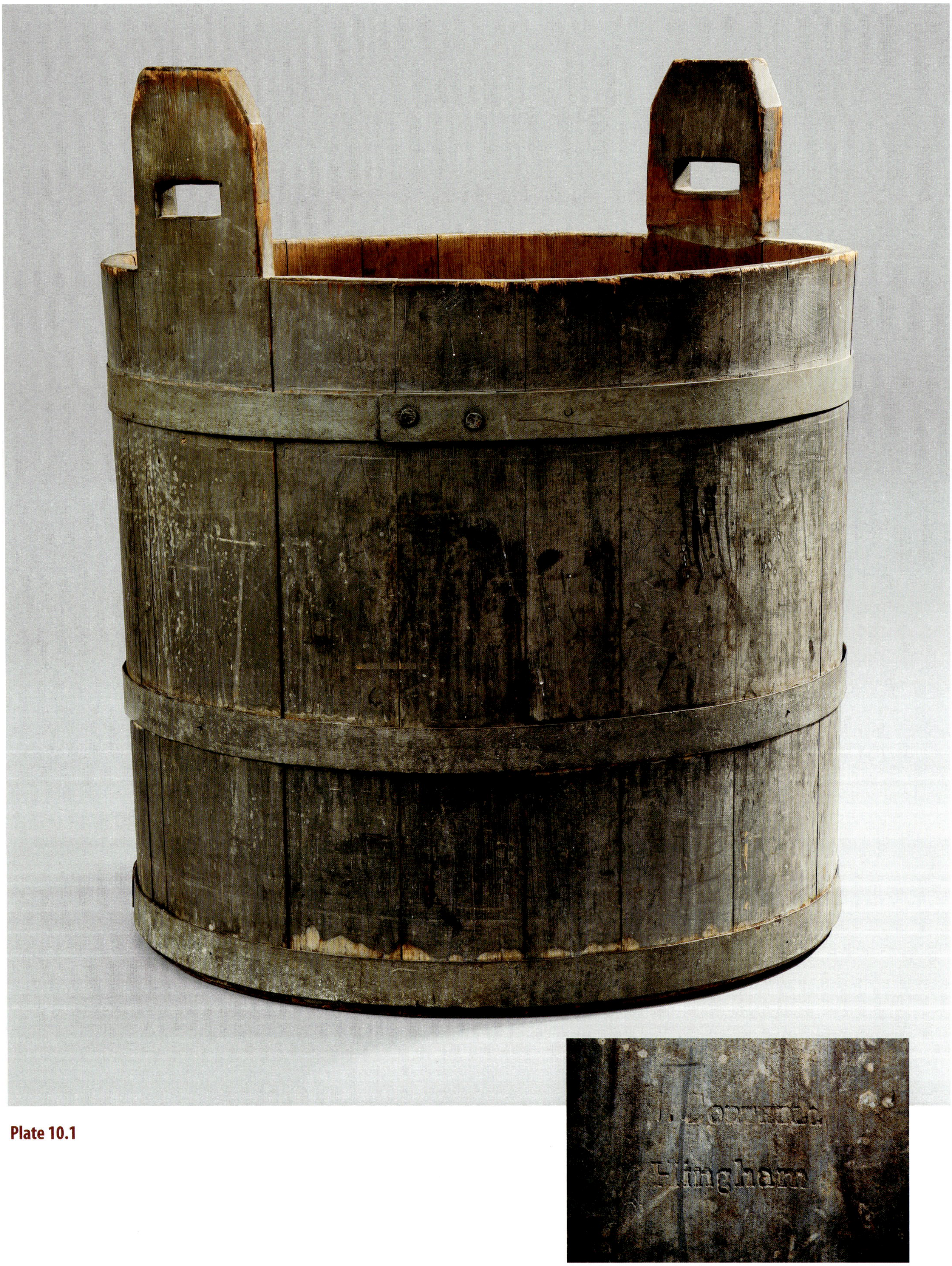

Plate 10.1

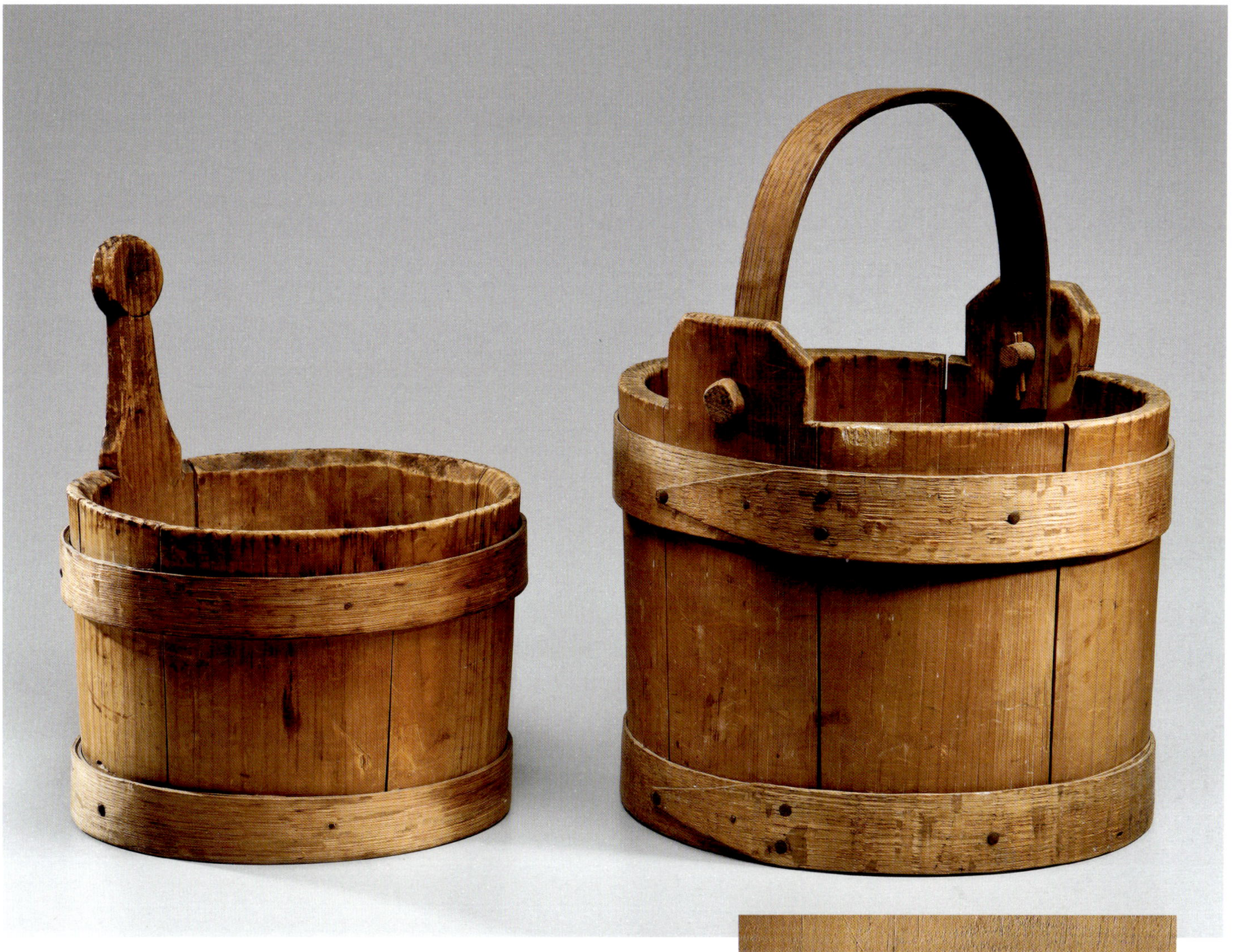

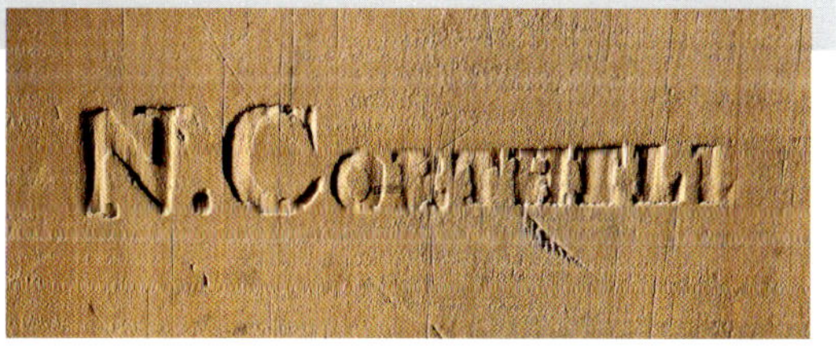

Plate 10.2

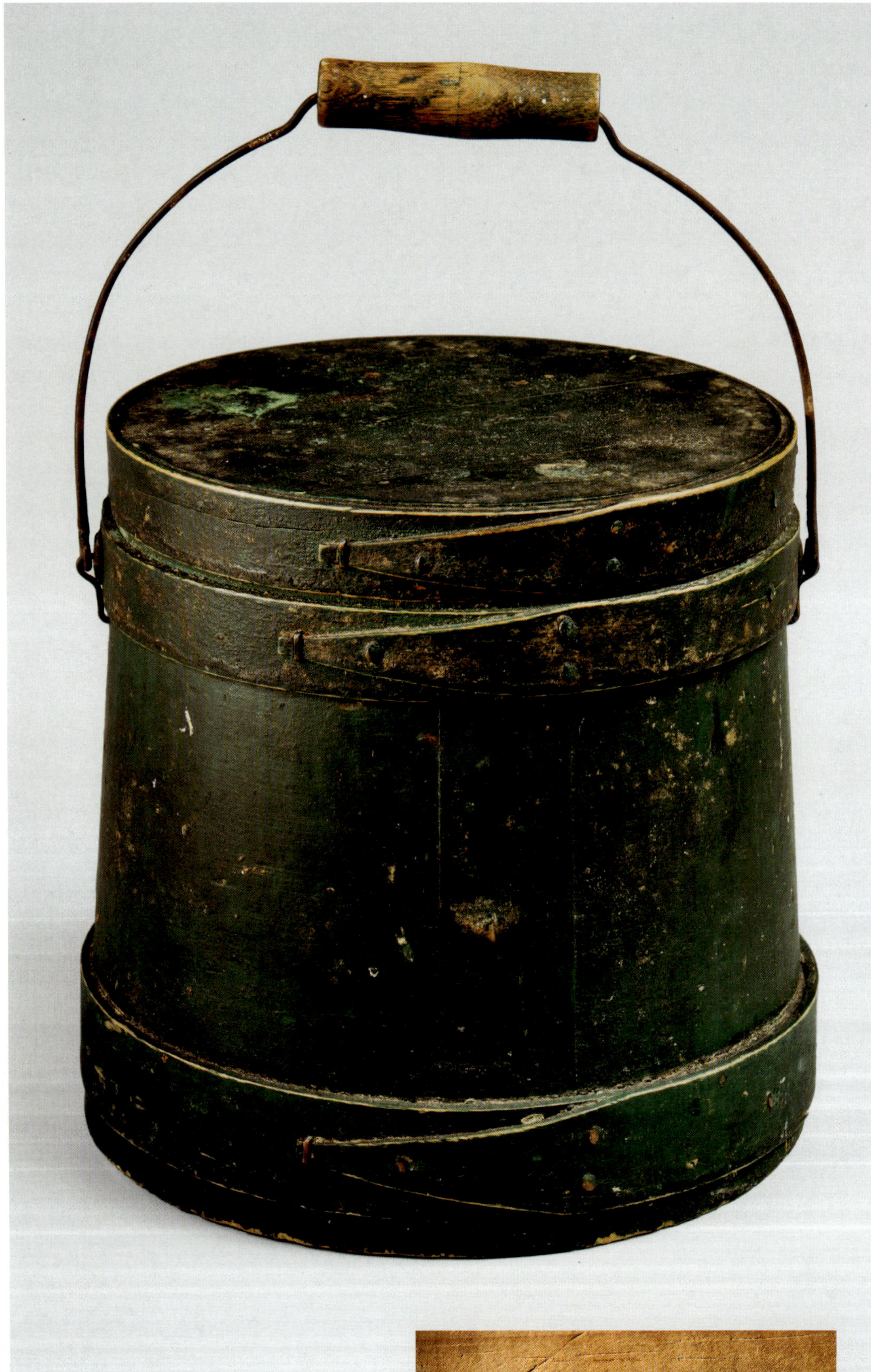

Plate 10.3

Plate 10.1

Tub, Nelson Corthell, 1830–60
Materials: White pine staves and bottom; iron hoops
Dimensions: H 25, Diam 22
Inscription: "N. CORTHELL / Hingham" stamped on bottom
Provenance: Courtesy of Larry Corthell

Plate 10.2

Pail and piggin, Nelson Corthell, 1840–1900
Materials: White pine staves, bottoms, and buttons; oak hoops and handle; copper nails
Dimensions: Pail: H 10 ¾, Diam 11 ⅛; piggin: H 11 ⅜, Diam 9
Inscription: "N. CORTHELL" stamped on bottoms
Provenance
Pail: Courtesy of Hanover Historical Society
Piggin: Courtesy of Peter and Diana Bennett

Plate 10.3

Bucket, 1890–1900
Materials: White pine staves, heads, and buttons; ash hoops; birch and wire handle; copper nails and staples; iron bail plates
Dimensions: H 7 ¾, Diam 8 3⁄16
Inscription: "N. CORTHELL" stamped on bottom
Provenance: Burns family, Emerald Street, Hingham, Massachusetts; acquired by John P. Richardson, Hingham; purchased by current owner at Willis Henry Auctions, November 19, 2011, lot 39. Courtesy of the Hersey family

11
COTTON HERSEY (1792–1881)

HERSEY STREET IN HINGHAM TEEMED WITH COOPERS' SHOPS IN THE EIGHTEENTH and nineteenth centuries.[1] Of these quaint buildings, the one that probably attracted the most attention belonged to Reuben Hersey (1757–1845), a fifer in the American Revolution and a third-generation craftsman.[2] He made the same durable goods as his competitors—everything from pails to piggins—but set himself apart by specializing in boxes and miniature woodenware for children.[3] These charming hand-sized containers became a steady side business for him and other enterprising members of the Hersey family.

Few objects crafted by Reuben are known to survive. His children, however, left behind a substantial material record (see chapter 3). Reuben's son Cotton was among the most prolific and long-lived coopers and toymakers of the nineteenth century. From an early age he balanced the heavy work of farming with coopering—through busy times, seasonal lulls, and in the evenings; in decades of unceasing pursuit of both livelihoods Cotton Hersey learned to make the full range of traditional woodenware, such as this classic six-inch oval box (pl. 11.1) painted blue and stamped "C.H." The same bold type is found on goods marked by other Herseys, including Cotton's nephew Samuel and son Bela French Hersey.[4]

In 1814, at age twenty-two, Cotton Hersey built a small Cape on a lot carved out from the family farm.[5] Although official records continued to identify him as a cooper and a boxmaker, his new shop, located just across the street, became the town's leading producer of toys and novelty woodenware. At the time toymaking was barely acknowledged as a profession anywhere in America, let alone in a rural town like Hingham. By midcentury, however, a new generation of ambitious craftsmen—mostly nephews and cousins of Hersey—were advertising themselves as toymakers.[6]

These small boxes (pl. 11.2) are typical of the keepsakes made in his shop. All three descended locally; the example in yellow paint has the most poignant history. Cotton Hersey made it in 1829 for his youngest daughter, Lydia, as a token of his affection.[7] The "C. HERSEY" stamped on the lid offers further evidence of its maker. Hersey also used this mark on a diminutive canteen (pl. 11.3) made around the same time. Local boys would have immediately recognized the form; military canteens had been a chief export for Hingham's coopers for over half a century.In fact, Cotton Hersey would have worn a similar, full-sized model during his service in the volunteer militia.[8]

Miniature buckets appear to have been Hersey's best seller—dozens of examples are extant today. They are nearly identical to their larger counterparts in construction, with sides of butt-joined staves and copper tacks and tiny wood pegs securing the hoops and heads. This pristine model (pl. 11.4) is stamped "C.H" on the top of the lid and is inscribed on the underside: "Made by / Cotton Hersey / 1878 / aged 86 years / Born 1792." Although many of his items were left in a natural finish, several are known with original painted surfaces in vibrant blues, greens, and reds (pls. 11.5–11.7). In fact, more of his goods are embellished with paint than any other Hingham maker's, an indication that Hersey had a hand in their decoration. This tradition was continued a generation later when his nephew Reuben Hersey also painted toys in his shop.

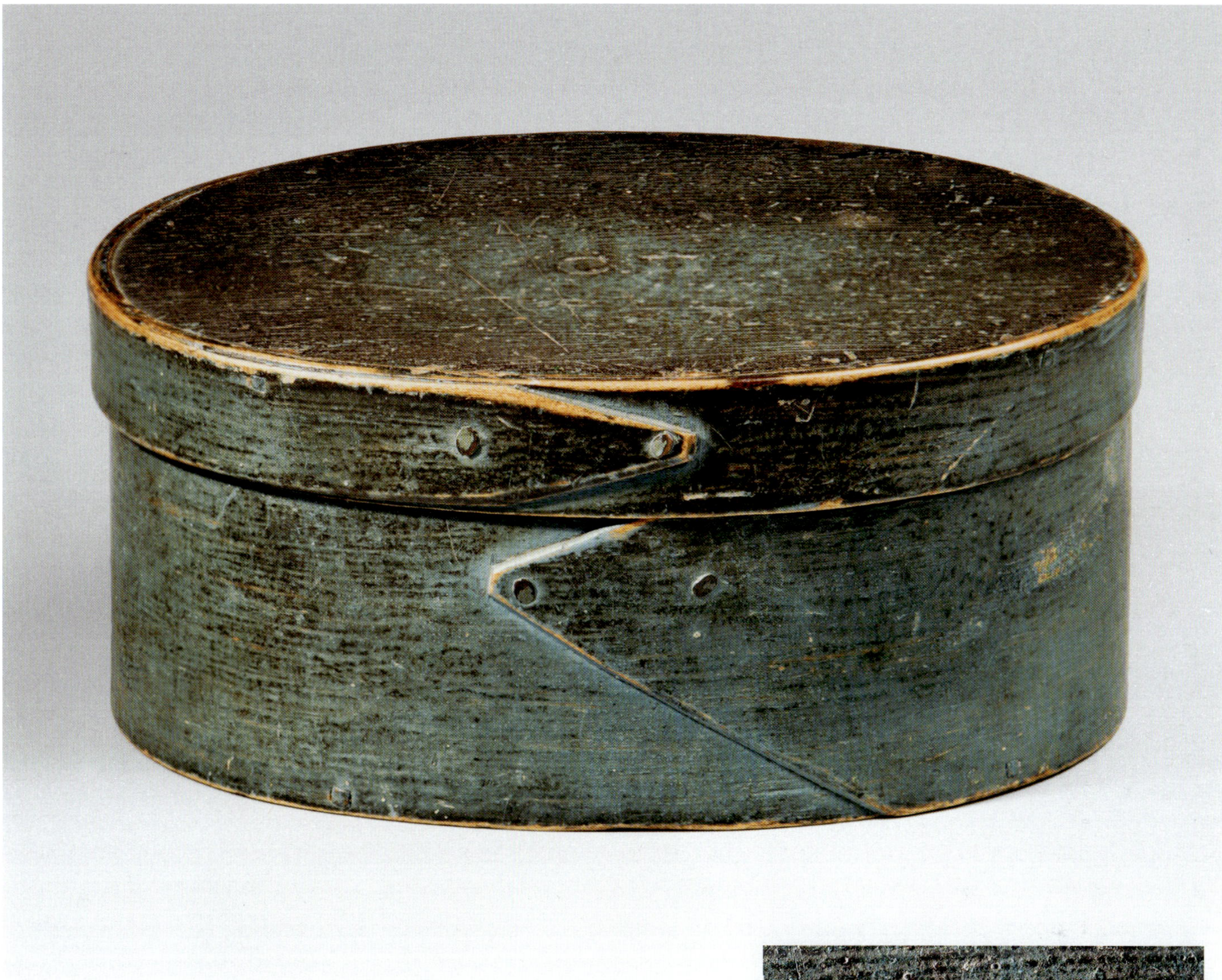

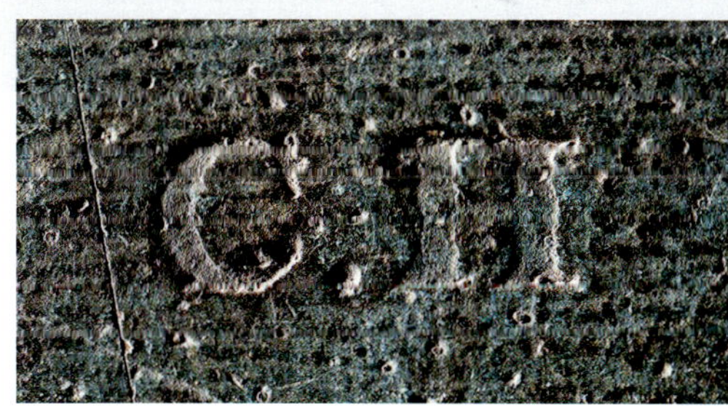

Plate 11.1

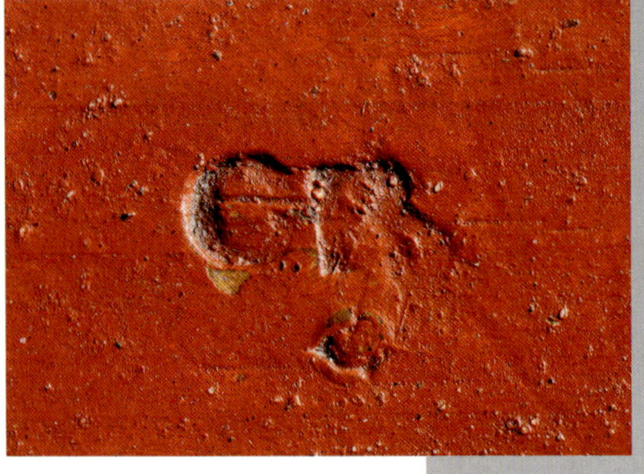

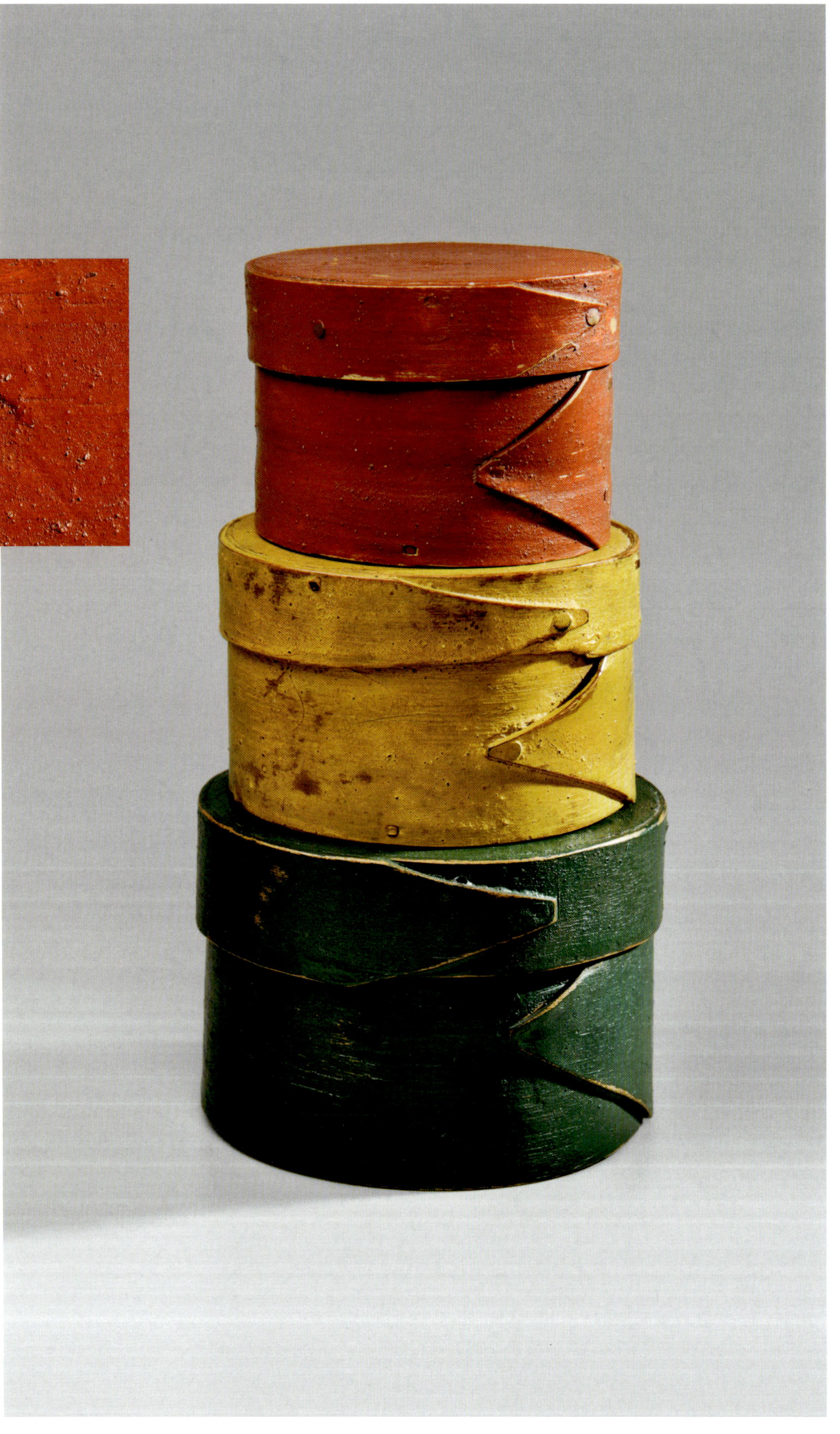

Plate 11.2

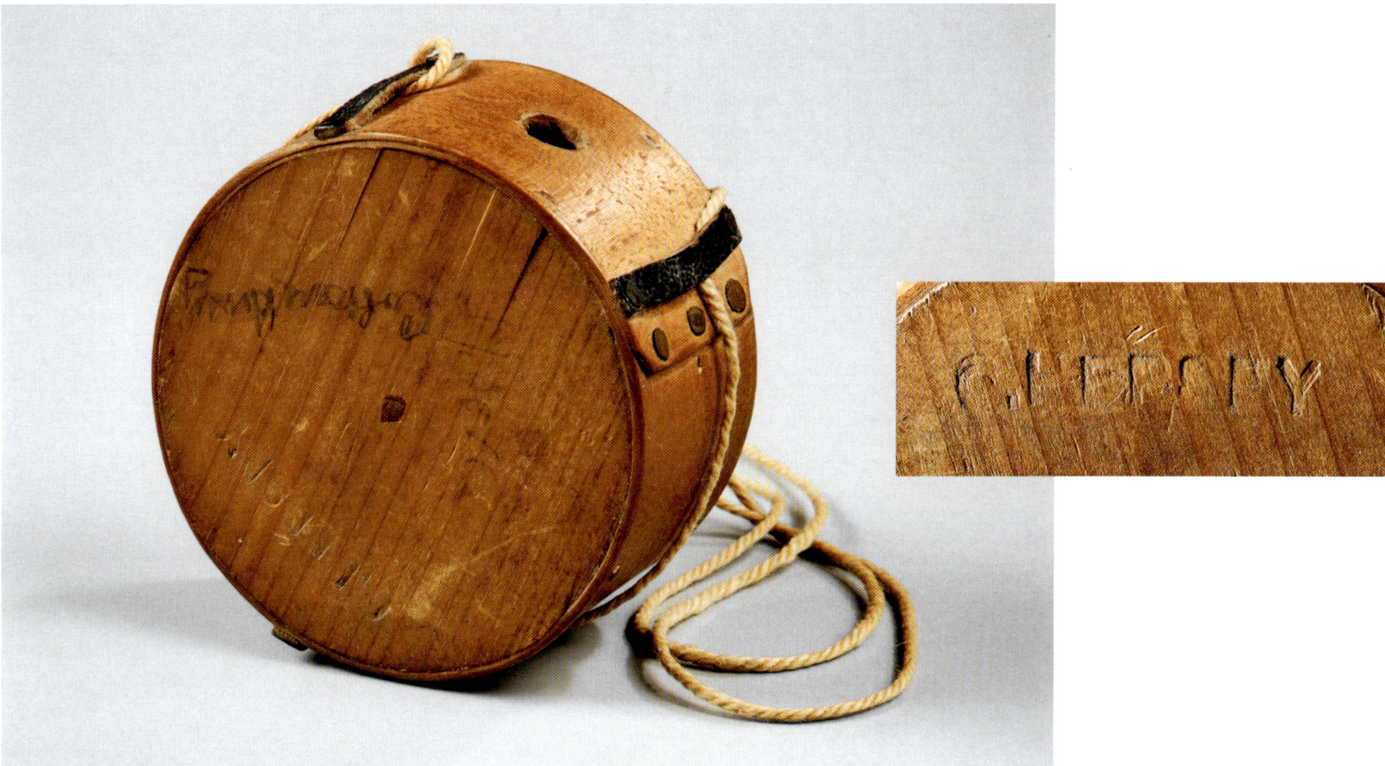

Plate 11.3

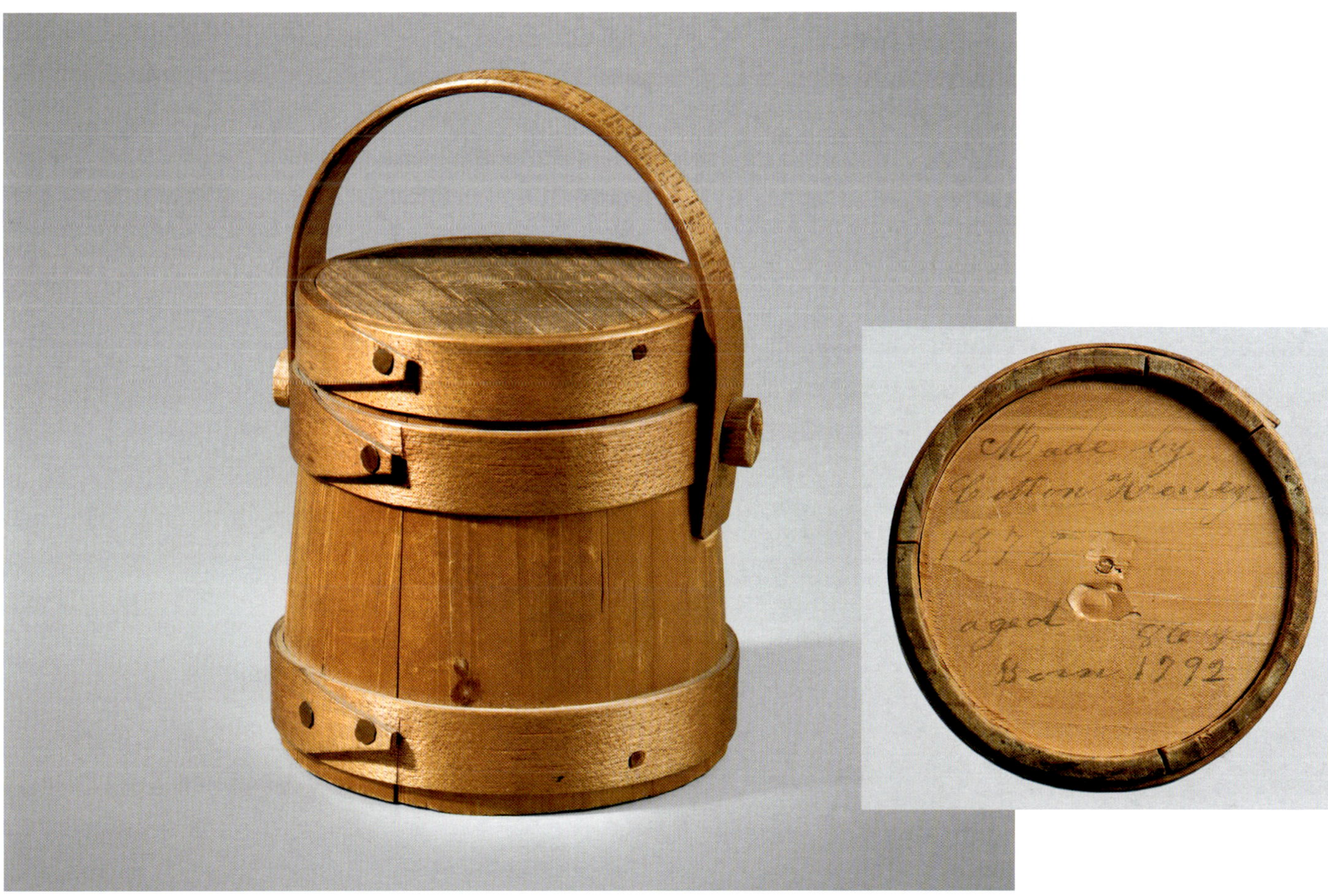

Plate 11.4

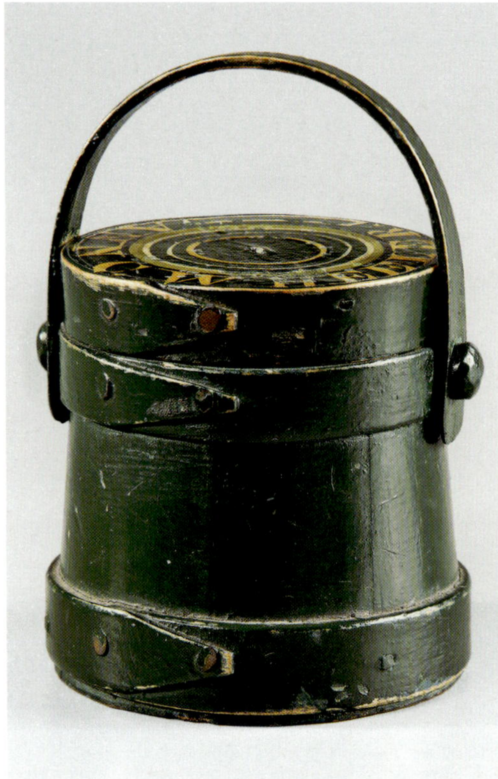

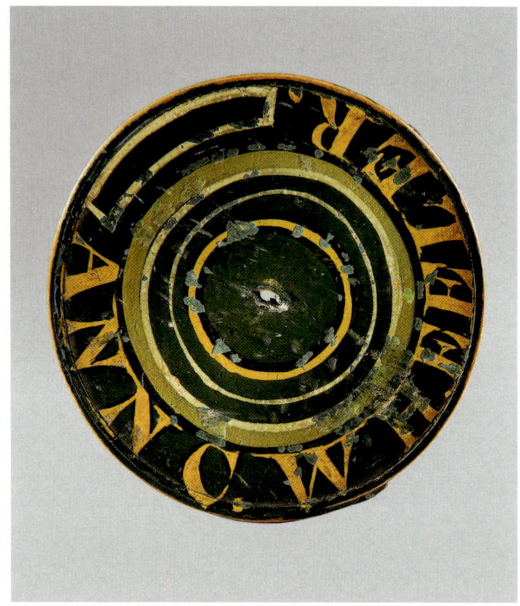

Plate 11.5

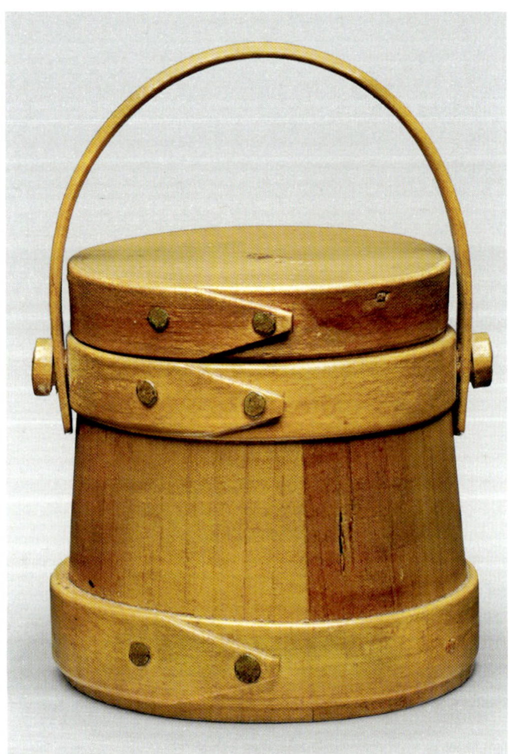

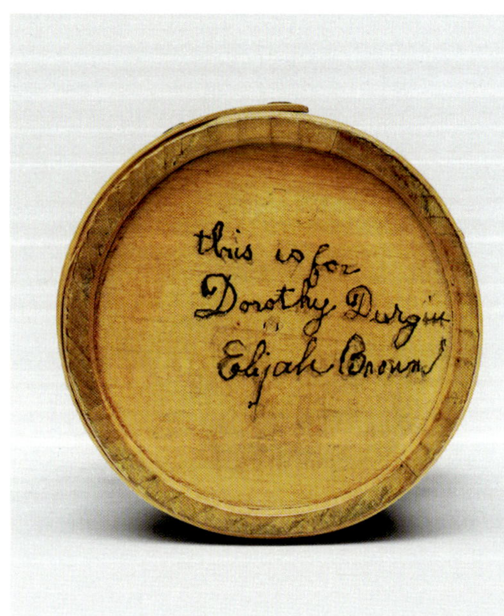

Plate 11.6

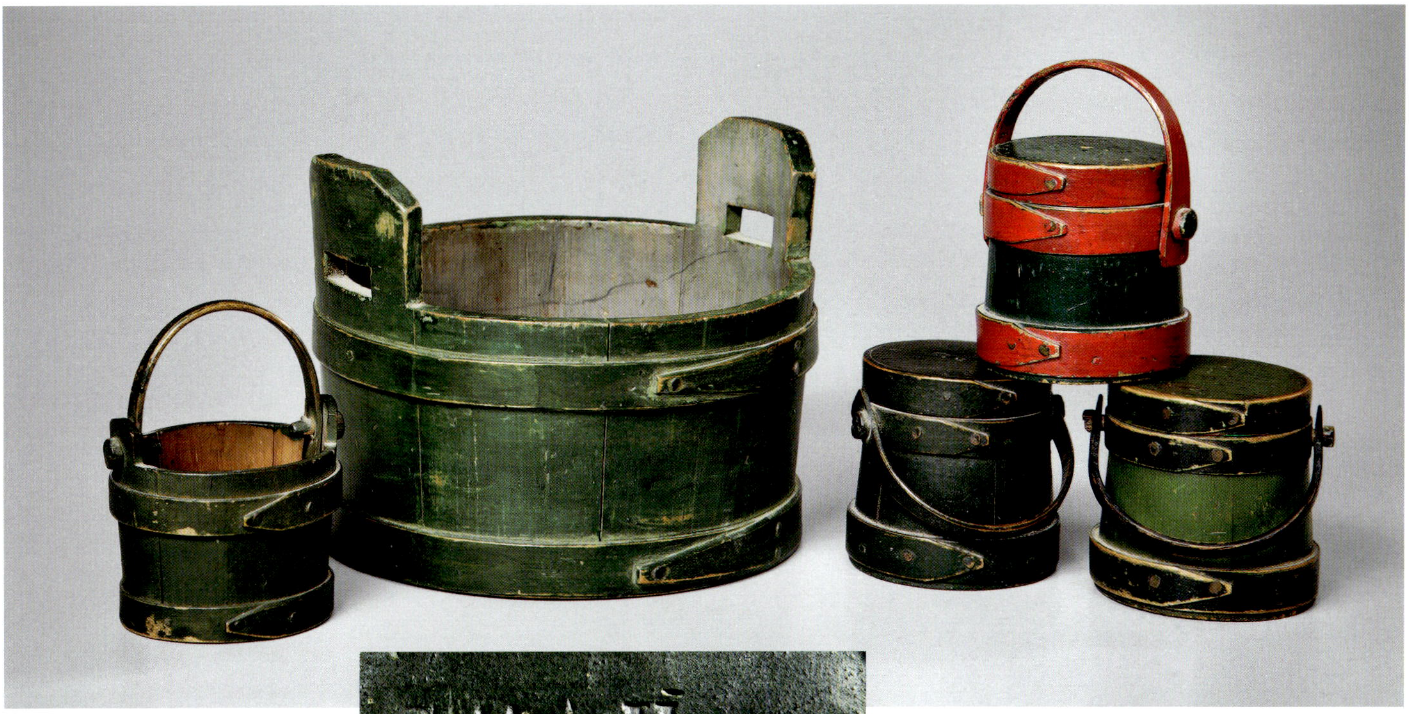

Plate 11.7

Plate 11.1

Oval box, Cotton Hersey, 1820–60

Materials: White pine heads; maple sides; copper tacks

Dimensions: H 2 5/8, W 6 1/4, D 4 9/16

Inscription: "C.H" stamped on lid

Provenance: Courtesy of the Hersey family

Plate 11.2

Miniature round boxes, Cotton Hersey, 1820–80

Materials: white pine heads; maple hoops; copper tacks

Dimensions: Red box: H 1 7/8, Diam 2 11/16; yellow box: H 1 7/8, Diam 3; green box: H 2 1/4, Diam 2 3/8

Inscriptions

Red box: "CH" stamped on lid and bottom; "MAF 4 / WOL / Randal / Lincoln" in graphite on underside of lid

Yellow box: "C.HERSEY" stamped on lid; "Made by / Cotton Hersey for / Lydia Hersey, Sept. 4 1829, BH" in graphite on underside of lid

Green box: "C.HERSEY" stamped on lid; "1832" in graphite on underside of lid

Provenance

Red box: William Otis Lincoln (b. 1838), Hingham, Massachusetts; to Selwyn Randall Lincoln (b. 1879), Hingham; to Mason A. Foley, Hingham; to John P. Richardson, Hingham; to current owner. Courtesy of the Hersey family

Yellow box: John P. Richardson, Hingham; to current owner. Courtesy of the Hersey family

Green box: John P. Richardson, Hingham; to current owner. Courtesy of the Hersey family

Plate 11.3

Miniature canteen, Cotton Hersey, 1820–80

Materials: White pine heads; maple hoop; copper tacks; leather strap loops

Dimensions: H 2 ¾, D 1 ⁷⁄₁₆

Inscriptions: "C.HERSEY" stamped on front and back; "Cotton Hersey" in graphite on front

Provenance: Courtesy of Bob and Peggy Mosher

Plate 11.4

Miniature bucket, Cotton Hersey, 1878

Materials: White pine staves and heads; maple hoops; copper tacks

Dimensions: H 2 ⅜, Diam 2 ⅜

Inscriptions: "C.H" stamped on bottom; "Made by / Cotton Hersey / 1878 / aged 86 years / Born 1792" in graphite on bottom

Provenance: Courtesy of the Hersey family

Plate 11.5

Miniature bucket, Cotton Hersey, 1820–60

Materials: White pine staves and heads; maple hoops; copper tacks

Dimensions: H 2 ⅜, Diam 2 ⅜

Inscriptions: "C.H" stamped on bottom; "ANN C. WHEELER" painted on lid

Provenance: Purchased by David A. Schorsch at Skinner Auctions, November 1, 1986, lot 67; sold to Alastair and Edith Martin, the Guennol Collection; purchased by David A. Schorsch at Christie's, September 29, 2010, lot 44. Courtesy of David A. Schorsch

Plate 11.6

Miniature bucket, Cotton Hersey, 1830–50

Materials: White pine staves and heads; maple hoops; copper tacks

Dimensions: H 2 ⅜, Diam 2 ⅜

Inscriptions: "C.H" stamped on lid; "this is for / Dorothy Durgin / Elijah Brown" in ink on bottom

Provenance: Given to Shaker Sister Dorothy Durgin (1825–98) by Brother Elijah Brown (1772–1851), Canterbury, New Hampshire; acquired from Canterbury Sister Marguerite Frost (1892–1971). Courtesy of Philadelphia Museum of Art, gift of Mr. and Mrs. Julius Zieget

Plate 11.7

Miniature buckets, miniature pail, and small tub, Cotton Hersey, 1820–60

Materials: White pine staves and heads; maple hoops and handles; copper tacks

Dimensions: buckets: H 2 ⅜, Diam 2 ⅜; pail: H 2 ¼, Diam 2 ¼; tub: H 4 ½, Diam 5 ⅝

Inscriptions

Buckets and pail: "C.H" stamped on lids or bottoms; tub: "Cotton Hersey / Hingham" stamped on bottom

Provenance

Green bucket: purchased by current owners at Willis Henry Auctions, April 19, 2008, lot 56. Courtesy of Maryellen and Brendan O'Toole

Red-and-green bucket, two-tone green bucket, green pail, and tub: Courtesy of David A. Schorsch

12

SAMUEL HERSEY (1820–1909)

IN THE AUTUMN OF 1873 THOUSANDS PACKED LORING HALL FOR THE ANNUAL Exhibition of the Hingham Agricultural and Horticultural Society. They came to eat, play games, and admire the prizewinning goods of local farmers and craftsmen. A judge for the Manufacturers' Department noted a crowd drawn to one display in particular: "The case of wooden toys . . . by Samuel Hersey attracted more attention than anything else. . . . They were looked upon with delight by the younger portion of our community, and with great interest by adults, who could not but express their gratification at the perfect finish and skill displayed in the construction of every article, even the smallest."[1]

By this time Samuel Hersey and his handcrafted toys were already well known in Hingham. In fact, Hersey marked the third generation of his family to sell miniature woodenware to a regional audience; his paternal grandfather, Reuben (1757–1845), began producing such goods in the late eighteenth century at a cooper's shop on Hersey Street.[2] By the 1820s Reuben's son and Samuel's uncle Cotton (1792–1881) had become the most prominent toymaker in town, offering many of his tiny buckets and boxes in colorful painted surfaces. Father and son lived only a few steps from Samuel, and they must have had a profound influence on the young craftsman's career.[3]

Samuel Hersey's earliest work is virtually identical to the wooden containers crafted by Cotton Hersey. Examples include this diminutive six-inch box and a small bucket decorated with light and dark green paint (pl. 12.1). The initials "S.H" crisply stamped on each identify him as the maker. Hersey crafted the box in 1839, probably shortly after completing his apprenticeship. His father, a successful cooper in his own right, proudly inscribed the bottom "made by Samuel Hersey when he was 19 years old."[4]

After a few years of quietly operating as the Hersey craftsmen had for generations, Samuel Hersey rebranded and expanded his business. Unlike his forebears, he dropped the title of cooper and, in 1850, began advertising himself exclusively as a toy and box manufacturer.[5] His listings, which appeared regularly in newspapers and directories throughout the Boston area until about 1894, attracted day-trippers and wealthy summer residents in Hingham, Hull, Cohasset, and other resort towns whose populations swelled during the warmer months.[6] Improved methods of transportation made it easier than ever for him to connect with distant customers; in the 1830s large passenger steamships had begun service between Boston and Hingham and nearby Nantasket Beach three times a day.[7]

Miniature pails, tubs, churns, and other woodenware (pl. 12.2; also see figs. 2.6, 2.7) remained a staple of Hersey's production. Among the most impressive of these goods is a masterfully coopered bucket measuring a mere one inch in height. As is the case with most of Hersey's toys, the exterior was left in a natural finish that has since oxidized to a golden brown. Samuel was also the first Hingham toymaker to market his goods with printed paper labels; one affixed here on the lid's underside reads, "Manufactured by / SAM'L HERSEY, / *Hersey Street,* / Hingham, Mass." A second label identifying him as a "Wholesale and Retail / Toy Ware and Box / MANUFACTURER" is found on a circular box fashioned on a similarly tiny scale.

Samuel Hersey further distinguished himself from his predecessors by offering miniature furniture and household accessories in his general line of toys, crafting everything from bedroom sets (see fig. 2.9) to cradles and card tables (pls. 12.3–12.5), all modeled after the cottage- and Empire-style furniture popular at the time. A miniature rocking chair (pl. 12.4) made around 1860 originally belonged to two sisters, Clara and Elizabeth Hoar of Concord. Their father, Ebenezer Rockwood Hoar (1816–95), was a state senator, judge of the Boston Court of Common Pleas, and attorney general of the United States.[8] Given the prominence of his clientele, it comes as no surprise that Hersey sold upwards of $1,200 in toys a year.[9]

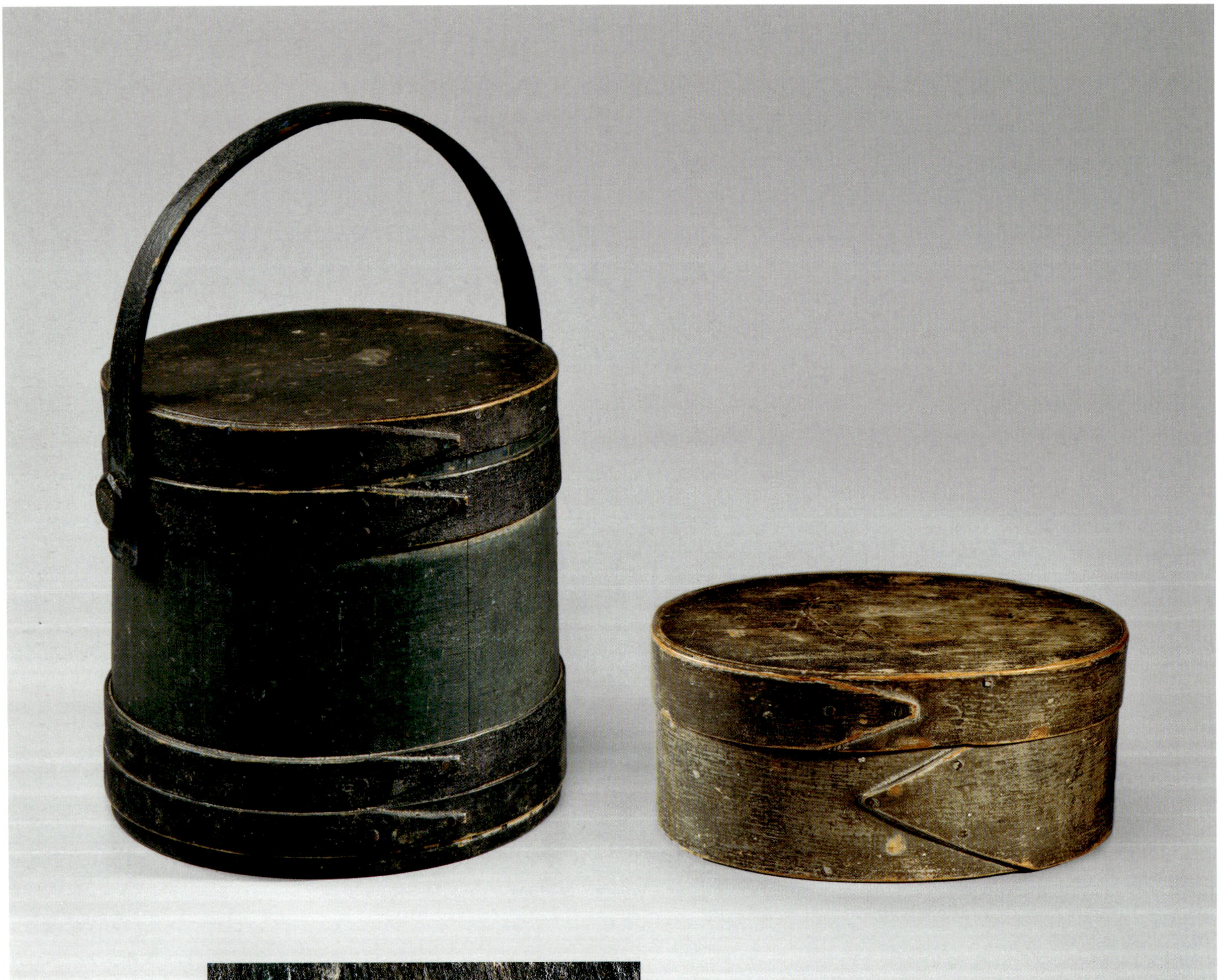

Plate 12.1

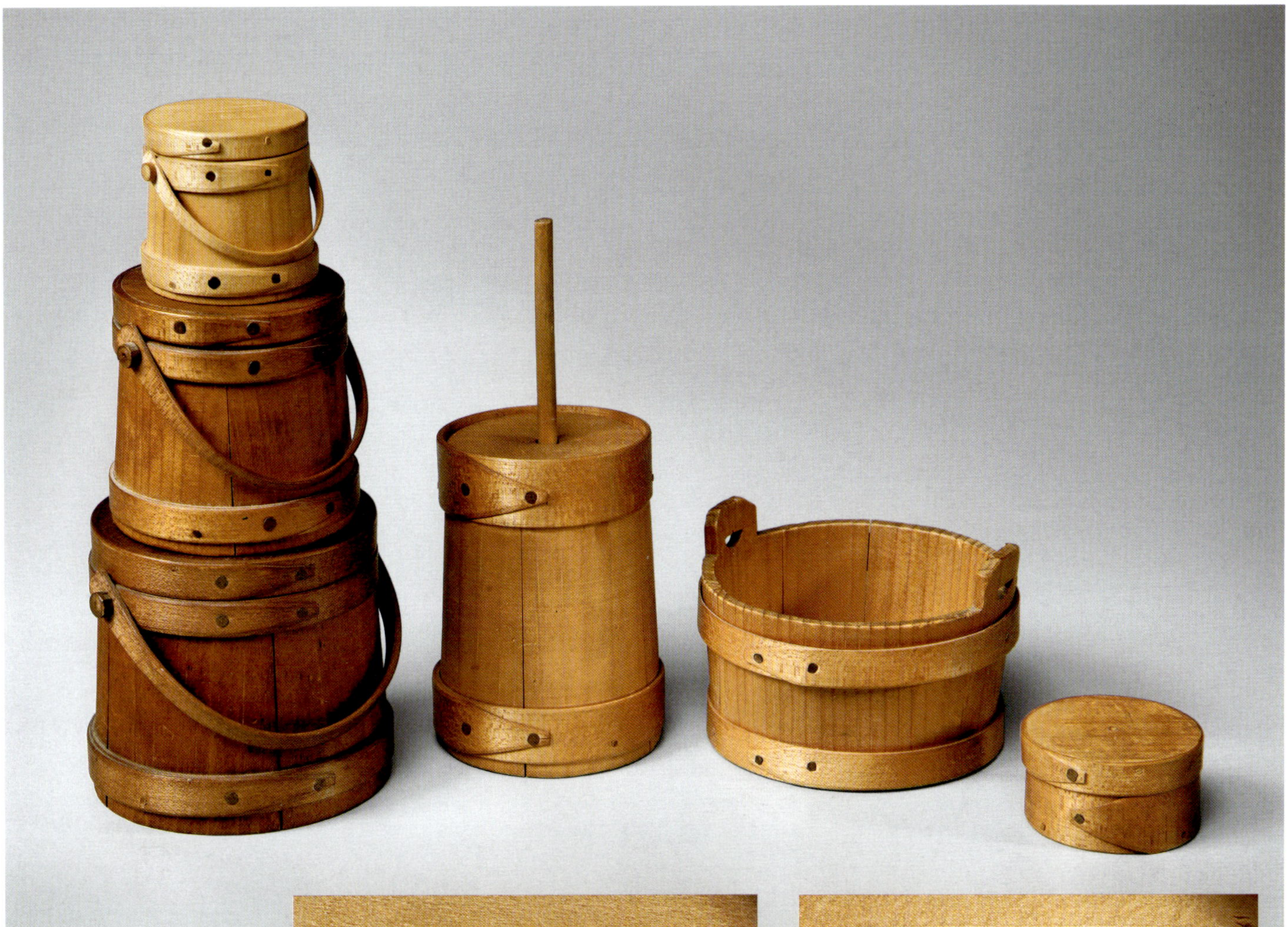

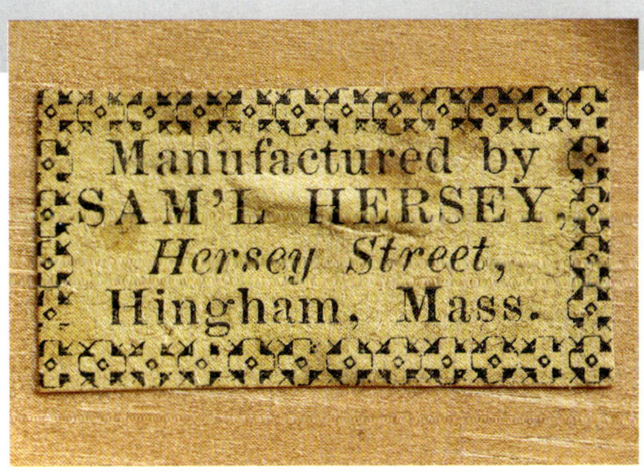
Manufactured by
SAM'L HERSEY,
Hersey Street,
Hingham, Mass.

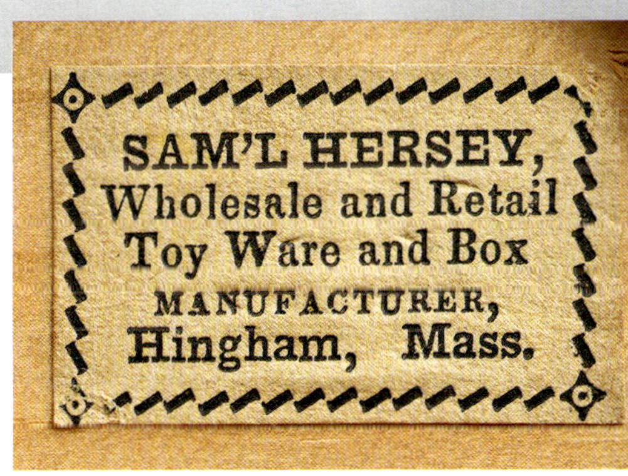
SAM'L HERSEY,
Wholesale and Retail
Toy Ware and Box
MANUFACTURER,
Hingham, Mass.

Plate 12.2

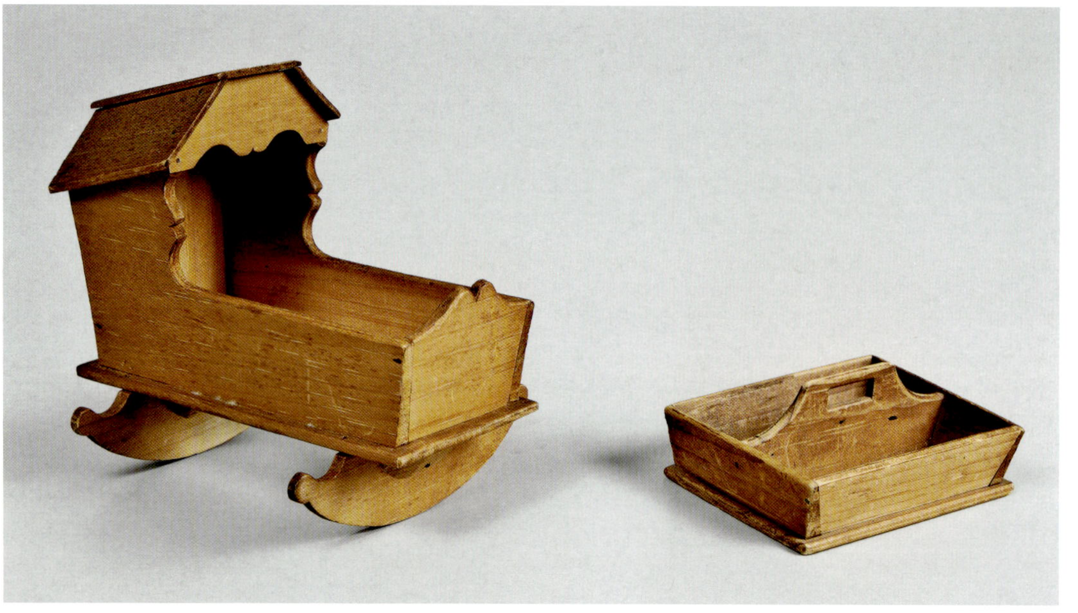

Plate 12.3

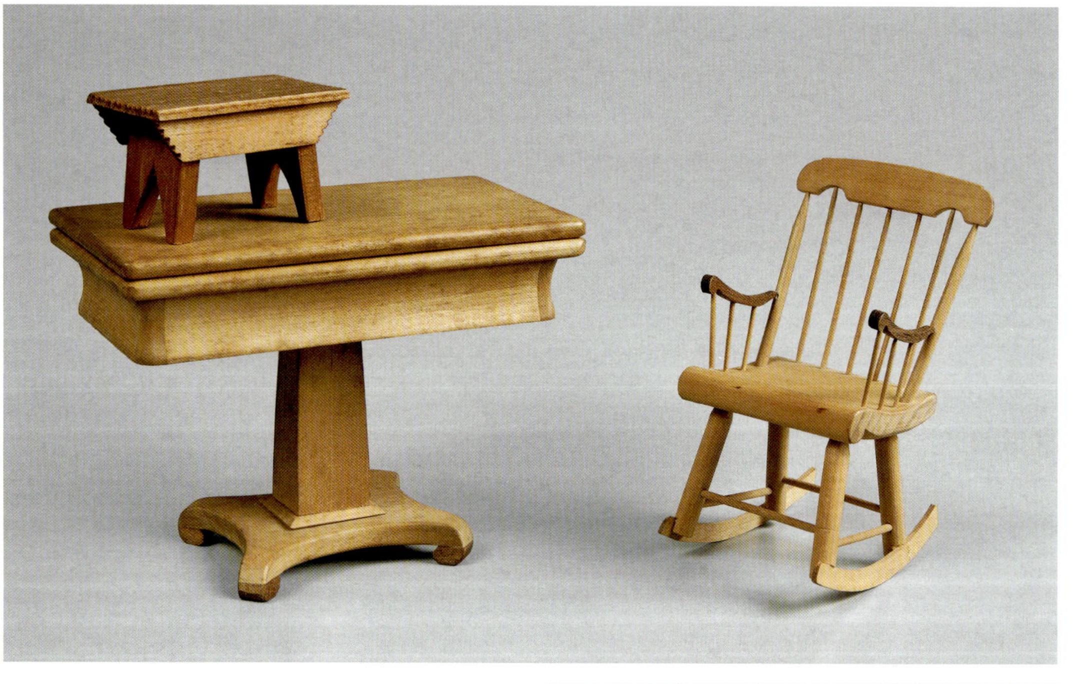

Plate 12.4

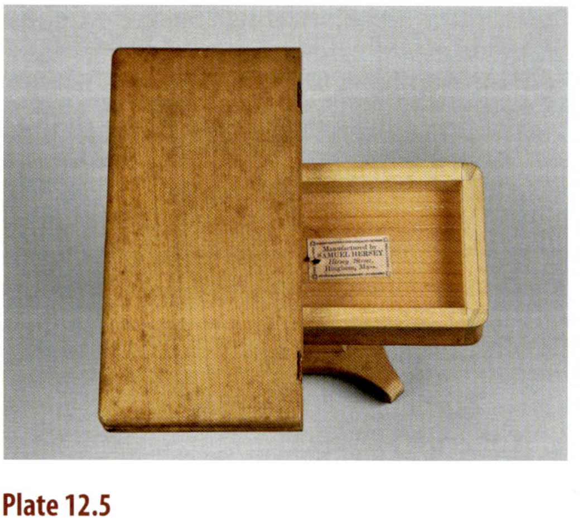

Plate 12.5

Plate 12.1

Small bucket and oval box , Samuel Hersey, 1839–60

Materials: White pine heads; maple hoops; copper tacks

Dimensions: Bucket: H 5 3/8, Diam 5 9/16; box: H 2 3/4, W 6, D 4 1/2

Inscriptions: "S.H" stamped on lid of bucket and bottom of box; "made by Samuel Hersey / when he was 19 years old / S.F.H." written in graphite on underside of box lid; "Mason A. Foley / 140 Hull Street / Hingham, Mass." written in ink on white sticker on underside of box lid

Provenance

Bucket: Courtesy of the Hersey family

Box: Mason A. Foley, Hingham, Massachusetts; to John P. Richardson, Hingham; purchased by current owner at Willis Henry Auctions, November 19, 2011, lot 1. Courtesy of David A. Schorsch

Plate 12.2

Miniature buckets, churn, tub, and box, Samuel Hersey, 1850–70

Materials: White pine staves, heads and dasher; maple hoops and handles; copper tacks

Dimensions: Top bucket: H 1 1/2, Diam 1 1/2; middle bucket: H 2 1/4, Diam 2 1/4; bottom bucket: H 2 1/2, Diam 2 13/16; churn: H 3 1/8, Diam 1 3/16; tub: H 2 1/4, Diam 3 1/4; box: H 13/16, Diam 1 1/2

Inscriptions: "Manufactured by / SAM'L HERSEY, / *Hersey Street,* / Hingham, Mass." printed on paper label on underside of churn and bucket lids; SAM'L HERSEY, / Wholesale and Retail / Toy Ware and Box / MANUFACTURER, / Hingham, Mass." printed on paper label on underside of tub and box lid

Provenance

Top bucket, bottom bucket, and churn: Courtesy of the Hersey family

Middle bucket, tub, and box: Courtesy of David A. Schorsch

Plate 12.3

Miniature cradle and knife tray, Samuel Hersey, 1850–70

Materials: White pine; iron nails

Dimensions: Cradle: H 6, W 5 1/8, D 6 3/16; knife tray: H 1 3/4, W 4 1/4, D 3 1/8

Inscriptions: "Manufactured by / SAMUEL HERSEY, / *Hersey Street,* / Hingham, Mass." printed on paper label on underside of cradle and knife tray

Provenance

Cradle: Courtesy of Hingham Historical Society

Knife tray: Courtesy of David A. Schorsch

Plate 12.4

Miniature card table, rocking chair, and footstool, Samuel Hersey, 1850–70

Materials: Card table: maple top, case, and pedestal; white pine case bottom; black walnut feet; iron nails, leather hinges; rocking chair: maple crest, spindles, and stretchers; white pine seat; black walnut arms; footstool: white pine

Dimensions: Card table: H 3 7/8, W 5 3/8, D (top closed) 2 11/16; rocking chair: H 4 1/2, W (seat) 2 3/8, D (seat) 1 15/16; footstool: H 1 1/4, W 2 7/16, D 1 1/4

Inscriptions: "Manufactured by / SAMUEL HERSEY, / Hersey Street, / Hingham, Mass." printed on paper label inside of card table and on underside of footstool. "Manufactured by / SAM'L HERSEY, / Hersey Street, / Hingham, Mass." printed on paper label on underside of rocking chair

Provenance

Card table: Collection of Robert Skinner; purchased by current owner at Skinner Auctions, October 28, 2012, lot 7. Courtesy of the Hersey family

Rocking chair: Clara and Elizabeth Hoar, Concord, Massachusetts; to Derin Bray, Portsmouth, New Hampshire; to current owner. Courtesy of the Hersey family

Footstool: Courtesy of Hingham Historical Society

Plate 12.5

Detail of label and swivel top on card table in pl. 12.4

13

EDWARD LEWIS HERSEY (1822–1853)

THE CAREER OF EDWARD LEWIS HERSEY WAS MARKED BY GREAT SUCCESS AND EVEN greater misfortune. In 1850, at the age of twenty-eight, he was one of only two people in Hingham—and probably all of Massachusetts—to advertise himself as a toymaker.[1] Local coopers had made "small ware for children" for generations, but until then none had taken the steps to professionalize it. Tragically, Hersey's efforts to pioneer the trade were short-lived. In March 1852 his wife and only child died of tuberculosis.[2] Within several months he too succumbed to the relentless disease.[3] An inventory of his estate taken shortly thereafter shows a craftsman in his prime: his small workshop contained tools, timber, rough stock, and finished and unfinished "toy-ware" appraised at $18.50.[4]

Hersey's rise to prominence, albeit brief, was built on the support of a large and established family network. His grandfather Reuben Hersey (1757–1845) was the patriarch of Hersey Street, a major center of domestic cooperage since 1730.[5] He was also a toymaker—the first known to have worked in Hingham. By 1850 Edward lived in the old homestead and may have worked in Reuben's shop.[6] During this period he also began to market his services aggressively. He advertised in regional and state directories and started branding some of his goods—such as small boxes and buckets (pl. 13.1)—with an elaborate circular stamp identifying him by name and location: "EDWARD L. HERSEY / H / ING / HAM."

Other items made in the shop, notably miniature woodenware, were more simply marked with Hersey's initials, "E.L.H." Examples include a piggin, butter churn, and washing machine (pl. 13.2) that survive in near-perfect condition. They were collected and lovingly preserved by a Hingham resident, Aurelia Fearing (1857–1913), who lent them for an exhibition held by the Society of Arts and Crafts, Hingham, in the early twentieth century.[7] The washing machine in particular is a rare, if not unique, survivor. No other miniature versions of this laundering device, called a dumb betty, are known to exist. Hersey also crafted tiny dippers with delicately turned handles. Though most were left unpainted, at least one retains an original yellow stain (pl. 13.3) similar to the finish found on boxes and other woodenware made in Shaker communities around the same time.[8]

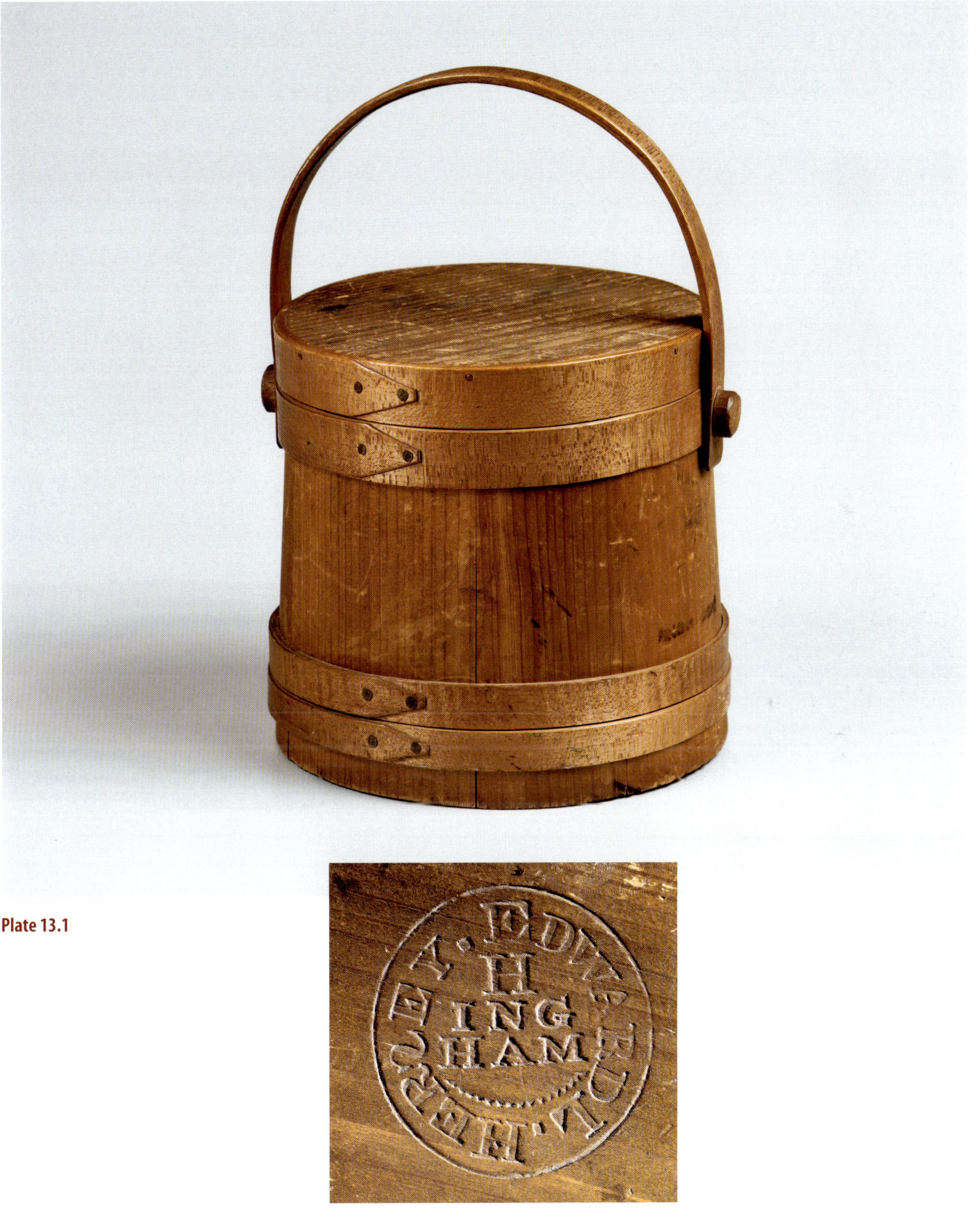

Plate 13.1

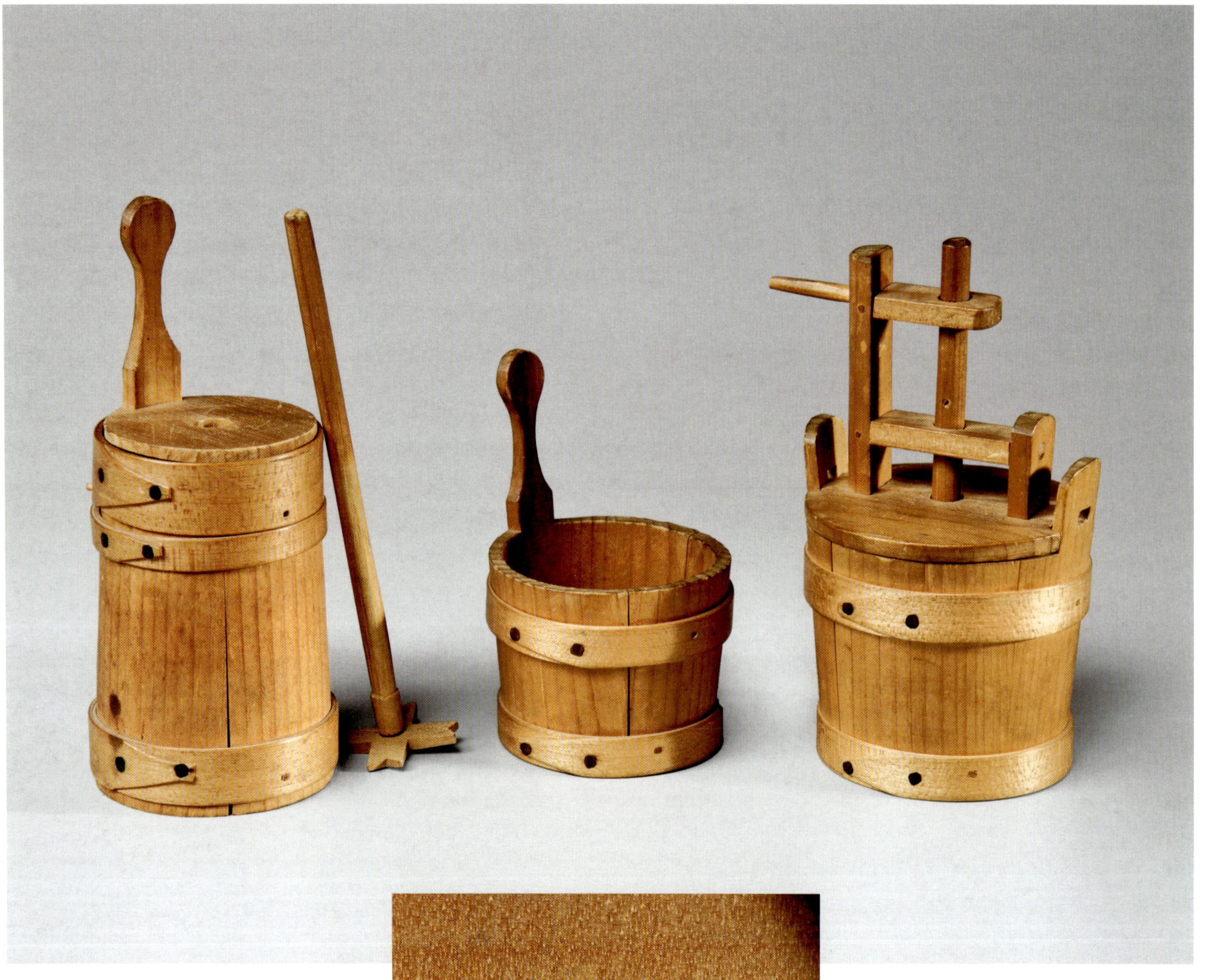

Plate 13.2

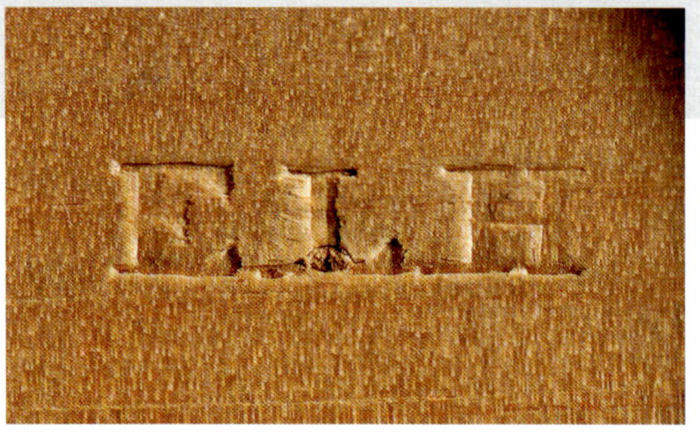

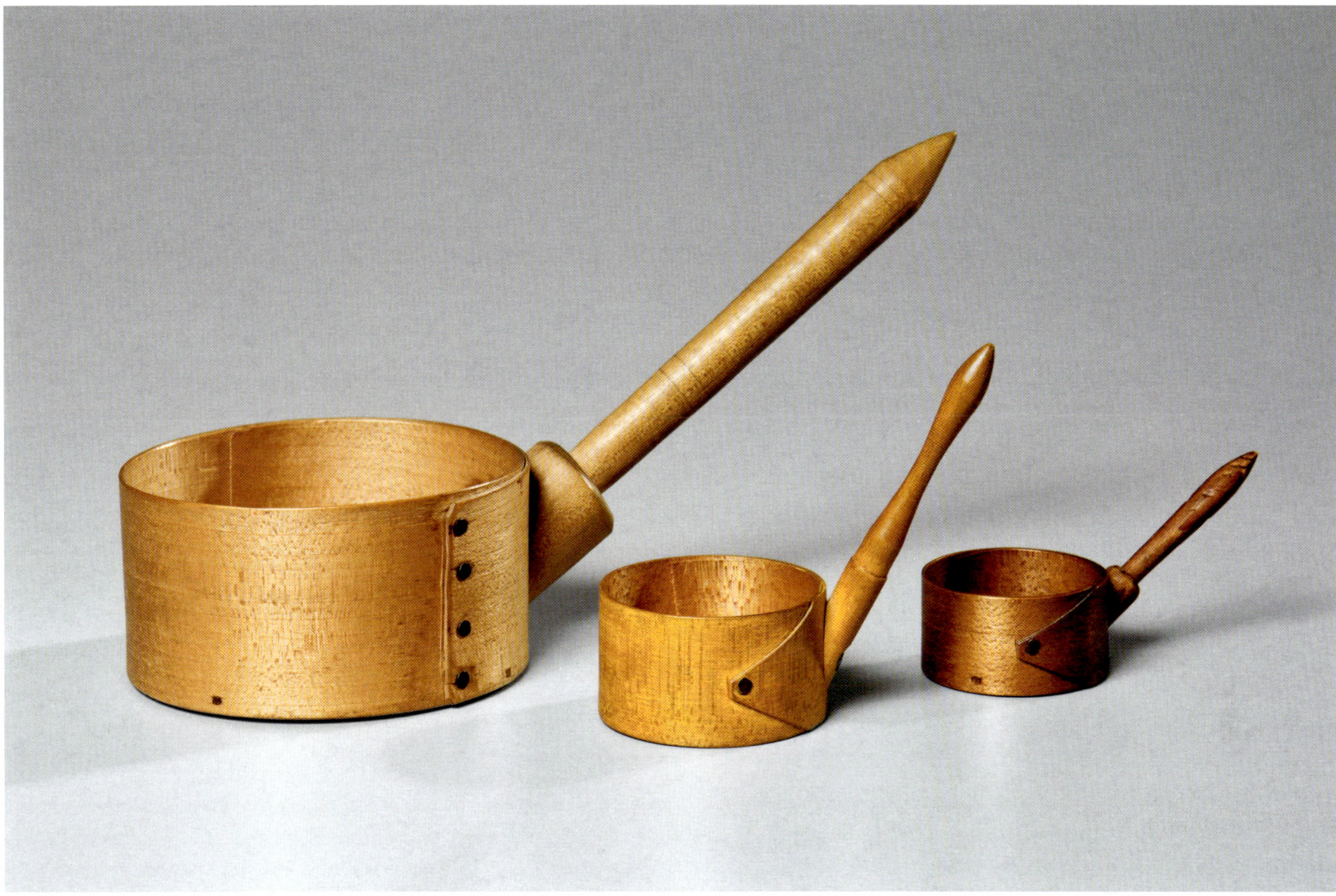

Plate 13.3

Plate 13.1
Small bucket, Edward L. Hersey, 1850–53
Materials: White pine staves, heads, and buttons; maple hoops and handle; copper tacks
Dimensions: H 5 3/8, Diam 5 7/8
Inscription: "EDWARD L. HERSEY. / H / ING / HAM" stamped on lid
Provenance: Courtesy of the Hersey family

Plate 13.2
Miniature churn, piggin, and dumb betty, Edward L. Hersey, 1850–53
Materials: White pine staves, heads, and dasher rod; maple hoops and dasher bottom; white pine and maple washing mechanism; copper tacks
Dimensions: Churn: H 5 1/8, Diam 2 1/8; piggin: H 3 3/8, Diam 2 1/8; dumb betty: H (overall) 4 3/4, Diam 2 3/8
Inscriptions: "E.L.H" stamped on bottoms; "Loaned by / Aurelia / Fearing" written in ink on white sticker on bottom of piggin
Provenance: Purchased by current owner at Willis Henry Auctions, Marshfield, Massachusetts, November 21, 2009, lots 91, 92, 94. Courtesy of the Hersey family

Plate 13.3
Miniature dippers, Edward L. Hersey, 1850–53
Materials: White pine bottoms; maple hoops; maple or white pine handles; copper tacks
Dimensions: Left dipper: H 3 5/8, Diam 3 1/8; middle dipper: H 2 3/4, Diam 1 11/16; right dipper: H 1 1/2, Diam 1 7/16
Inscriptions: "E.L.H" stamped on bottoms
Provenance
Left dipper: Purchased by current owner at Willis Henry Auctions, Marshfield, Massachusetts, November 21, 2009, lot 93; courtesy of the Hersey family
Middle and right dippers: Courtesy of David A. Schorsch

14

WILLIAM SEWALL TOWER (1826–1898)

WILLIAM S. TOWER (pl. 14.1) HAS LONG BEEN CELEBRATED AS A PIONEER OF INDUSTRY. The seminal book *Toys in America*, published in 1961, identified him as the founder of one of the earliest toy manufacturing businesses in the country: "[It] was originally a kind of guild, an association of craftsmen in different fields who, by pooling their efforts and products, could offer a more varied and extensive line of toys than any single craftsman. . . . Joseph Jacob, who had a hatchet factory, turned out toy tools and other novelties. A boat builder made toy boats, cabinet makers produced doll's furniture. Altogether a score of workmen contributed to the line of Tower Toys."[1]

For decades this romanticized account has been cited in studies of American manufacturing and repeated in dozens of books and articles on children's playthings—yet it is largely incorrect. New information about Tower and other Hingham craftsmen has come to light, offering fresh insights into the town's extraordinary past as a toymaking center and Tower's role within it.

Like many Hingham youngsters, William S. Tower began his career at the cooper's bench. He worked for a number of years at Crocker Wilder's large bucket factory near Cushing Pond, where he mastered the use of machinery and excelled as an innovator of fancy woodenware and other novelty items.[2] Among his best work was "a very curious pail" made of 47,568 pieces of red and white cedar and inlaid around with stars (see fig. 2.10). Tower displayed this elaborate container at regional fairs; tens of thousands of people would have seen it in 1865 at the exhibition of the Massachusetts Charitable Mechanic Association held in Boston.[3]

Having demonstrated his prowess as a craftsman, Tower sought to make his mark on the business world, too. In 1869 the *Hingham Journal* announced his latest venture:

> *The toy ware manufactory of W. S. Tower & Co., at South Hingham, recently started in the bucket factory of the Messrs. Wilders, is now fairly under way, and turning out some fine specimens of toy ware, croquet sets, etc. Most of their work is done by machinery, they having purchased that formerly used in the factory of the Hingham Wooden Ware Co., thereby they are enabled to compete with any similar establishment in the country. The gentlemen comprising the firm are excellent mechanics, industrious and energetic and we predict for them good success—which they certainly deserve in their undertaking. They also attend to orders for wood sawing, turning, & c. Our readers are referred to their advertisement in other columns.*[4]

Tower's new two-story building—rented from his former employer—was the first toy factory in town (see fig. 2.11).[5] Similar enterprises had existed elsewhere in New England since midcentury, but they were new to Hingham, where small handcraft shops still dominated the trade.[6] Contrary to popular belief, Tower's business was not, in fact, supported by a guild. His only partner was his brother Charles (and possibly his brother George, who worked at the factory during its first years); no other craftsmen had a stake in the firm. Early scholars seem to have conflated the history of Tower's toy factory with that of the Society of Arts and Crafts, Hingham.[7] Founded in 1901, the society, a guildlike organization, was made up of local artists and artisans from wide-ranging backgrounds; some were toymakers who maintained independent shops. They contributed their wares to a central storefront and to traveling exhibitions, but never to William S. Tower & Company.[8]

Tower's factory was an immediate success. Upon opening, it employed about ten full-time hands, including young women, who produced a broad range of wooden toys and novelties. In the following decades the workforce more than doubled; by 1893 its new owner—Tower's son-in-law Frank Burrell—predicted employing as many as forty people.[9] Items were sold directly from the factory and locally at the region's major summer destinations, such as the Pavilion at Nantasket Beach, Rockland Café, and Downer Landing, a sprawling pleasure palace of Victorian entertainment that attracted thousands of visitors to Hingham's scenic waterfront.[10] Large shipments of toys were also sent to Boston and New York, where agents distributed them to a world market; Tower filled wholesale orders for clients throughout the country—including the western United States—as well as Montreal, Liverpool, London, and other places in Europe.[11] In 1878 a selection of Tower's toys was awarded a bronze medal at the Exposition Universelle in Paris.[12]

Not surprisingly, Tower's earliest toys included a range of coopered items similar to the fancy woodenware he had perfected years earlier. His small tubs, pails, and buckets often featured alternating light and dark wood staves (pl. 14.2) for a dazzling effect. Like the goods produced at Wilder's factory, their parts were machine-cut and machine-turned, and they were held together with riveted tinned iron hoops. Most examples also have wire handles with diamond-shaped bale plates and, occasionally, lids with turned knobs. The tops and undersides are often stamped in tiny lettering: "WM. S. TOWER / SO HINGHAM, MASS."

By the 1880s Tower boasted over a hundred different toys, from tenpin sets and game boards to carts (pl. 14.3), nests of pails (pl. 14.4), and miniature gristmills (see fig. 2.15).[13] A former employee once recalled the "wild hustle" to keep up with the craze for number puzzles—still known today as 15 puzzles—which the company turned out by the thousands.[14] The firm developed new goods as well, including educational play sets like Tower's Log Cabin, a miniature old-fashioned farmstead complete with an interlocking fence and well sweep with a tiny bucket (pl. 14.5). The set—which could be ordered through popular children's magazines—was delivered in a small box with an instructional label pasted on the front.

While boys could play at farming and milling with Tower's toys, his line of miniature furniture (pl. 14.6) and household accessories let girls play at the domestic arts. His chests, tables, and washstands, modeled after contemporary forms, seem to have sold well; they are frequently found among the furnishings of dollhouses from the period. Though strikingly similar to the works of Samuel Hersey—the first Hingham toymaker to produce miniature furniture—Tower's pieces lack Hersey's attention to detail and construction. He also favored turned chestnut-shaped drawer pulls, rather than the more stylistically accurate round knobs used by Hersey. For a premium Tower's miniature furniture was available in black walnut—a modest upgrade from the more common and less expensive white pine.

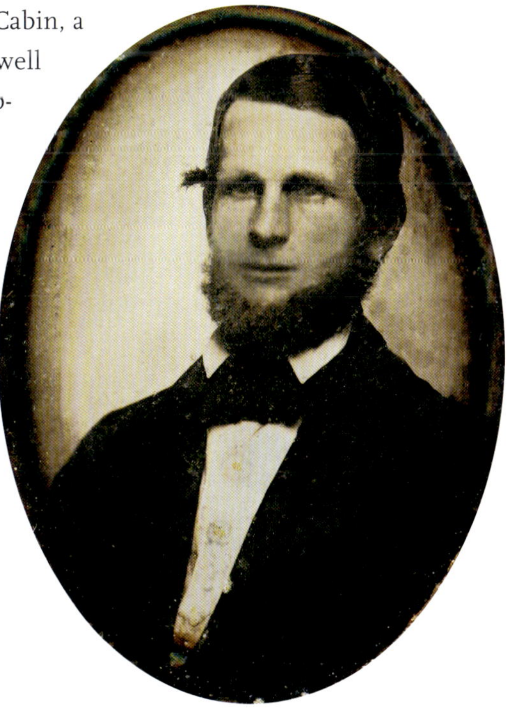

Plate 14.1

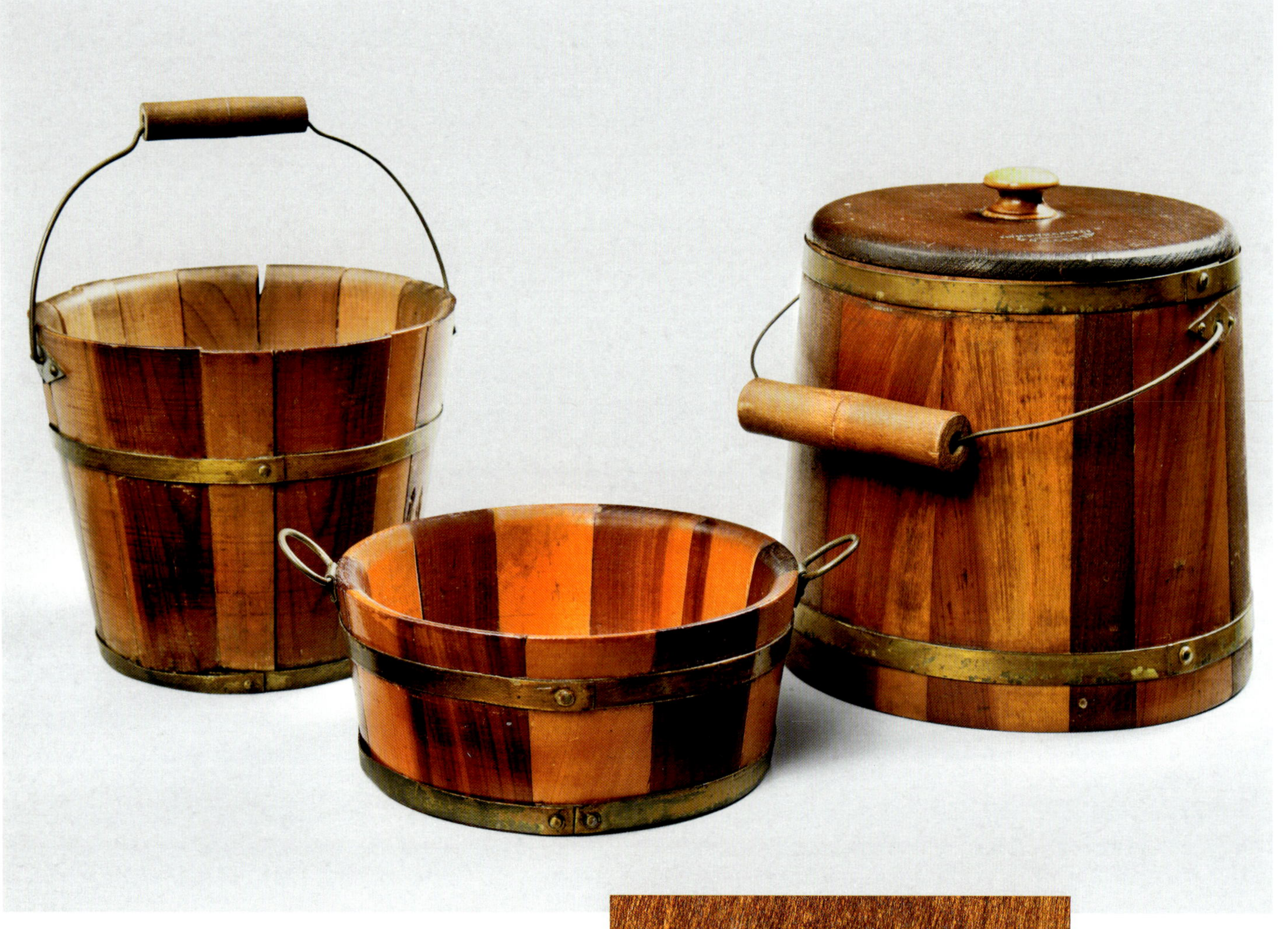

Plate 14.2

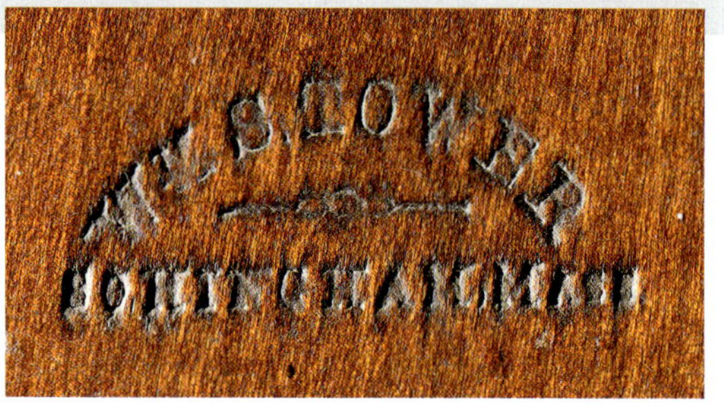

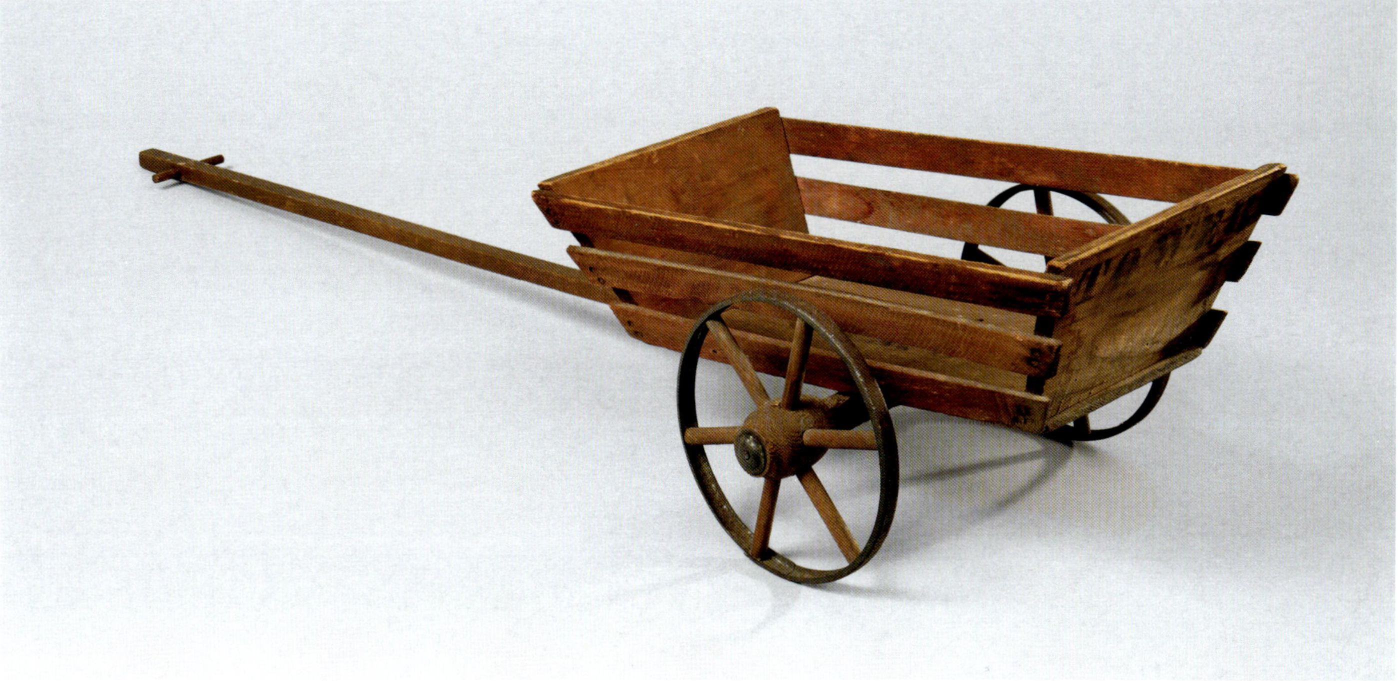

Plate 14.3

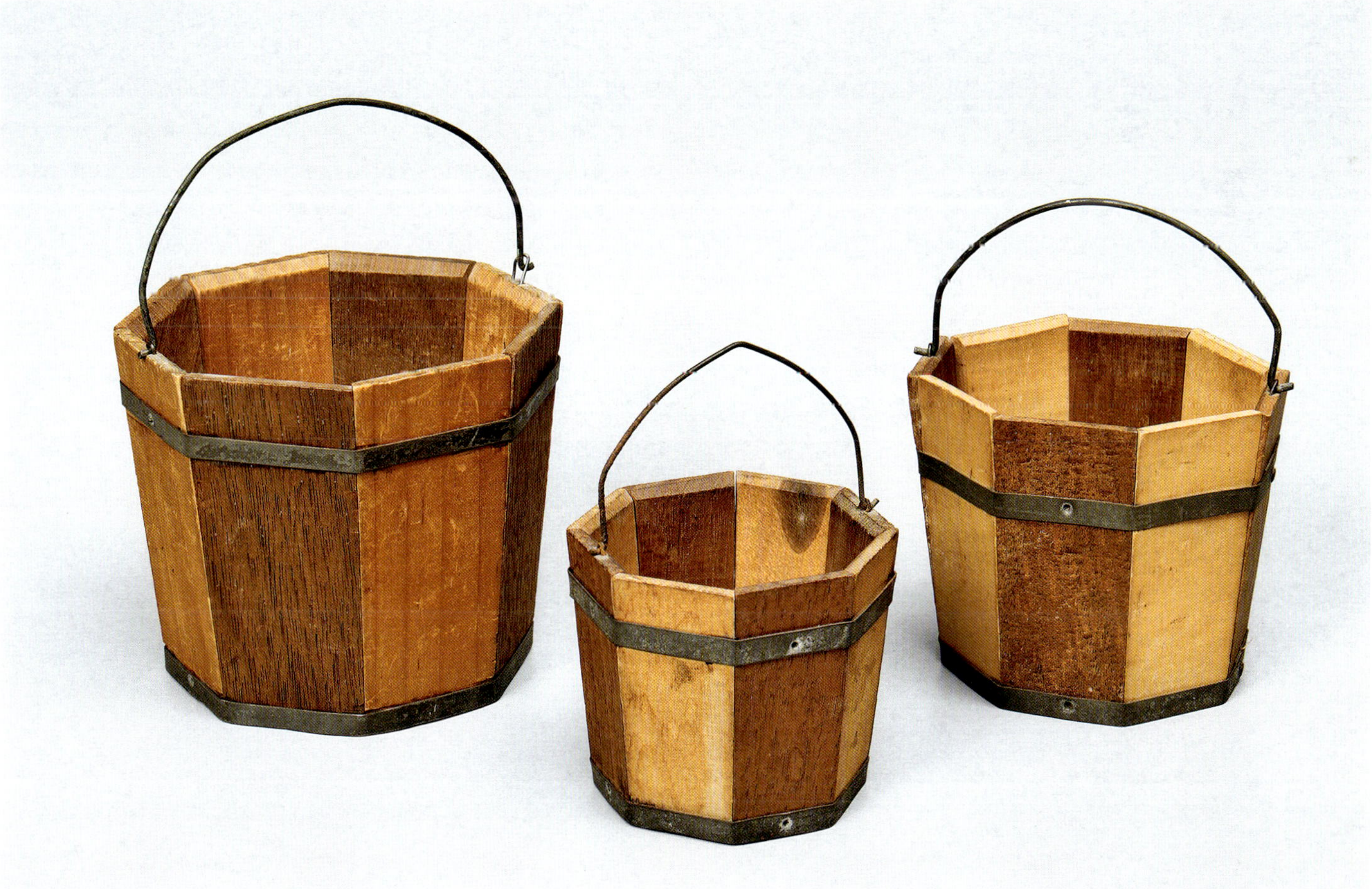

Plate 14.4

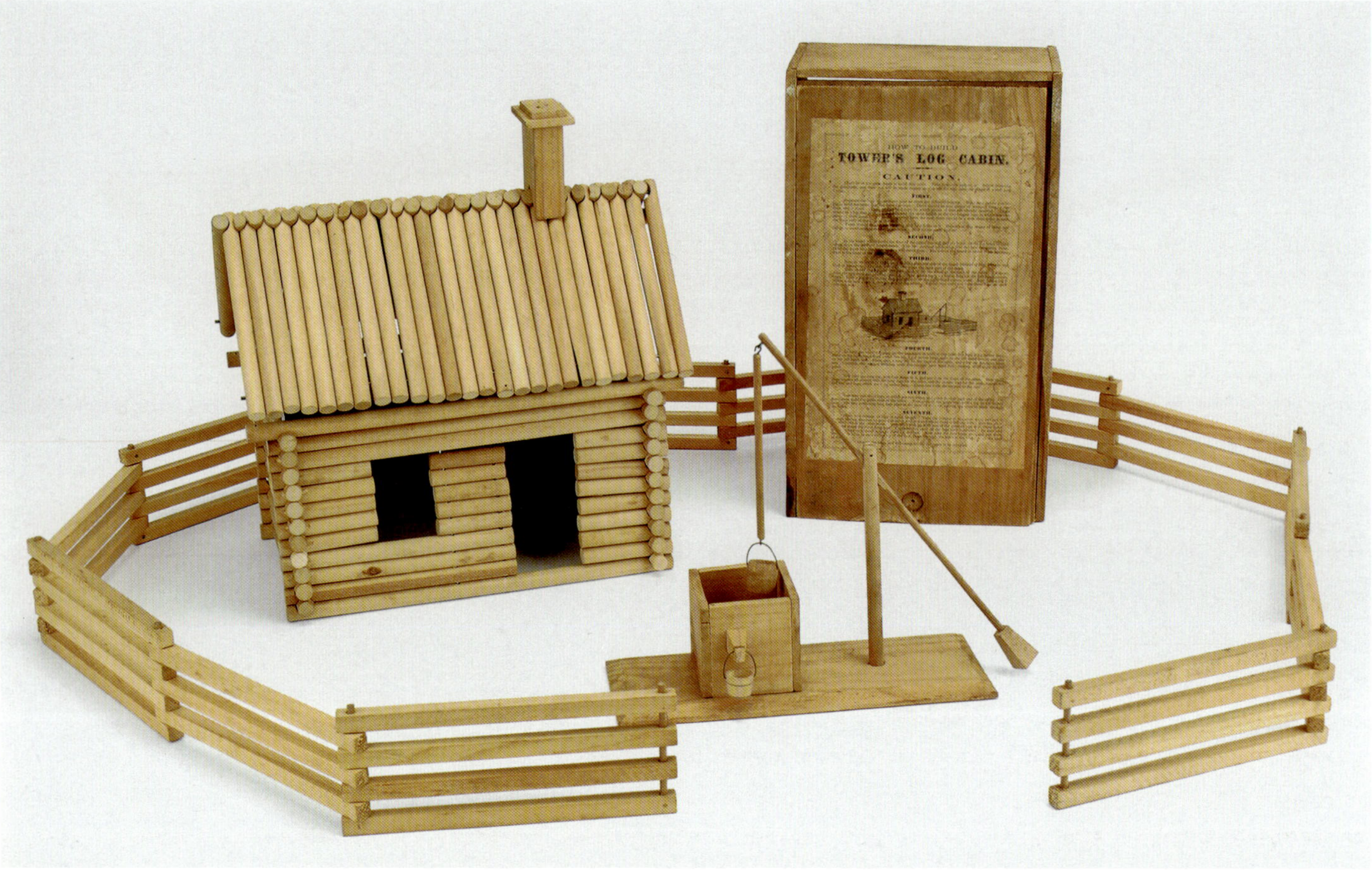

Plate 14.5

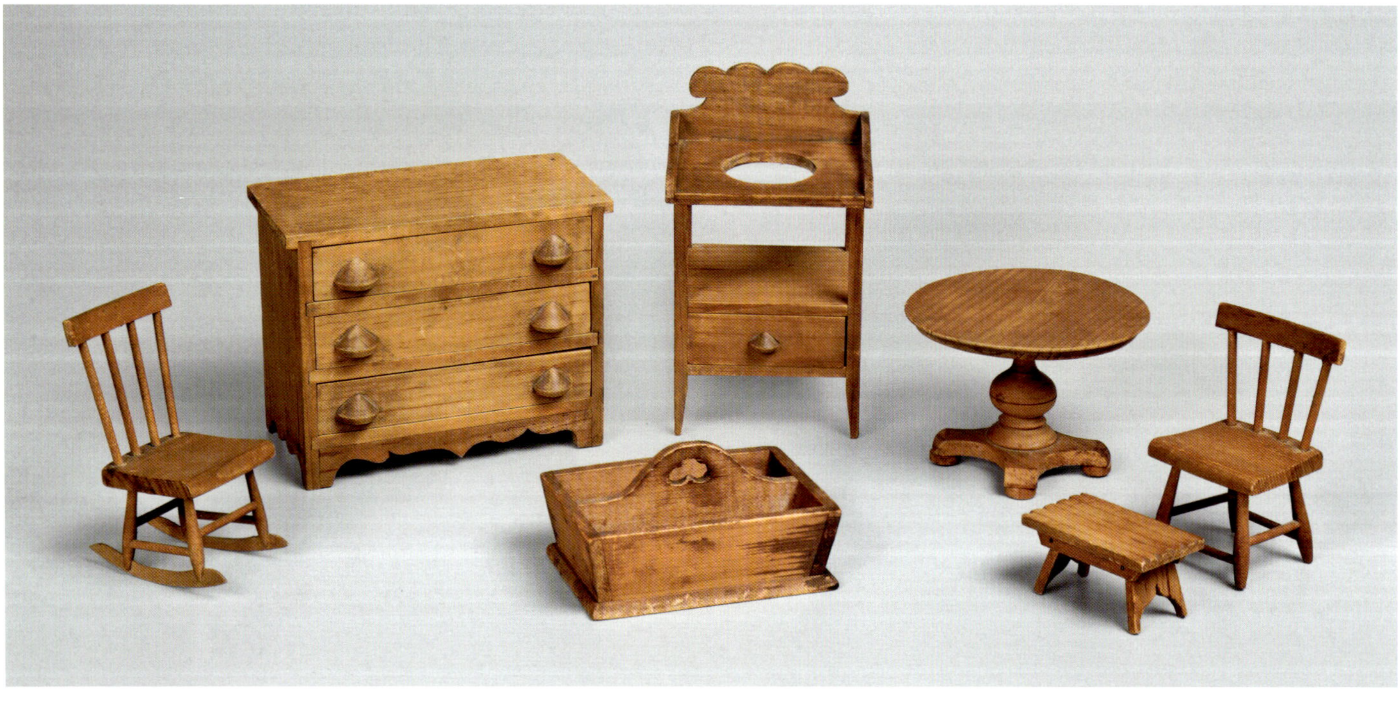

Plate 14.6

Plate 14.1

Daguerreotype portrait of William S. Tower, 1850–60
Provenance: Courtesy of the Hersey family

Plate 14.2

Small pail, tub, and bucket, William S. Tower, 1869–85
Materials: Red cedar staves; white cedar staves; white pine heads; birch handles; maple knob; tinned iron hoops; wire handles
Dimensions: Pail: H 4 ½, Diam 5 ½; tub: H 2 ¼, Diam 5 9⁄16; bucket: H 6 ¼, Diam 6 ¼
Inscriptions: "WM. S. TOWER / SO. HINGHAM, MASS." stamped on bottoms
Provenance
Pail and bucket: Courtesy of the Hersey family
Tub: Courtesy of Bob and Peggy Mosher

Plate 14.3

Miniature cart, William S. Tower, 1869–85
Materials: White pine body; tulip-poplar tongue; possibly birch spokes; tin wheels; iron nails
Dimensions: H 6 ¼, L (overall) 31 ½, W 9 ¾
Inscriptions: "TOWER" stenciled on back, front, and inside bottom; "REWOT" stamped twice on bottom
Provenance: Selwyn Randall Lincoln (b. 1879), Hingham, Massachusetts; to John P. Richardson, Hingham; purchased by current owner at Willis Henry Auctions, November 19, 2011, lot 110F. Courtesy of the Hersey family

Plate 14.4

Nest of miniature pails, William S. Tower, 1869–85
Materials: White pine staves and bottoms; holly or beech staves; black walnut staves; tinned iron bands; wire handles
Dimensions: Largest: H 3 ¼; W 4; smallest: H 2 ¼, W 2 ½
Inscriptions: "WM. S. TOWER / SO. HINGHAM, MASS." stamped on bottoms
Provenance: Discovered by current owner among the multigenerational furnishings of the Murray-Marbury-Dodge dollhouse, originally crafted for Juliet A. Murray (1827–1916) of Mount Lubentia in Prince Georges County, Maryland. The dollhouse descended to her daughter Meta Marbury Dodge (1861–1923), and then to her son Frances Chapman Dodge (1889–1975); to current owner. Courtesy of David A. Schorsch

Plate 14.5

Tower's Log Cabin, William S. Tower, ca. 1879
Photograph courtesy of Noel Barrett Auctions

Plate 14.6

Miniature furniture and knife tray, William S. Tower, 1869–85
Materials: White pine; maple pedestal, spindles, and stretchers; tulip-poplar rockers; iron nails
Dimensions: Rocking chair: H 4 ⅜, W (seat) 2 ¼, D (seat) 2; chest: H 4 ⅝, W 6, D 3; washstand: H 6 ¼, W 3 ¾, D 3; table: H 3, Diam 4 ⅜; side chair: H 3 ¾, W (seat) 2 ⅛, D (seat) 2; footstool: H 1 ¼, W 2 ½, D 1 ½; knife tray: H 2 ¼, W 2 ⅞, D 4 ¼
Inscriptions: "WM. S. TOWER / SO. HINGHAM, MASS." stamped on back of chest and bottoms of chairs, washstand, and knife tray
Provenance
Miniature furniture: Courtesy of Bonnie Benson Hanson
Knife tray: Courtesy of Boston Children's Museum

15

ELIJAH LORING CORTHELL (1836–1917)

THOUGH MORE THAN A DOZEN TOYMAKERS OPERATED SMALL SHOPS AND FACTORIES in nineteenth-century Hingham, only a single price list for their work is known to survive. In 1877 Elijah L. Corthell printed a modest wholesale catalogue (pl. 15.1) for his new toy manufactory on Hersey Street. This tattered sheet, discovered in Corthell's home, documents a breadth of wooden playthings not typically associated with local craftsmen. In addition to traditional coopered items, Corthell made and sold watchman's rattles, birdhouses, swings, rakes, sleds, and a variety of puzzles and games, including Fox and Geese. No labeled example of these goods has come to light, though a group of miniature churns, pails, and tubs—sold individually and in nests—raises an intriguing possibility (see fig. 2.14, pl. 14.4). They match Corthell's description for "Octagon Ware," made of alternating pine and walnut staves and metal hoops—but these toys commonly bear the impressed stamp of William S. Tower, which suggests that Tower produced similar items, or purchased the toys from Corthell and sold them under the Tower brand.[1]

Corthell spent most of his life at the center of Hingham's woodenware industry. His father, also named Elijah, established a small boxmaking shop on Hersey Street around 1835, shortly after he married into the prominent Hersey family of coopers and toymakers.[2] The senior Corthell worked in partnership with his two sons in the early 1860s, but by 1875 his son Elijah had struck out on his own, opening the Hingham Toy Manufactory.[3] His new venture continued to offer traditional oval boxes made by hand, but it also produced toys and other items with the aid of steam-powered saws and lathes. The 1880 manufacturer's census shows that Corthell employed upwards of six people, half of whom were over the age of sixteen.[4] Skilled craftsmen in the shop made $1.50 a day and worked ten-hour shifts between the months of May and November, producing approximately $4,000 in goods a year.[5] By 1885 Corthell had left the toy business and begun specializing in picture frames "of any description."[6] At the turn of the century and for the remainder of his life, he was listed in town directories as a peddler.[7]

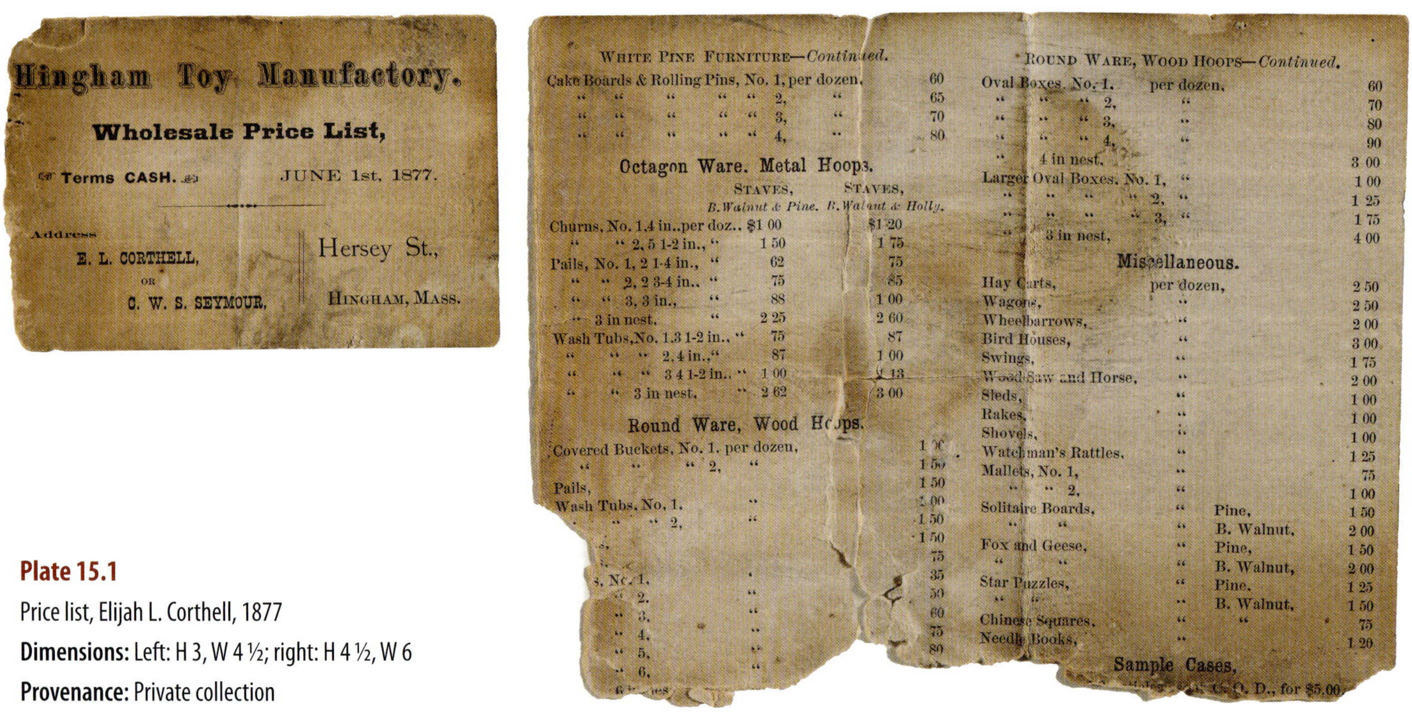
Hingham Toy Manufactory.
Wholesale Price List,
Terms CASH. JUNE 1st, 1877.
Address
E. L. CORTHELL,
OR
C. W. S. SEYMOUR,
Hersey St.,
HINGHAM, MASS.

WHITE PINE FURNITURE—*Continued.*

Cake Boards & Rolling Pins, No. 1, per dozen,	60
" " " " " 2, "	65
" " " " " 3, "	70
" " " " " 4, "	80

Octagon Ware. Metal Hoops.

	STAVES, *B. Walnut & Pine.*	STAVES, *B. Walnut & Holly.*
Churns, No. 1, 4 in., per doz.,	$1 00	$1 20
" " 2, 5 1-2 in., "	1 50	1 75
Pails, No. 1, 2 1-4 in., "	62	75
" " 2, 2 3-4 in., "	75	85
" " 3, 3 in., "	88	1 00
" 3 in nest, "	2 25	2 60
Wash Tubs, No. 1, 3 1-2 in., "	75	87
" " " 2, 4 in., "	87	1 00
" " " 3 4 1-2 in., "	1 00	1 13
" " 3 in nest, "	2 62	3 00

Round Ware, Wood Hoops.

Covered Buckets, No. 1, per dozen,	1 00
" " " 2, "	1 50
Pails,	1 50
Wash Tubs, No. 1, "	2 00
" " 2, "	1 50
[illegible]	1 50
[illegible]	75
[illegible], No. 1, "	35
" 2, "	50
" 3, "	60
" 4, "	75
" 5, "	80
" 6, "	[illegible]
6 [illegible]	

ROUND WARE, WOOD HOOPS—*Continued.*

Oval Boxes, No. 1,	per dozen,	60
" " " 2,	"	70
" " " 3,	"	80
" " " 4,	"	90
" 4 in nest,		3 00
Larger Oval Boxes, No. 1,	"	1 00
" " " " 2,	"	1 25
" " " " 3,	"	1 75
" 3 in nest,		4 00

Miscellaneous.

Hay Carts,	per dozen,		2 50
Wagons,	"		2 50
Wheelbarrows,	"		2 00
Bird Houses,	"		3 00
Swings,	"		1 75
Wood Saw and Horse,	"		2 00
Sleds,	"		1 00
Rakes,	"		1 00
Shovels,	"		1 00
Watchman's Rattles,	"		1 25
Mallets, No. 1,	"		75
" " 2,	"		1 00
Solitaire Boards,	"	Pine,	1 50
" "	"	B. Walnut,	2 00
Fox and Geese,	"	Pine,	1 50
" "	"	B. Walnut,	2 00
Star Puzzles,	"	Pine,	1 25
" "	"	B. Walnut,	1 50
Chinese Squares,	"	"	75
Needle Books,	"		1 20

Sample Cases,
[illegible] C. O. D., for $5.00

Plate 15.1
Price list, Elijah L. Corthell, 1877
Dimensions: Left: H 3, W 4 ½; right: H 4 ½, W 6
Provenance: Private collection

16
GEORGE WILLIAM FEARING (1838–1918)

BY 1903, LESS THAN TWO YEARS AFTER THE SOCIETY OF ARTS AND CRAFTS, HINGHAM, had formed, one of its members, George Fearing, had emerged as the central figure in the group's campaign to promote and preserve traditional handcrafts in Hingham.[1] As part of this effort, William Luce, a local artist, staged a series of publicity photographs of the sixty-five-year-old cooper and toymaker at work in his cluttered shop near Main Street (pl. 16.1; see fig. 2.16).[2] These striking images, remarkable for their detail, appeared in major print media such as *Good Housekeeping*, the *Southern Workman*, and the *Boston Globe*.[3] They were usually accompanied by short biographies describing Fearing as the "sole survivor" of Hingham's coopering industry and the owner and user of a set of ancient tools that could no longer be duplicated.[4]

This depiction, though romanticized for marketing purposes, was mostly accurate. Fearing learned the trade of coopering from his father, Perez, who operated a sizable water-powered factory in Weymouth, near the Hingham border.[5] By 1870 George was working at the Wilder bucket mill,[6] where he probably met William S. Tower—his eventual employer at the toy factory next door.[7] Unlike Loring H. Cushing, his close friend and colleague, Fearing left Tower's firm after only a few years to open a small toy and novelty shop of his own. An 1888 announcement in the local newspaper declared his whirligigs "very unique, made in the style of a windmill with a miniature wooden man turning the crank; just the thing to place on the roof of a small building to please the children."[8]

Fearing worked in anonymity for most of his career; he finally gained national attention for his handcrafted woodenware when he joined the Society of Arts and Crafts. Two small butter churns (pl. 16.2) that were probably made during this early society period feature butt-joined staves, a hallmark of the small-shop cooper. Ironically, Fearing spent most of his adult life in factories, where he produced goods that were similar but assembled with machine-cut parts, like staves with rabbeted, rather than butted, joints. The round-headed copper tacks holding the churns' and piggin's hoops in place are a product of the machine age; rose- or flat-head nails are typical of woodenware made before and through the mid-nineteenth century. Fearing stamped the undersides of each piece "FEARING / HINGHAM."[9]

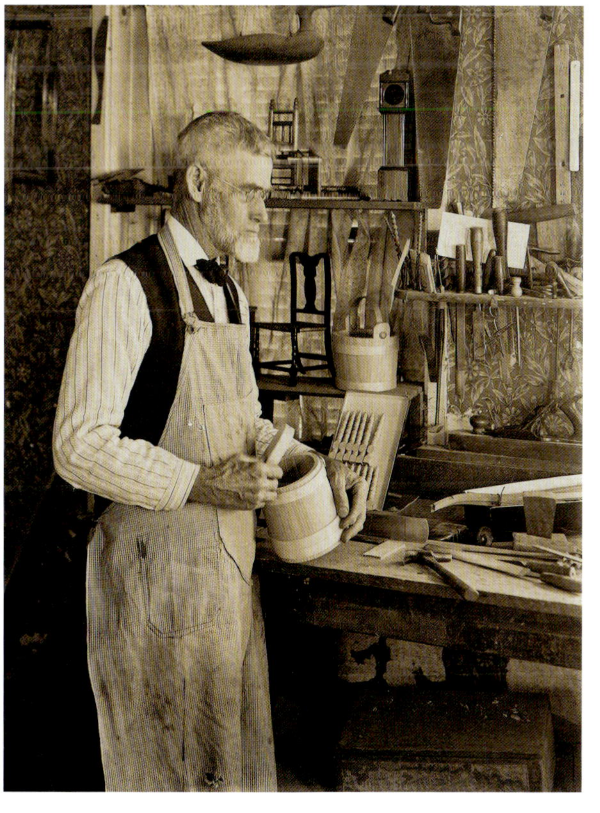

Plate 16.1

Miniature desks, chests, ladder-back chairs, and other toy furniture bearing Fearing's printed paper label (pls. 16.3–16.5) survive in far greater numbers than his coopered goods. An article published in 1903, in which the author likened Fearing to a modern-day Santa Claus, claimed that many of these items were copied from Massachusetts furniture of the colonial era.[10] These include a group of furniture, some pieces unlabeled, that descended through the Hyde family of Glens Falls, New York. A Queen Anne–style side chair and tall clock—complete with an alarm clock from the Waterbury Clock Company—can be seen in one of Luce's photographs. Whereas these pieces are generic in design, the Pilgrim Century armchair is not; it is modeled after a great chair long thought to have been owned by John Carver, the first governor of Plymouth Colony (pl. 16.5).[11]

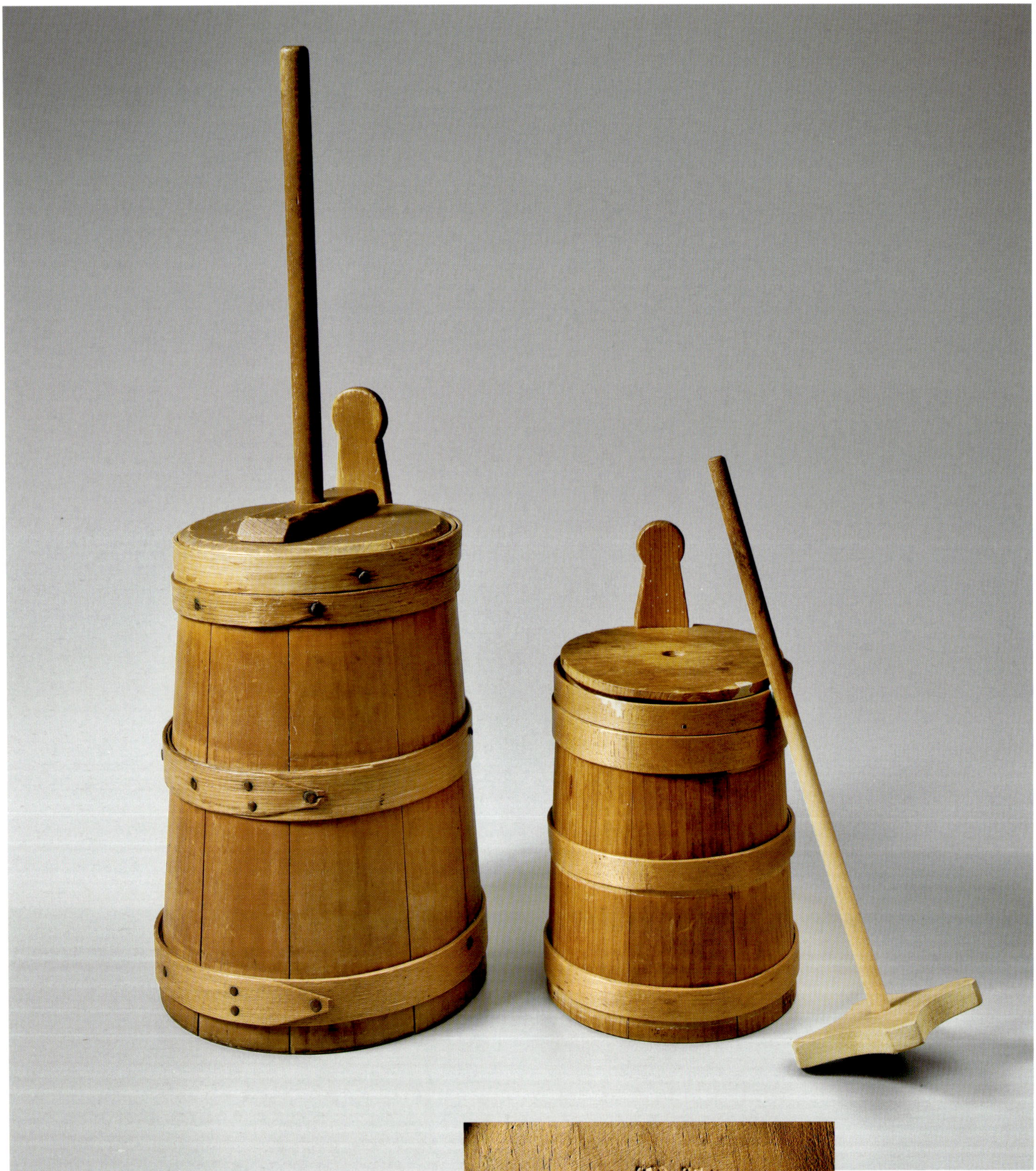

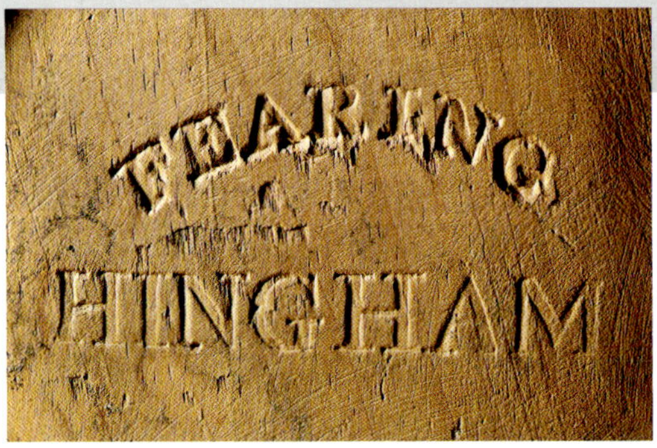

Plate 16.2

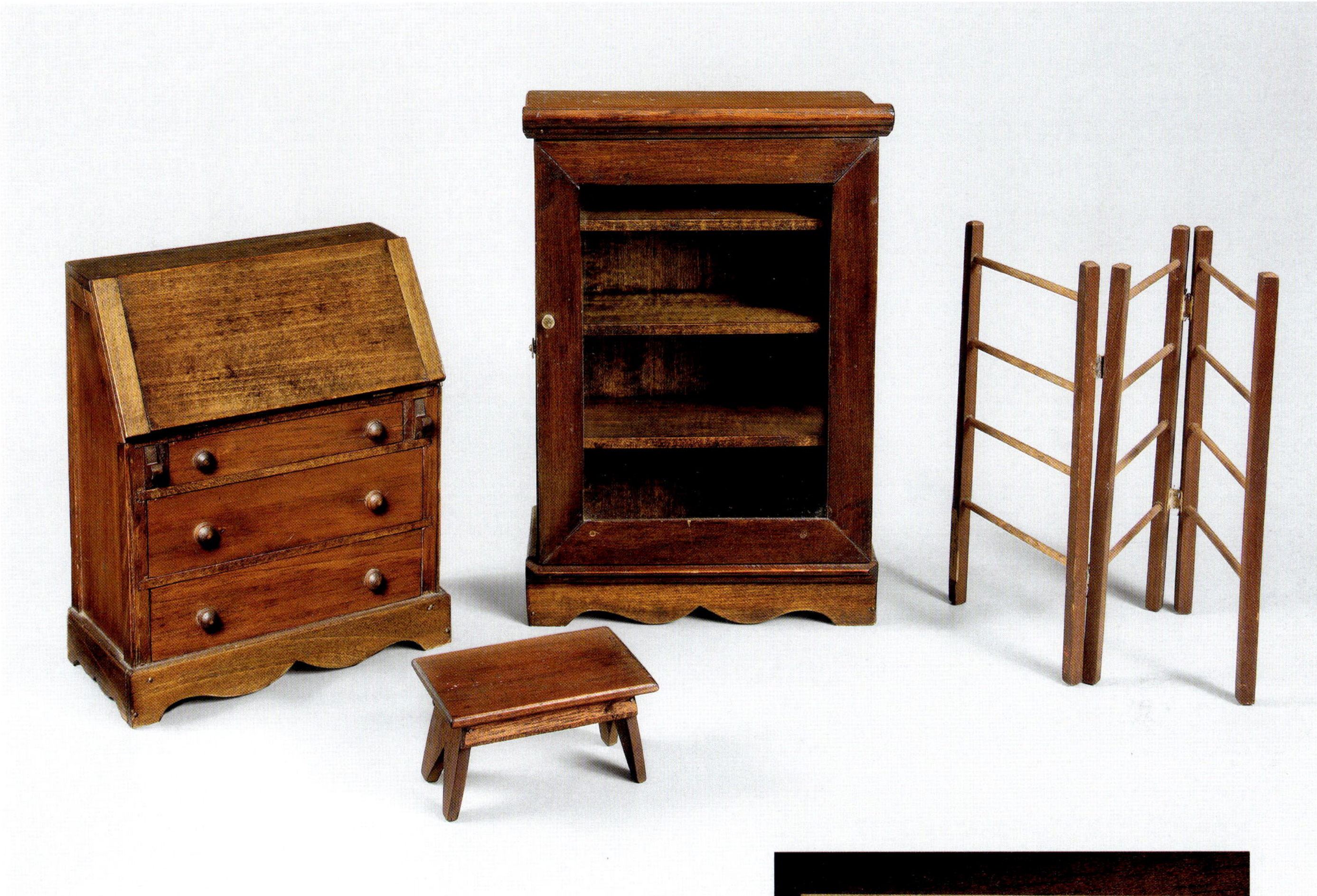

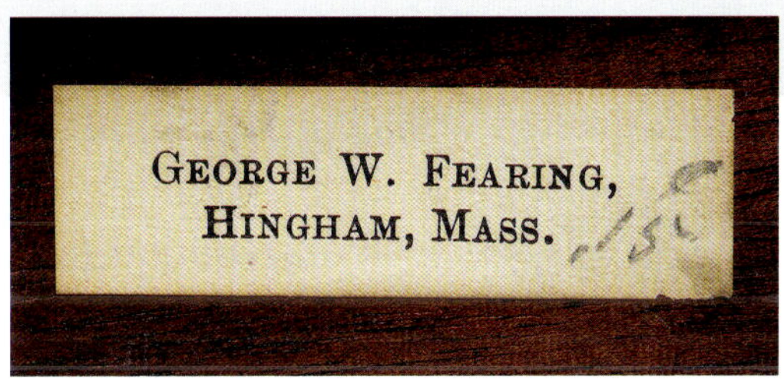

Plate 16.3

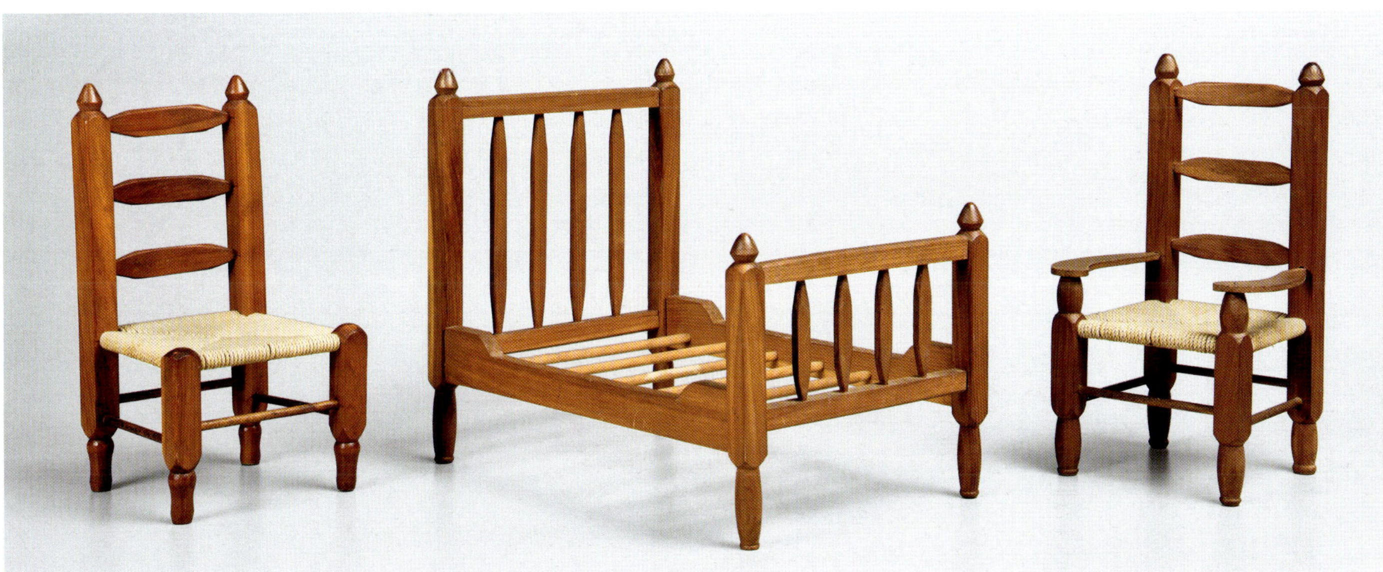

Plate 16.4

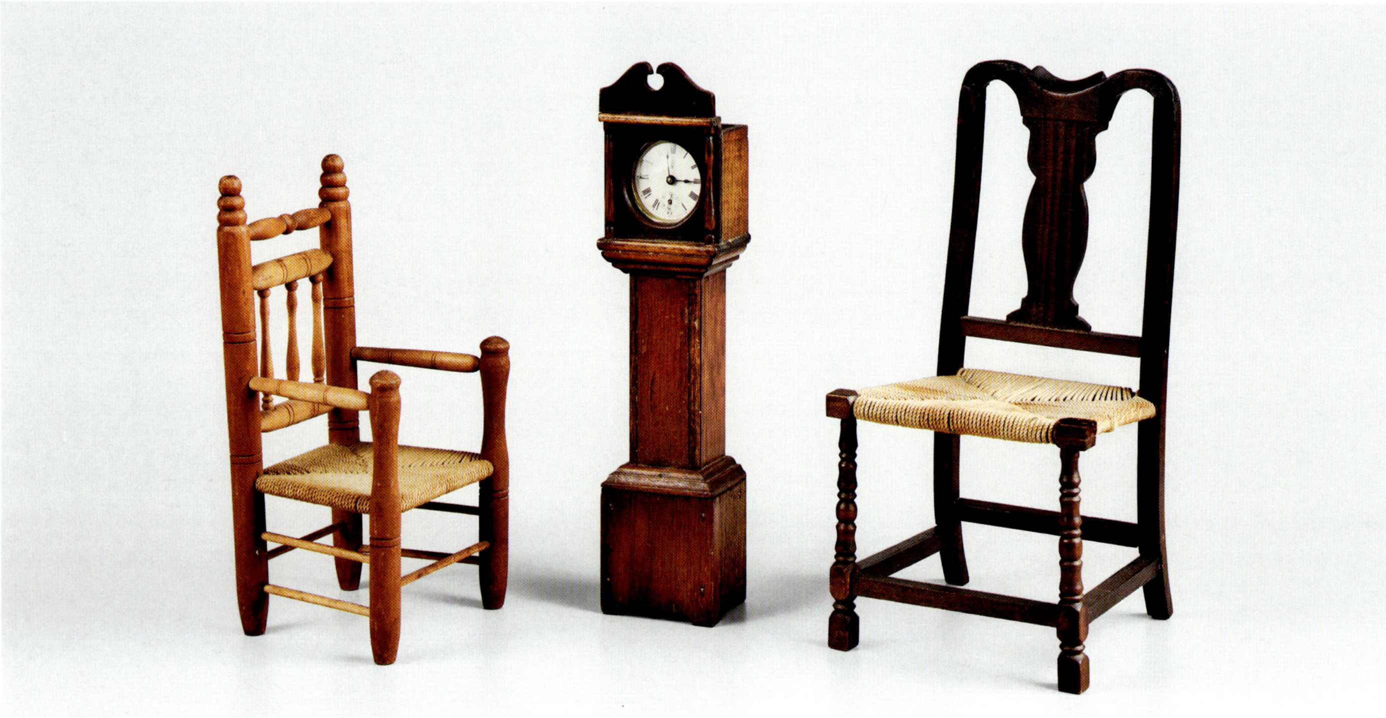

Plate 16.5

Plate 16.1
George Fearing, by William Blake Luce, 1904
Provenance: John P. Richardson Photograph Collection. Courtesy of Hingham Historical Society

Plate 16.2
Small churns, George Fearing, 1901–18
Materials: White pine staves and bottoms; ash, maple, oak hoops; copper nails
Dimensions: Left: H 12 ½, Diam 7 ¼; right: H 9 ⅜, Diam 5 ⅝
Inscriptions: "FEARING / HINGHAM" stamped on bottoms
Provenance: Courtesy of the Hersey family

Plate 16.3
Miniature furniture, George Fearing, 1901–18
Materials: Desk: white pine; footstool: black walnut; china cabinet: tulip-poplar, glass door; drying rack: tulip-poplar
Dimensions: Desk: H 9 ¼, W 7 ⅜, D 4 ¼; footstool: H 2 ⅜, W 4 ½, D 2 ⅞; china cabinet: H 12 ½, W 7 ¾, D 4 ¼; drying rack (each section): H 9 ½, W 5 ¼
Inscriptions: "GEORGE W. FEARING, / HINGHAM, MASS." printed on paper label on inside of desk, underside of footstool, back of china cabinet, and side of drying rack
Provenance:
Desk and drying rack: Courtesy of Hingham Historical Society
Footstool: Courtesy of the Hersey family
China cabinet: Collection of Flora Gill Jacobs, Chevy Chase, Maryland; Ron Rhoads Auctioneer, Nov. 21, 2009, lot 111. Courtesy of Judith and Richard Armitstead

Plate 16.4
Miniature side chair, armchair, and bedstead, George Fearing, 1901–18
Materials: Possibly red cedar; twine seats
Dimensions: Side chair: H 8 ¾, W (seat) 4 ⅝, D (seat) 4 9⁄16; armchair: H 9 ¼, W (seat) 5 7⁄16, D (seat) 5 ½; bedstead: H 9, W 7 ¼, D 9
Inscriptions: "GEORGE W. FEARING, / HINGHAM, MASS." printed on paper label on underside of side chair; remnant of printed paper labels on slat of bedstead and bottom of armchair
Provenance: Collection of Flora Gill Jacobs, Chevy Chase, Maryland; Ron Rhoads Auctioneer, Nov. 21, 2009, lots 108, 109. Courtesy of Judith and Richard Armitstead

Plate 16.5
Miniature Carver chair, tall clock, and Queen Anne chair, George Fearing, 1901–18
Materials: Carver chair: possibly tulip-poplar; twine seat; tall clock: white pine, tulip-poplar; Queen Anne chair: tulip-poplar; twine seat
Dimensions: Carver chair (left): H 10, W (seat) 5, D (seat) 4 ⅞; tall clock: H 12 ⅜, W 2 15⁄16, D 2 ¼; Queen Anne chair (right): H 12 ¼, W (seat) 6, D (seat) 5 ⅜
Inscription: "GEORGE W. FEARING, / HINGHAM, MASS." printed on paper label inside hood of tall clock
Provenance: Descended through the Hyde family of New York; to Derin Bray, Portsmouth, New Hampshire; to current owner. Courtesy of the Hersey family

17
LORING HERSEY CUSHING (1840–1915)

ON AUGUST 17, 1881, THE *BOSTON EVENING JOURNAL* REPORTED AN ACCIDENT NEAR the Old Colony Railroad Station in Hingham. Loring H. Cushing, a toymaker, was returning from Nantasket Beach when an oncoming locomotive spooked his horse. He attempted to rein the animal in, but it raced wildly alongside the tracks until the carriage it was pulling collided with the moving train. The vehicle's wheels and dashboard broke apart, and everything and everyone aboard was thrown violently to the ground. Many wooden toys and other cargo were smashed to pieces, but the horse and passengers, remarkably, walked away unharmed.[1]

Cushing's brush with disaster was a minor setback in an otherwise long and fruitful career at the Tower toy factory in South Hingham. During his seventeen years there, he rose to the rank of general manager and helped the internationally known firm produce, exhibit, and sell goods at sites that included its summer stand in Nantasket.[2] His lasting contribution, however, was as a designer of toys, such as Tower's Log Cabin, an instructive miniature replica of an old homestead (see pl. 14.5).[3] After his employer, William S. Tower, retired from the business in the early 1890s, Cushing took up work in a shoe factory, where he spent his days at a sewing machine stitching leather.[4]

When a group of Hingham residents formed the local Society of Arts and Crafts at the turn of the century, Cushing seized the opportunity to reinvent himself as a small-shop toymaker. With a guiding mission to "promote artistic work in all branches of handi-craft"—many of which had nearly vanished locally—the society invited basketmakers, metalworkers, embroiderers, and other skilled craftsmen, both professional and domestic, to submit goods to its panel of judges. If their workmanship was deemed worthy, the artists gained membership in the group, and their items were sold at annual exhibitions in Hingham and eventually throughout the country.[5]

Cushing embraced this new enterprise with ambition, determination, and savvy business sense. He converted the loft of his stable into a shop and installed an electric motor to run a lathe and saw. The site, described by one contemporary as "very workmanlike and not widely picturesque," became a popular tourist attraction for children and adults eager to observe an "old-fashioned" toymaker in action. Those lucky enough to visit would have also seen "his old army coat, faded and soiled, and his time-worn knapsack" hanging prominently within.[6] Cushing had worn the garments (pls. 17.1, 17.2) in the Civil War as a member of Company K, 43rd Massachusetts Regiment.[7] Some time in the 1950s his niece donated the items—including his hat—to the Hingham Historical Society. An early tintype (pl. 17.3) shows the young soldier neatly dressed in the uniform.[8]

Seeing opportunity in the Colonial Revival craze to collect and preserve American antiques, Cushing developed a line of miniature colonial-style chests, chairs, and other forms (pls. 17.4, 17.5), some of which he illustrated in a 1912 price list (pl. 17.7). His larger case pieces, like this slant-lid desk (pl. 17.6), made from tulip-poplar, sold for two dollars—a price that kept it out of the reach of most families. It is constructed with the same simple lap joints and tiny iron nails used by Tower, though Cushing's work is generally more fastidious; he typically filled nail holes with putty before finishing them. His printed paper label "MADE BY / LORING H. CUSHING / 14 High Street / SO. HINGHAM MASS." is glued to the underside, as is the official seal of the Society of Arts and Crafts, Hingham: a bold graphic of a classic Hingham bucket with "APPROVED" printed below.

Cushing preferred to leave his toys in a "natural" state, but his clients often requested a brown finish. To achieve the perfect look—one in which the wood was dark enough to simulate mahogany without effacing the grain—he spent hours experimenting with a mixture of commercial stains.[9] This explains the great variety of finishes found on his surviving toys. Among the most unusual and pleasing is a red stain on a ladder-back rocking chair (see pl. 17.6); it also has the woven twine seat and distinctive beehive finials that characterize much of his seating furniture. The underside features a paper label that identifies the maker and his hometown but omits his address.

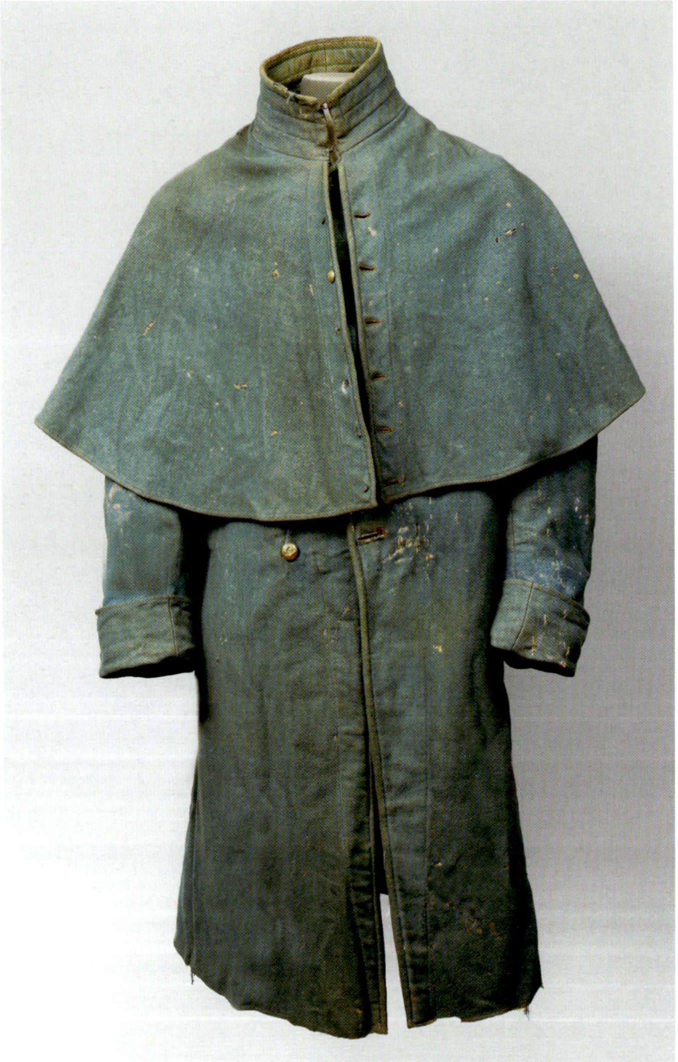

Plate 17.1

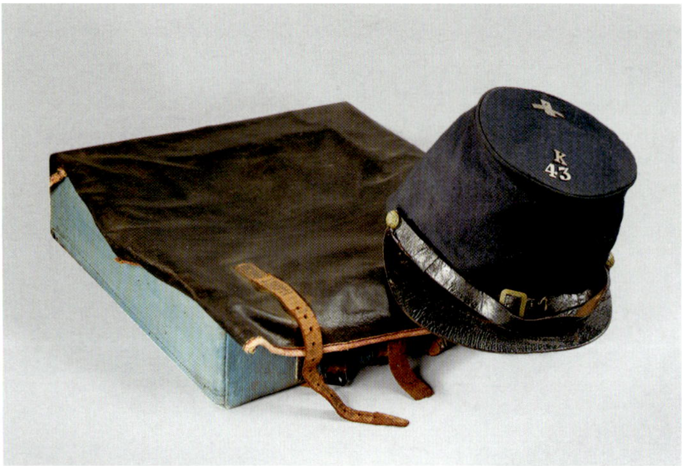

Plate 17.2

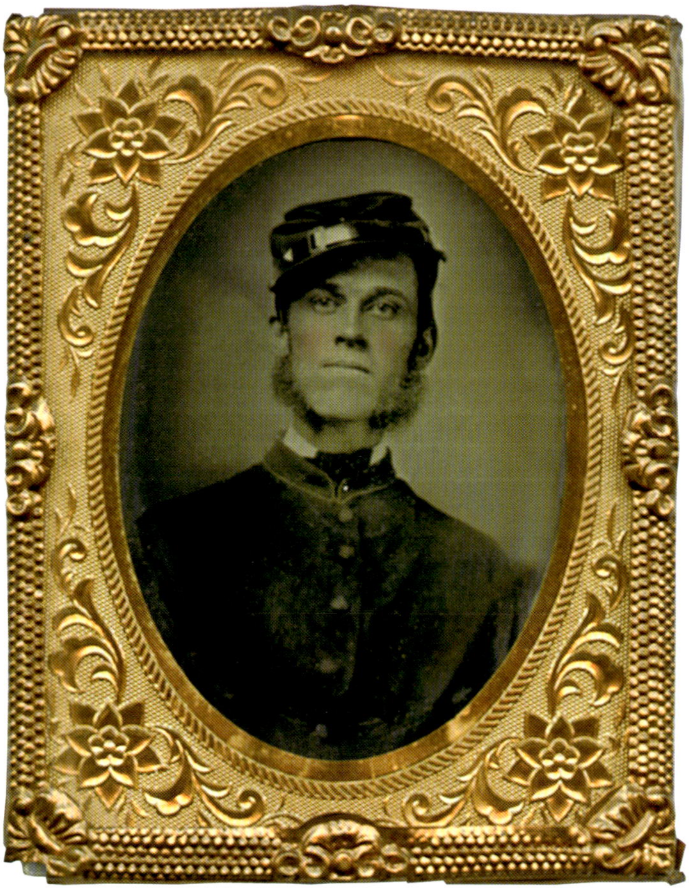

Plate 17.3

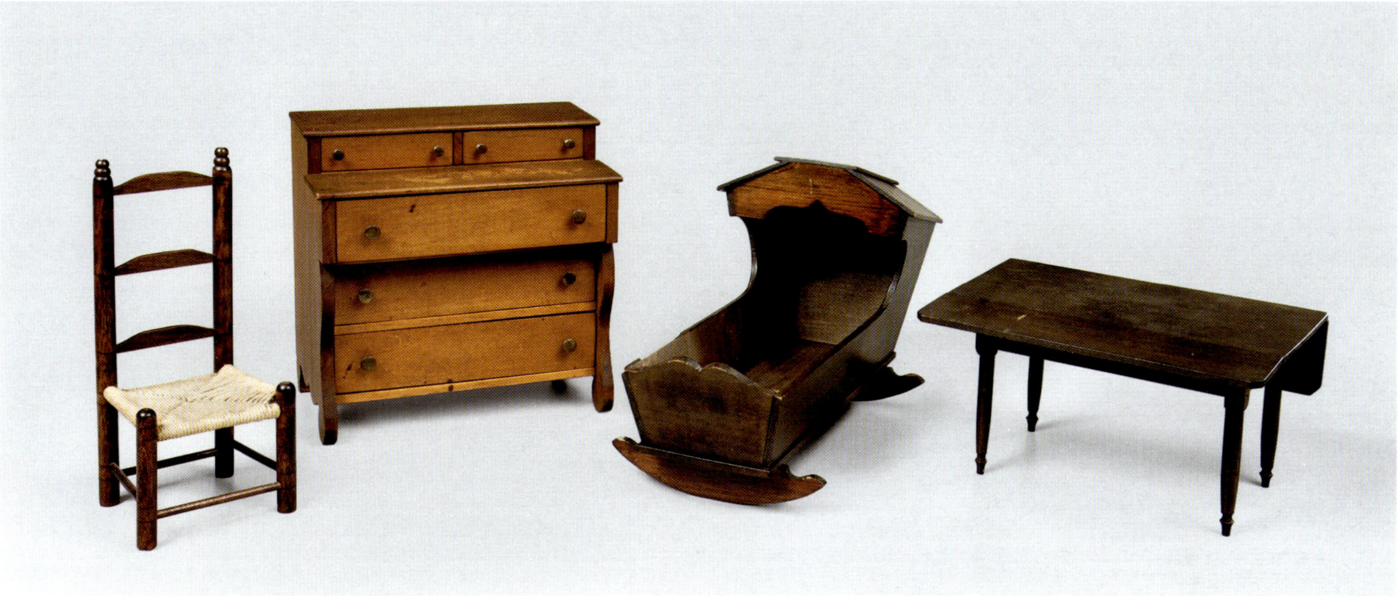

Plate 17.4

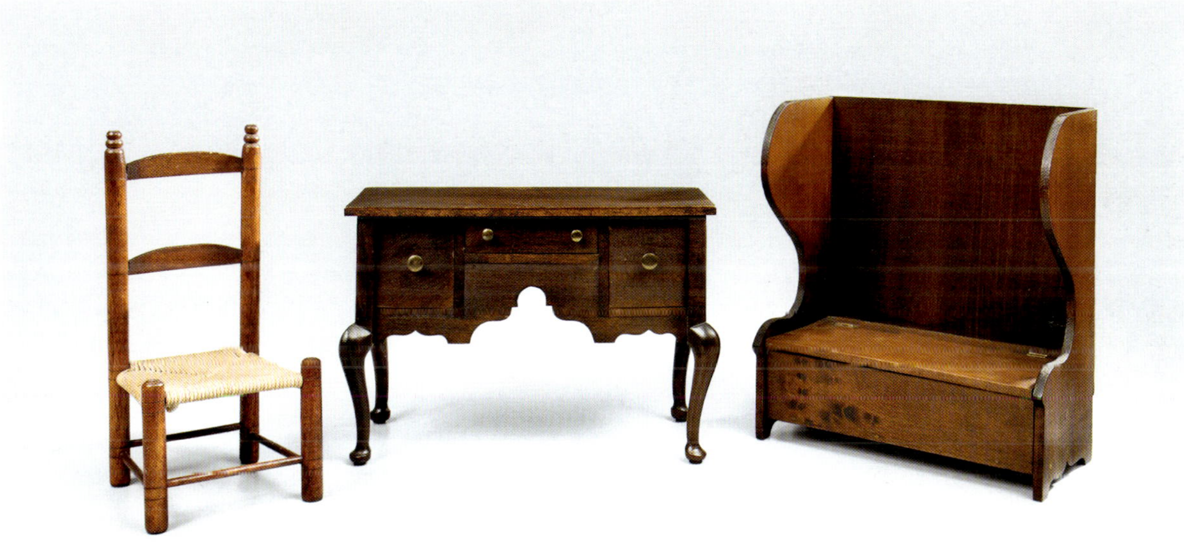

Plate 17.5

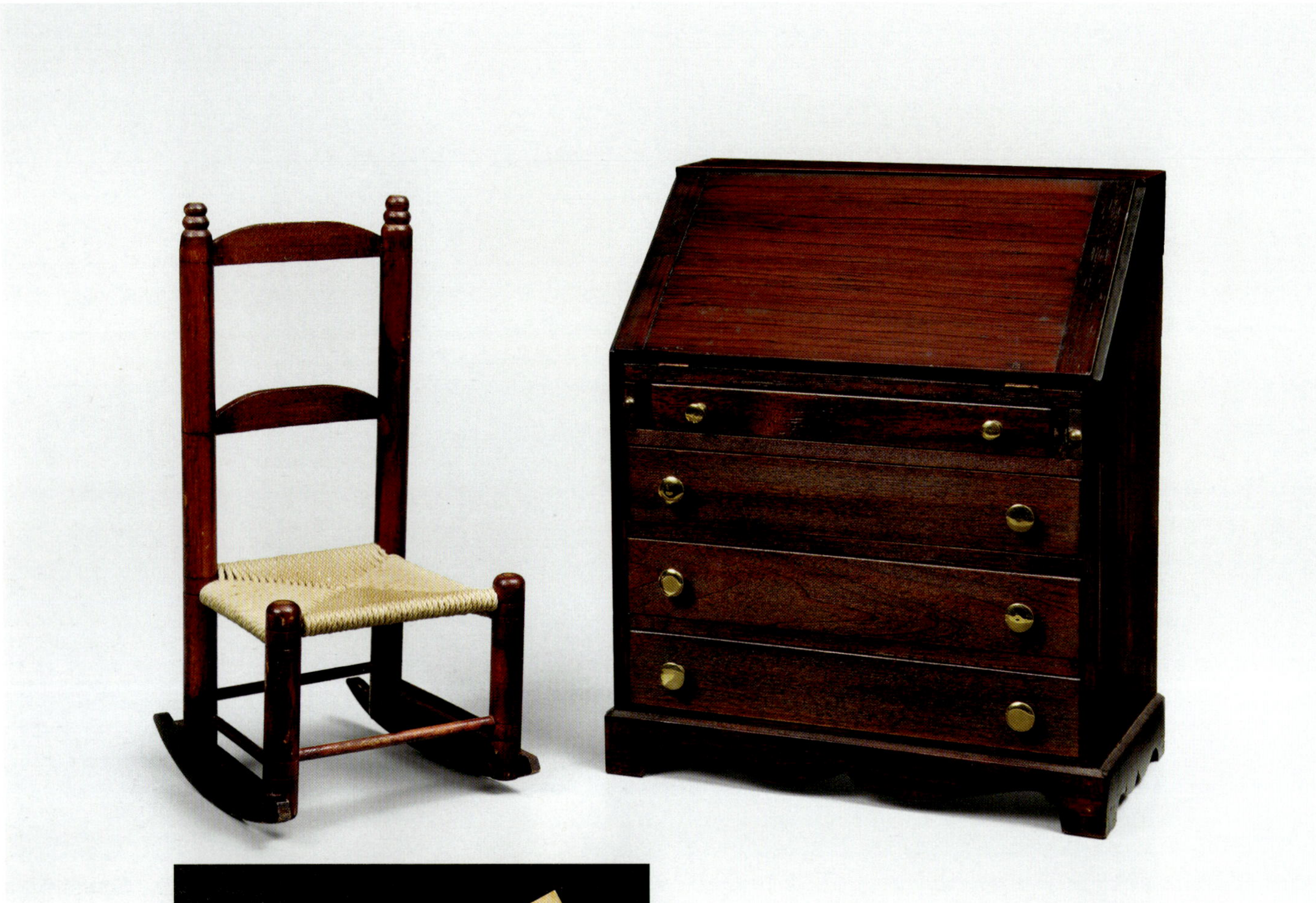

Plate 17.6

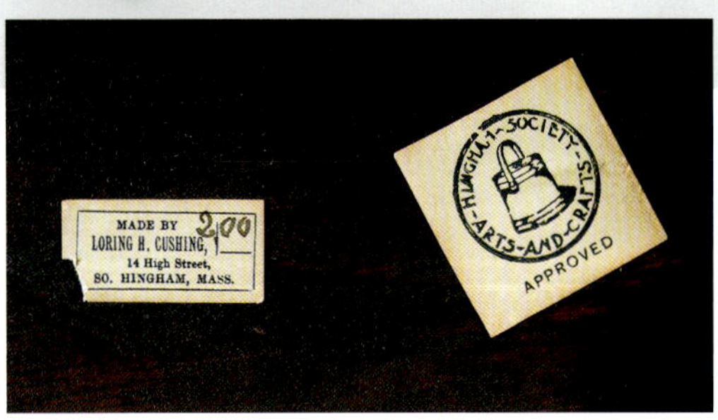

Plate 17.1

Greatcoat worn by Loring H. Cushing during the Civil War, 1861–64

Provenance: Loring H. Cushing; to his niece Mrs. Winthrop Jacob Cushing; given to the Hingham Historical Society in 1946. Courtesy of Hingham Historical Society

Plate 17.2

Hat and knapsack worn by Loring H. Cushing during the Civil War, 1861–64

Provenance: Loring H. Cushing; to his niece Mrs. Winthrop Jacob Cushing; given to the Hingham Historical Society in 1946. Courtesy of Hingham Historical Society

Plate 17.3

Tintype portrait of Loring H. Cushing, 1861–64

Provenance: Loring H. Cushing; to his niece Mrs. Winthrop Jacob Cushing; given to the Hingham Historical Society in 1946. Courtesy of Hingham Historical Society

Plate 17.4

Miniature chair, bureau, cradle, and table, Loring H. Cushing, 1901–15

Materials: Chair: tulip-poplar; rope seat; bureau: white pine and tulip-poplar; brass knobs; iron nails; cradle: tulip-poplar; iron nails; table: tulip-poplar; leather hinges

Dimensions: Chair: H 9 ⅛, W (seat) 4 ⅜, D (seat) 3 ⅞; bureau: H 8 ⅝, W 8 ¾, D 5 ⅜; cradle: H 8, W 6 ⅛, D 11 ⅞; table: H 5 ⅛, W 9, D 5

Inscriptions: "MADE BY / LORING H. CUSHING, / So. Hingham, Mass." printed on paper label on bottoms of chair and bureau; "MADE BY / LORING H. CUSHING, / 14 High Street, / SO. HINGHAM, MASS." printed on paper label on bottoms of cradle and table; "HINGHAM—SOCIETY / ARTS—AND—CRAFTS" surrounding a graphic of a bucket printed on a paper label inside cradle

Provenance: Courtesy of the Hersey family

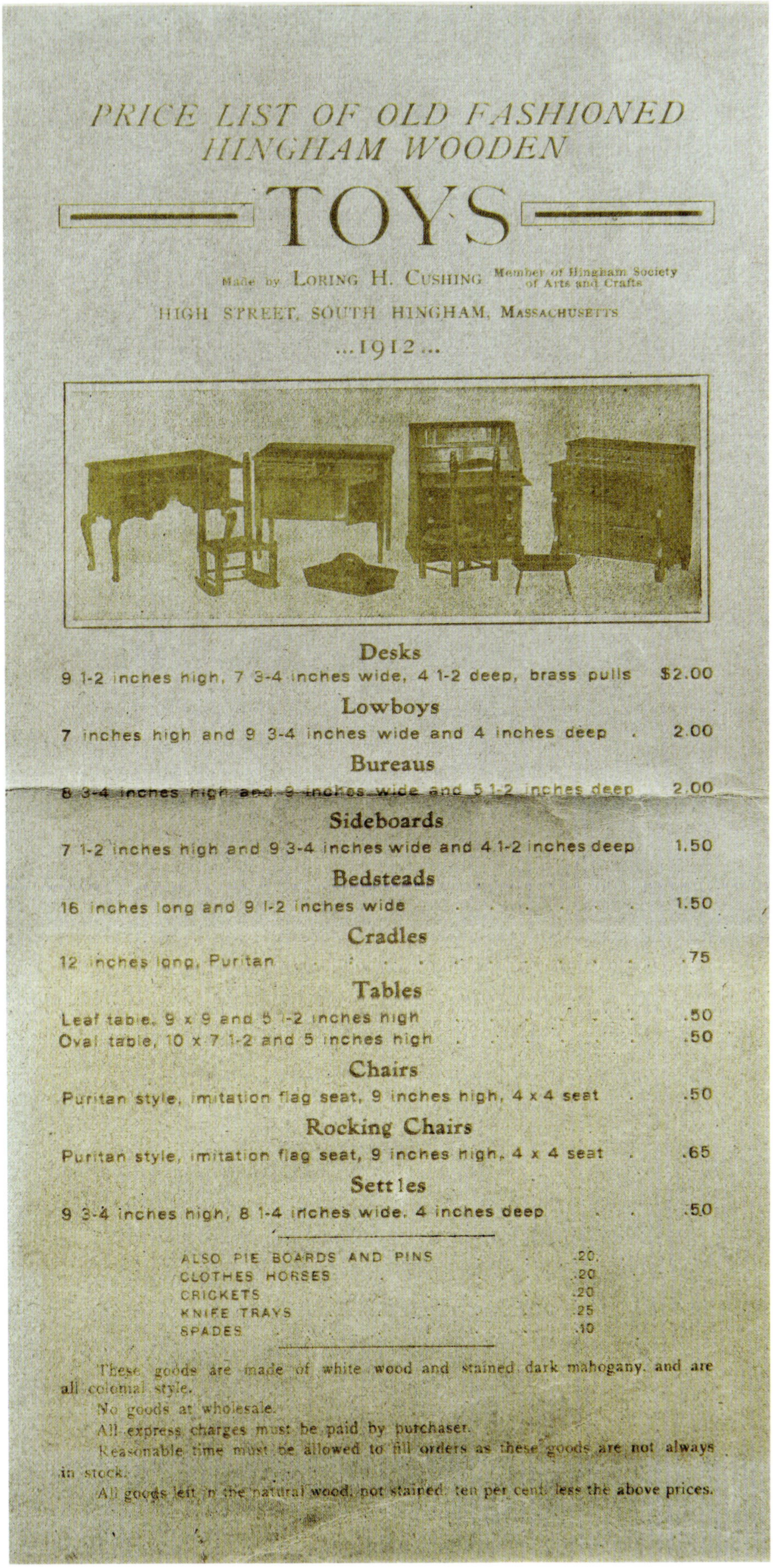

PRICE LIST OF OLD FASHIONED HINGHAM WOODEN

TOYS

Made by LORING H. CUSHING Member of Hingham Society of Arts and Crafts

HIGH STREET, SOUTH HINGHAM, MASSACHUSETTS

...1912...

Desks

9 1-2 inches high, 7 3-4 inches wide, 4 1-2 deep, brass pulls	$2.00

Lowboys

7 inches high and 9 3-4 inches wide and 4 inches deep	2.00

Bureaus

8 3-4 inches high and 9 inches wide and 5 1-2 inches deep	2.00

Sideboards

7 1-2 inches high and 9 3-4 inches wide and 4 1-2 inches deep	1.50

Bedsteads

16 inches long and 9 1-2 inches wide	1.50

Cradles

12 inches long, Puritan	.75

Tables

Leaf table, 9 x 9 and 5 1-2 inches high	.50
Oval table, 10 x 7 1-2 and 5 inches high	.50

Chairs

Puritan style, imitation flag seat, 9 inches high, 4 x 4 seat	.50

Rocking Chairs

Puritan style, imitation flag seat, 9 inches high, 4 x 4 seat	.65

Settles

9 3-4 inches high, 8 1-4 inches wide, 4 inches deep	.50

ALSO PIE BOARDS AND PINS	.20
CLOTHES HORSES	.20
CRICKETS	.20
KNIFE TRAYS	.25
SPADES	.10

These goods are made of white wood and stained dark mahogany, and are all colonial style.

No goods at wholesale.

All express charges must be paid by purchaser.

Reasonable time must be allowed to fill orders as these goods are not always in stock.

All goods left in the natural wood, not stained, ten per cent. less the above prices.

Plate 17.7

Plate 17.5

Miniature chair, lowboy, and settle, Loring H. Cushing, 1901–15

Materials: Chair: tulip-poplar; rope seat; lowboy: tulip-poplar; brass knobs; iron nails; settle: tulip-poplar; brass hinges; iron nails

Dimensions: Chair: H 9 ⅛, W (seat) 4 ⅜, D (seat) 3 ⅞; lowboy: H 7, W (legs) 10, D (legs) 4 ⅞; settle: H 9 ½, W 8 ¼, D 4 ¼

Inscriptions: "MADE BY / LORING H. CUSHING, / So. Hingham, Mass." printed on paper label on bottoms of lowboy and settle; "MADE BY / LORING H. CUSHING, / 14 High Street, / SO. HINGHAM, MASS." printed on paper label on bottom of chair

Provenance: Courtesy of the Hersey family

Plate 17.6

Rocking chair and desk, Loring H. Cushing, 1901–15

Materials: Rocking chair: tulip-poplar slats and rockers; possibly birch posts and stretchers; twine seat; desk: tulip-poplar case; brass knobs and hinges

Dimensions: Rocking chair: H 9, W (seat) 4 ¼, D (seat) 3 7⁄16; desk: H 9 ¼, W 8, D 4 5⁄16

Inscriptions: "MADE BY / LORING H. CUSHING, / So. Hingham, Mass." printed on paper label on underside of chair; "MADE BY / LORING H. CUSHING, / 14 High Street, / SO. HINGHAM, MASS." printed on paper label on bottom of desk; "HINGHAM—SOCIETY / ARTS—AND—CRAFTS" surrounding a graphic of a bucket printed on a paper label on bottom of desk

Provenance

Rocking chair: Courtesy of Hingham Historical Society

Desk: Courtesy of the Hersey family

Plate 17.7

Price list, Loring H. Cushing, 1912

Dimensions: H 12, W 6

Provenance: Courtesy of Hingham Historical Society

18
AUGUSTUS LINCOLN HUDSON (1830–1922)

WOOD CARVING IS AMONG HINGHAM'S MOST REVERED ARTISTIC TRADITIONS.[1] IN THE second half of the nineteenth century the best work in town came from the little-known shop of Augustus Hudson on North Street.[2] A cabinetmaker noted for his furniture embellished with carved animals and fruit, Hudson is best remembered today for the extraordinary sign he made for the Torrent Engine Company in West Hingham (pl. 18.1)—a work completed in 1860 at the substantial cost of sixty dollars.[3] That same year, before it was hung over the entrance to the firehouse, it was exhibited to great fanfare at the local Agricultural and Horticultural Society Fair.[4] Local legend holds that the lower half of the sign deteriorated long ago, but the dapper fireman that stood atop it, albeit weather-beaten, still survives, sporting a corded hose and tin helmet (pl. 18.2).[5]

The painted scene on Hudson's masterpiece, though altered over time, may have been executed—or at least influenced—by Augustus's older brother William, a prominent local artist known for his portraits, still life paintings, and views of the First Parish Church (see fig. 2).[6] The brothers worked together on a number of occasions. In 1865, after William had become the town's leading professional photographer, they copatented an improved sensitizing box for exposing developing plate negatives.[7] Between 1860 and 1890 Augustus also manufactured a variety of wooden handheld and tabletop stereoscopes.[8] These optical instruments allowed users to view, in three dimensions, the popular stereo cards produced and sold by photographers like William. A mahogany example (pl. 18.3) is marked "AH" on the hood and "A. L. HUDSON / WARRANTED" on the end of the narrow slide.

Near the end of his career, Hudson became a member of the Society of Arts and Crafts, Hingham. In 1911, at the age of eighty-one, he displayed picture frames and an elegant work table at the society's annual show and sale at Loring Hall.[9] His specialty, however, was miniature bedsteads and tilt-top tables (pl. 18.4). These items reveal the skill and trained eye of a master cabinetmaker and rank among the finest toys made in Hingham. Hudson's rope bedsteads, including a model with a scalloped headboard, feature narrow rails and tall mahogany posts with expertly turned baluster-and-ring decoration.[10] His miniature tables, though simpler in appearance, are equally graceful in form and construction; an unusual example with an incised border is finished with red and black grain painting. Much like its larger counterparts, this table's octagonal top is secured with small screws and tapered cleats, which in turn pivot on a small block atop the pedestal. A brass clasp ensures the table stays locked when in the closed position. Hudson's paper label, "MANUFACTURED BY / AUGUSTUS L. HUDSON / WEST HINGHAM, MASS.," is found on both pieces.

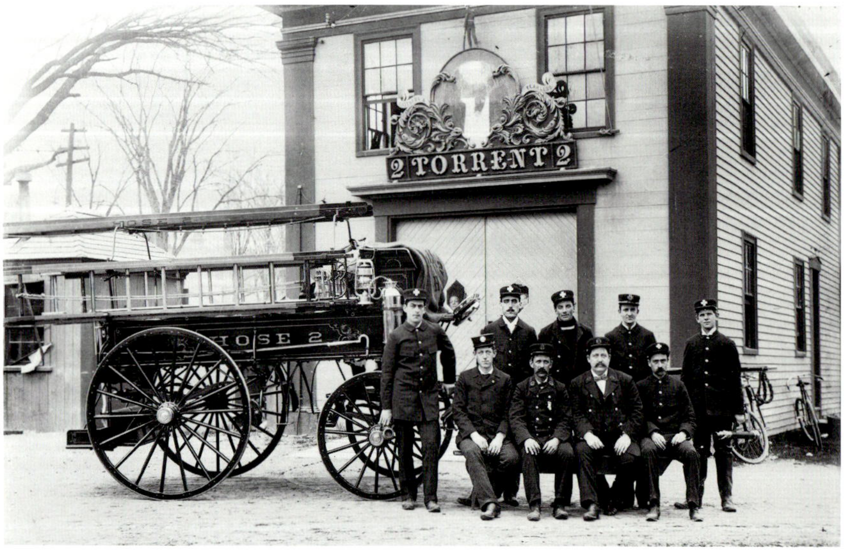

Plate 18.1

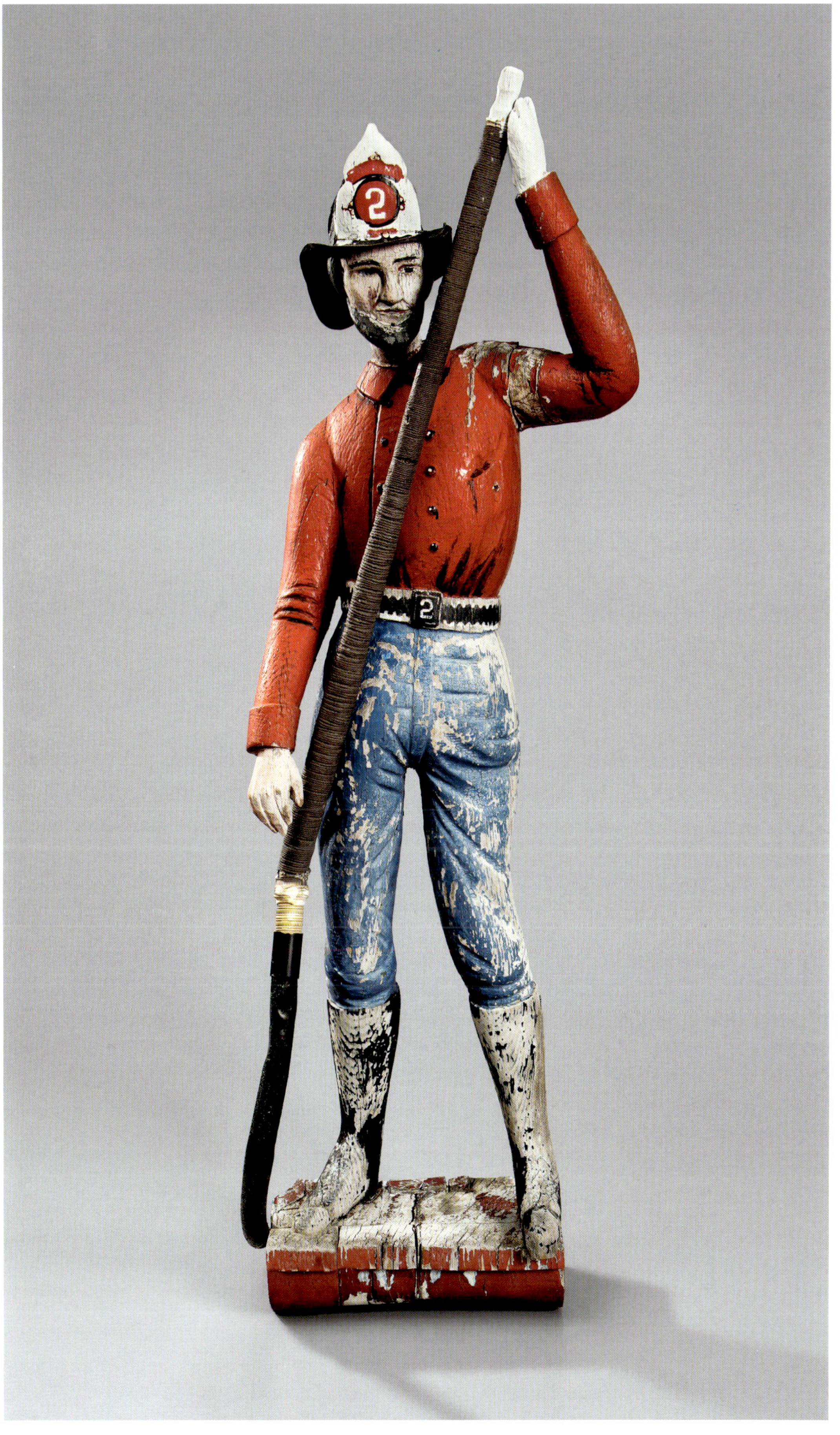

Plate 18.2

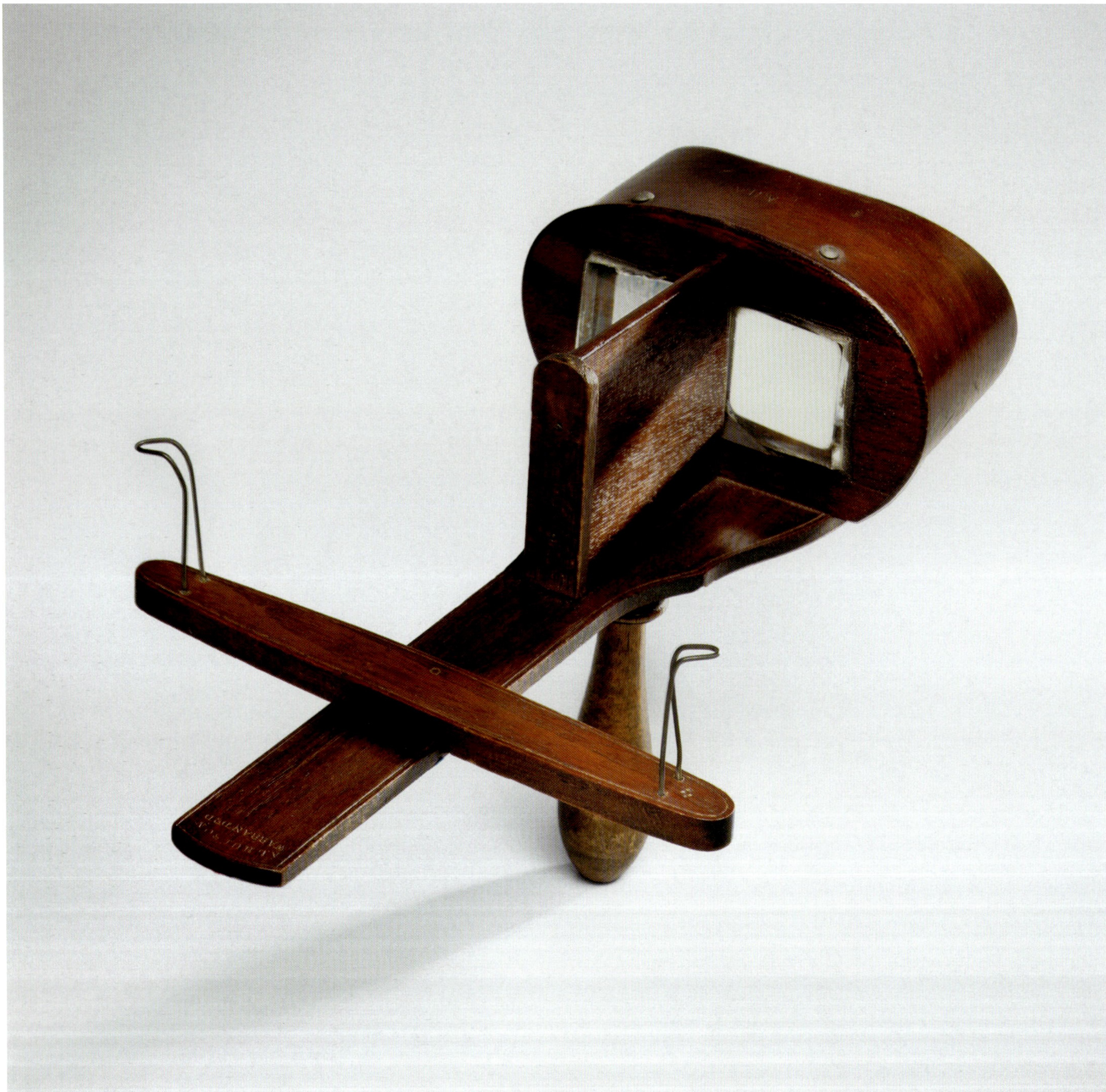

Plate 18.3

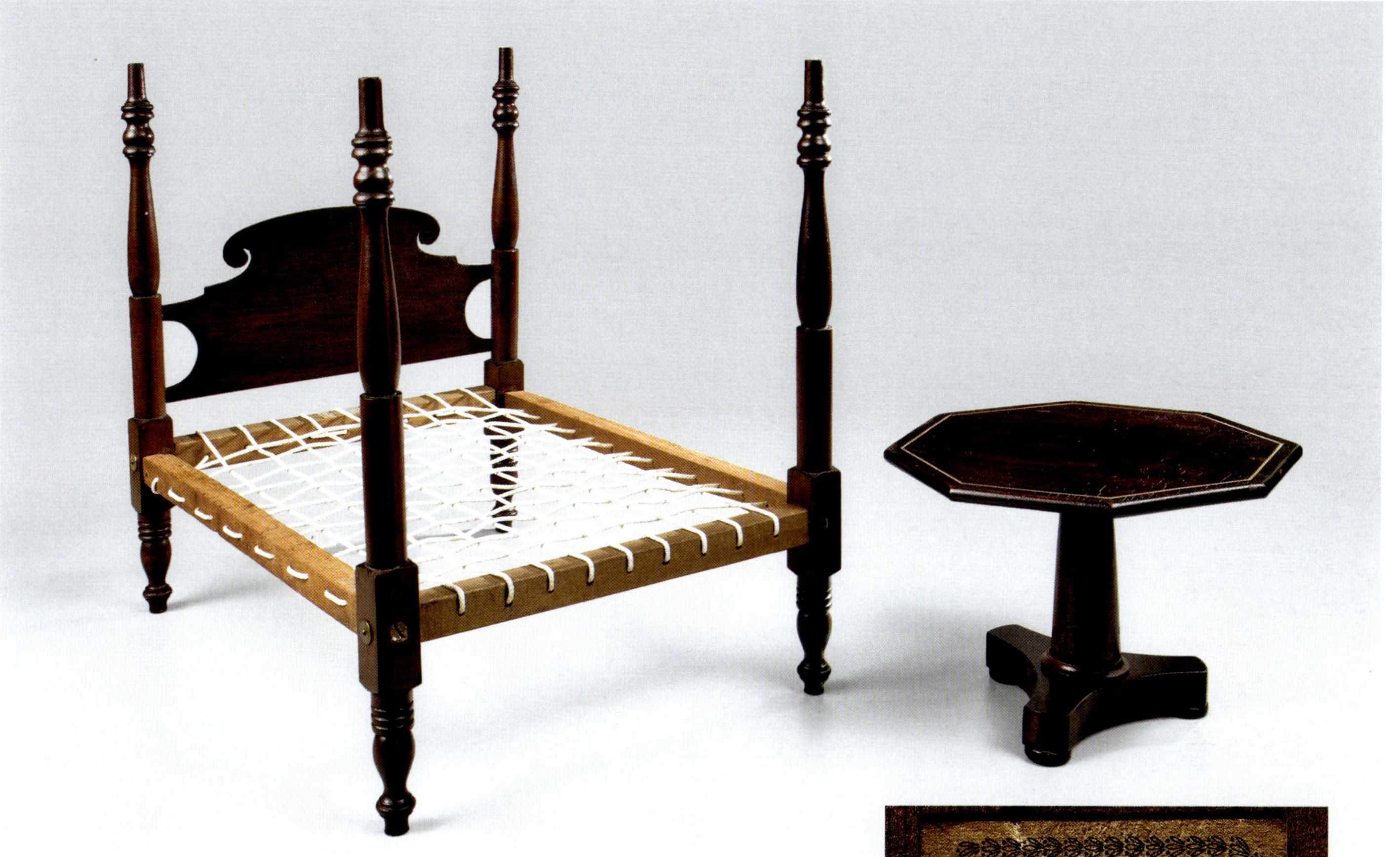

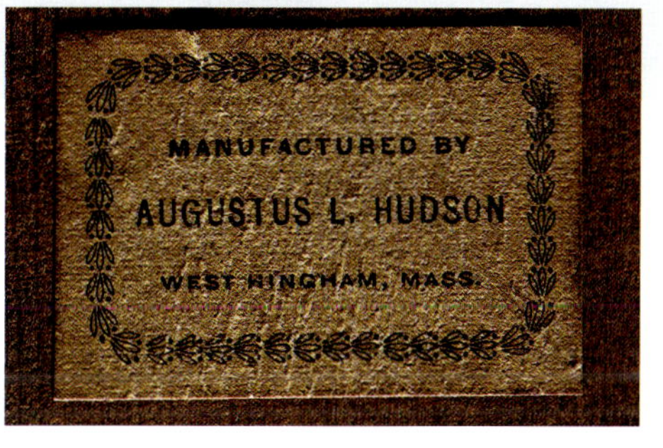

Plate 18.4

Plate 18.1

Torrent Engine Company, ca. 1881
Provenance: Courtesy of Bare Cove Fire Museum

Plate 18.2

Fireman, Augustus L. Hudson, 1860
Dimensions: H 28, W $7 \frac{1}{2}$, D 4
Provenance: Courtesy of the Town of Hingham

Plate 18.3

Stereoscope, August L. Hudson, 1865–85
Materials: Mahogany; glass; brass hardware
Dimensions: H $7 \frac{3}{8}$, W $7 \frac{1}{2}$, D 12
Inscriptions: "A. L. HUDSON / WARRANTED" stamped on slide; "AH" stamped on hood
Provenance: Courtesy of the Hersey family

Plate 18.4

Miniature bedstead and table, Augustus L. Hudson, ca. 1911
Materials: Bedstead: mahogany posts, headboard, and footboard; tulip-poplar front and back rails; walnut side rails; twine; table: tulip-poplar; brass catch; iron screws
Dimensions: Bedstead: H $15 \frac{3}{8}$, W $11 \frac{7}{8}$, D 16; table: H $6 \frac{1}{4}$, W $8 \frac{1}{2}$, D $8 \frac{1}{2}$
Inscriptions: "MANUFACTURED BY AUGUSTUS L. HUDSON / WEST HINGHAM, MASS." printed on paper label on bedstead rail and underneath table top
Provenance: Courtesy of the Hersey family

19
WILLIAM BLAKE LUCE (1860–1924)

THE HISTORIAN JOHN P. RICHARDSON (1935–2011) SPENT A LIFETIME RUMMAGING through attics, barns, and abandoned trash sites in search of early Hingham artifacts.[1] He hit the proverbial jackpot in 1977, when Mary Luce, a retired schoolteacher, invited him to purchase "odds and ends" from the studio of her late father-in-law.[2] Much to Richardson's amazement, the small building had stood largely undisturbed for over half a century, and it still contained thousands of items owned by William Blake Luce, a prominent photographer and toymaker (pl. 19.1). In addition to a large collection of finished and unfinished wooden toys (pl. 19.2), Richardson salvaged a group of glass-plate negatives, including more than twenty rare views of miniature furniture and dollhouses made and photographed by Luce (see Appendix B).[3] This extraordinary archive of images and objects—never before seen by the public—has shed much-deserved light on one of Hingham's most accomplished, yet largely forgotten, master craftsmen.

William Luce was a restless artist and inventor. Before moving to the South Shore in 1897, he lived in Boston, where he worked as an illustrator for the popular children's magazine the *Youth's Companion*.[4] In his spare time he practiced photography and tinkered with designs for electric fans, lamps, and other gadgets for which he received almost a dozen patents—only one of which achieved modest commercial success: a small snapshot "detective" camera manufactured by Perry Mason & Company in 1890.[5] A few years later Luce pioneered the art and science of aerial kite photography. Although he self-published an instructional book on the topic and promoted it in trade journals, it never gained widespread appeal.[6]

The Society of Arts and Crafts, Hingham, founded in 1901, offered the perfect venue to showcase Luce's many talents.[7] In 1904 he staged several publicity photographs of the society's leading members—most notably the veteran toymaker George W. Fearing (see fig. 2.16; pl. 16.1). Within a few years Luce began to make and exhibit toys of his own.[8] A miniature paneled ice chest (pl. 19.3), complete with a tin interior and drainage pipe, represents some of his earliest work; the back is signed and dated 1908. Beneath the inscription Luce pasted a small circular label printed with his name and an engraving of the Old Ship Church—an ancient landmark that stood only a few hundred feet from his home and studio on Main Street.

Miniature antique furniture accounted for most of Luce's output. Signed pieces, of which only a few are known, include a work table and an Empire bureau in a dark stain (see pl. 19.3). Luce stamped both items "W. B. LUCE, / HINGHAM MASS." in blue ink on the interior of the drawers. The bureau also bears the seal of the Society of Arts and Crafts, an indication that it was sold at one of the group's many exhibitions. Similar markings appear on Luce's small split-spindle mirrors with faux-gilded surfaces. According to one contemporary, these versatile toys suited "lady's chambers" as well as dollhouses. The upper tablets often featured hand-colored prints, probably of Luce's own manufacture. An example stamped "*LUCE & IVES CO. / Hingham, Mass.*" was made around 1910 (pl. 19.4), during a brief partnership between the toymaker and a local salesman, August P. Ives.[9]

Wealthy families from all over the country purchased Luce's extraordinary dollhouses. A timber-framed model with a fully furnished interior, as documented by a series of early photographs (pls. 19.5, 19.6), confirms Luce's status as a master artist and craftsman; it may be the same dollhouse he displayed at the society's annual exhibition in 1915. A local newsman enthusiastically reported:

> *A home in the Black Forest is the theme of this year's masterpiece. The picturesque wooden house with the long slanting roof is big enough nearly to cover the top of a small dining table. Despite the cute knocker and latch-string the exterior gives but little idea of the wonders of the interior. The marvels of an old German home, however, are easily seen, for sides and roof of the residence all lift to allow a peek within. In one of the living rooms is a bed with curtains, a vivid reminder of Jerome K. Jerome's adventures in sleeplessness. On a carved table are beer mugs and a plate of pretzels. Spinning wheel and flax wheel attest to the persistence of domestic thrift. The wall paper has an aspect of old Nuremberg. This home of Teuton conservatism, withal, is lighted by real electric lamps.*[10]

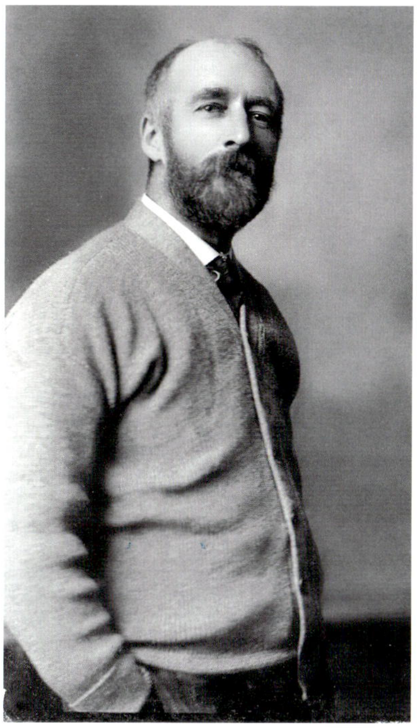

Plate 19.1

Following Luce's death in 1924, his son Stuart (b. 1888) took control of the studio, which had been rebranded "The Luce Specialty Shop: Makers of Old Colonial Toy Furniture."[11] It was probably during this period that the shop produced a line of simple, often brightly painted chairs, tables, mirrors, bedsteads, and chests, many with floral decoration (pl. 19.7).[12] Although reasonably well made, they lack the detail and workmanship found in Luce's earlier toy furniture and dollhouses. By 1934 Stuart closed the shop and took up work as a store clerk.[13] Without the creative guidance of his father and the built-in audience of the Society of Arts and Crafts, which had disbanded by this time, he would have found it difficult to earn a living making toys.

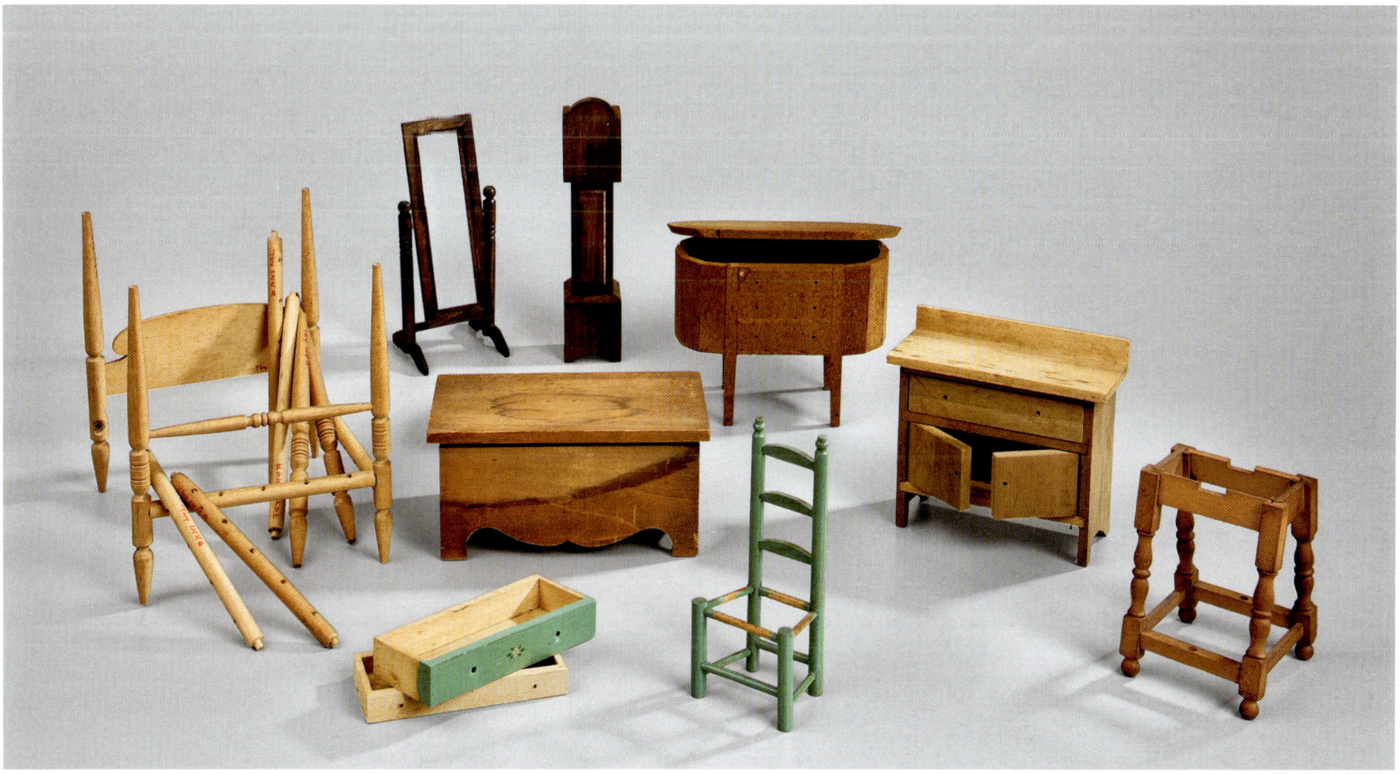

Plate 19.2

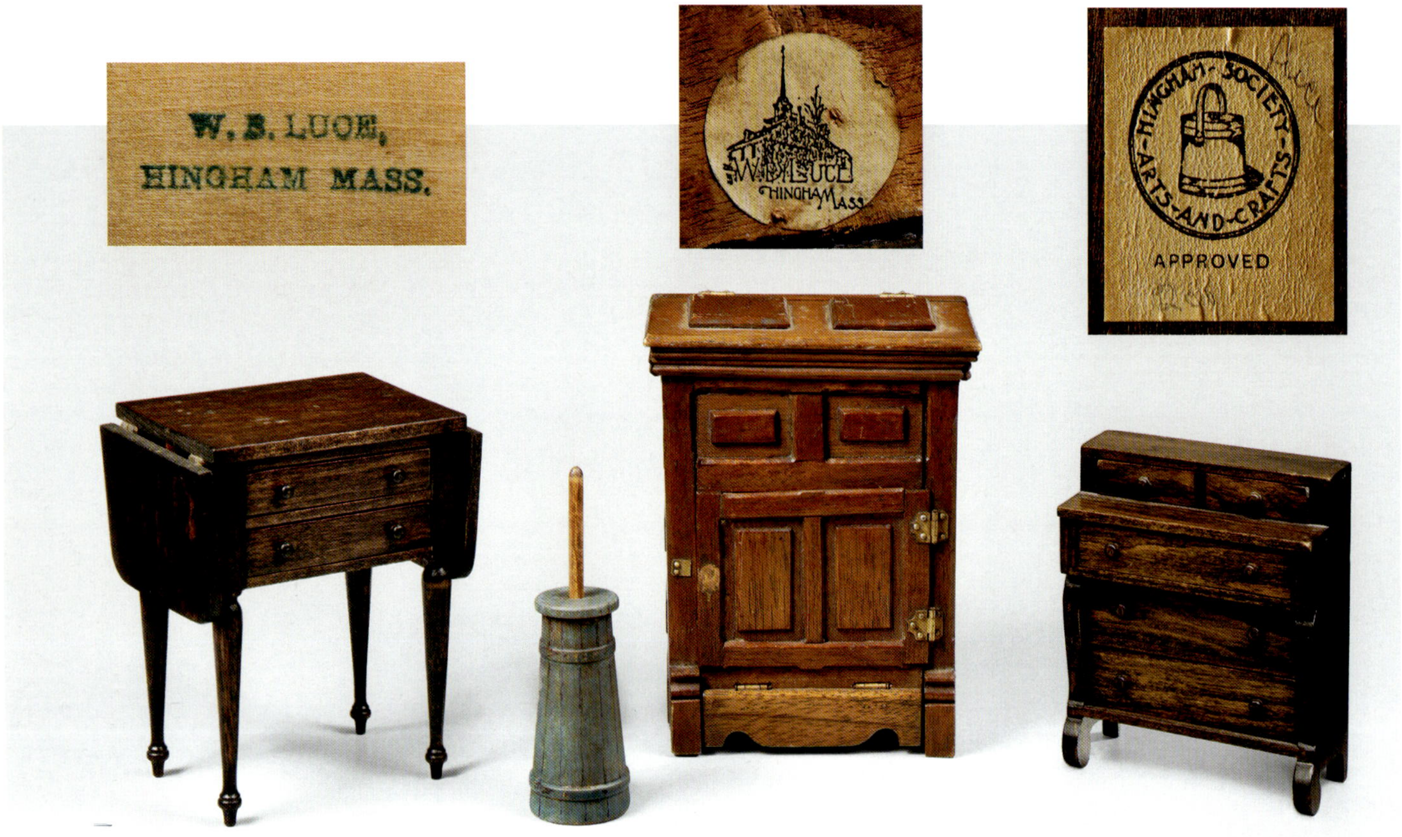

Plate 19.3

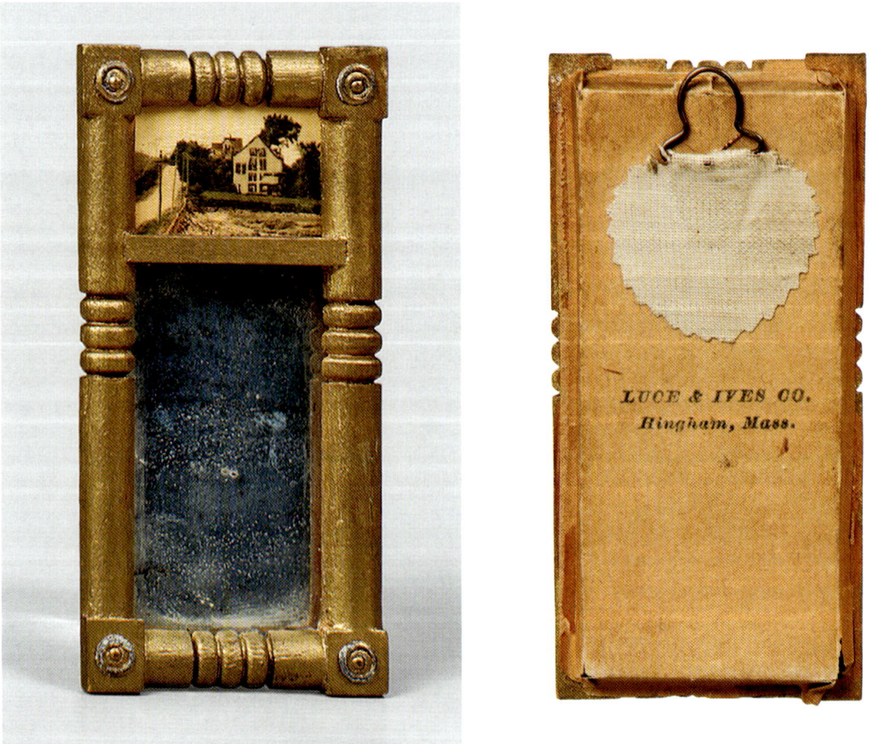

Plate 19.4

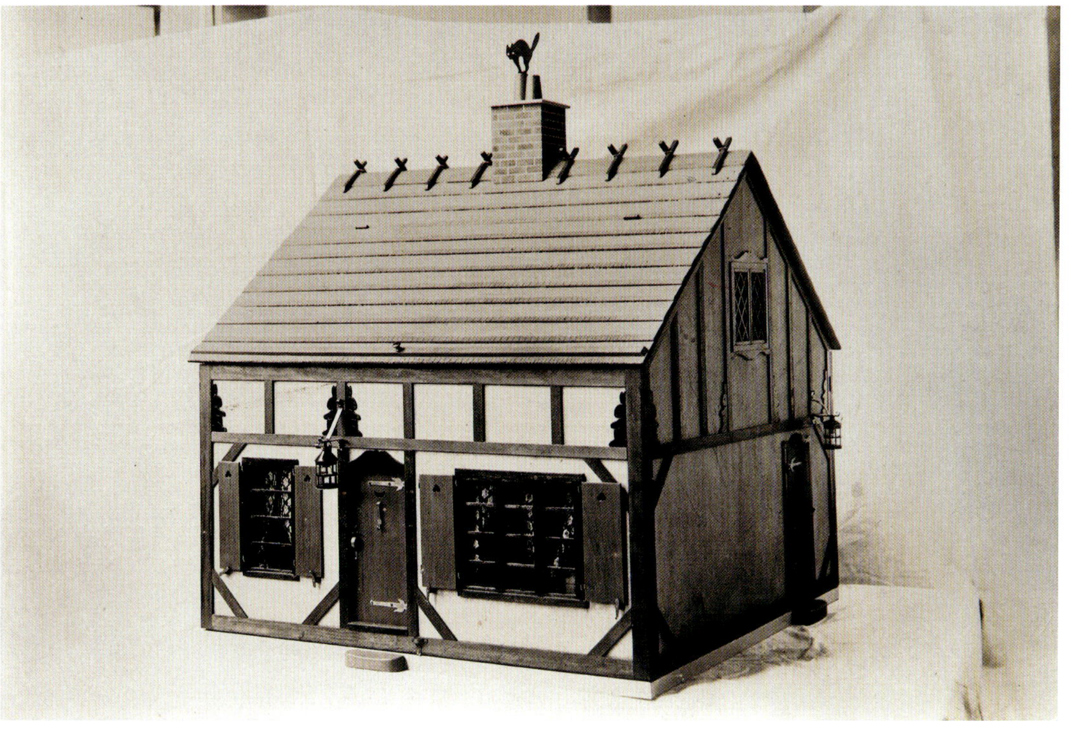

Plate 19.5

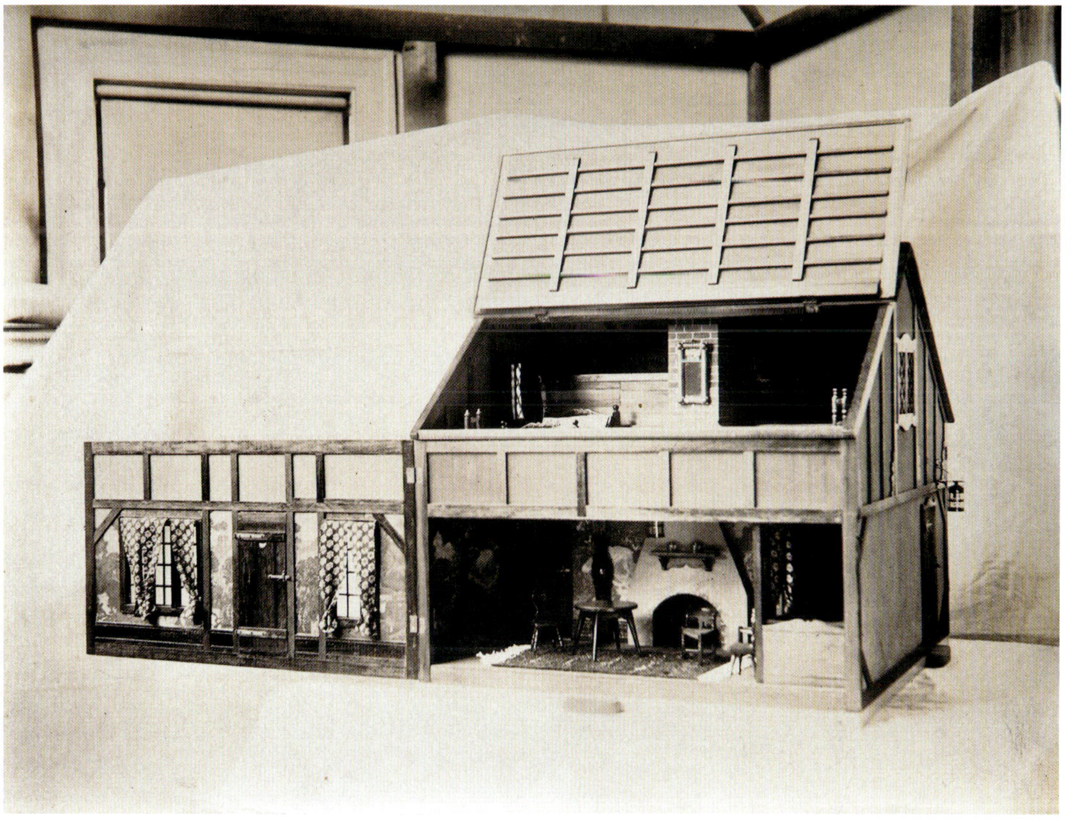

Plate 19.6

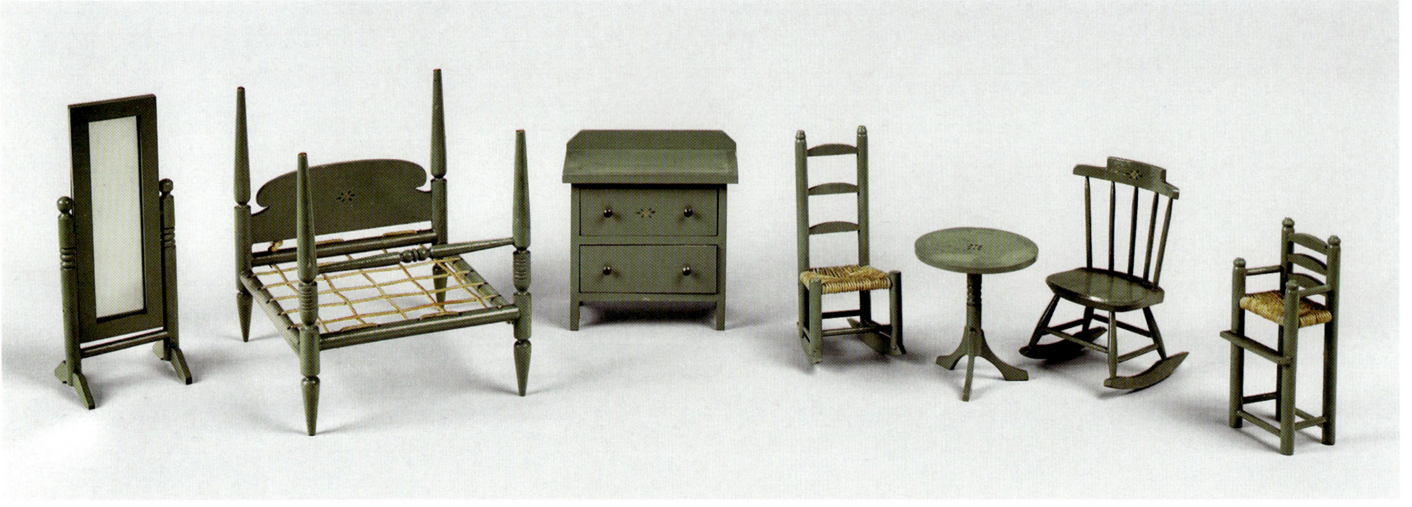

Plate 19.7

Plate 19.1

Self-portrait of William B. Luce, 1900–05

Provenance: John P. Richardson Photograph Collection. Courtesy of Hingham Historical Society

Plate 19.2

Unfinished miniature furniture, William B. Luce, 1904–34

Materials: Bedstead: tulip-poplar; standing looking glass: tulip-poplar; clock case: tulip-poplar; work table: mahogany; cabinet: tulip-poplar; butterfly table base: maple; drawers, tulip-poplar; blanket chest: tulip-poplar; chair: probably tulip-poplar

Dimensions: Bedstead: H 5 ⅝, W 4 ¾, D 5 ⅝; standing looking glass: H 5 ¾, W 2 ⅜, D 2; clock case: H 6 ⁷⁄₁₆, W 1 ⅜, D 1 ¹⁄₁₆; work table: H 4, W 5 ⅜, D 2 ⅝; cabinet: H 4, W 4 ½, D 2 ⁷⁄₁₆; butterfly table base: H 3 ⁷⁄₁₆, W 2 ¾, D 2 ⁷⁄₁₆; blanket chest, H 2 ⁷⁄₁₆, W 5 ⅜, D 3 ¼; chair: H 4 ⅜, W (seat) 2, D (seat) 1 ⁹⁄₁₆

Provenance: Courtesy of Hingham Historical Society

Plate 19.3

Miniature work table, churn, ice chest, and bureau, William B. Luce, 1904–24

Materials: Work table: tulip-poplar; churn: unknown wood; brass hinges; ice chest: walnut; tin liner; brass hinges; bureau: tulip-poplar

Dimensions: Work table: H 4 ¾, W 4, D 2 ¾; churn: 4 ¼, Diam 1 ¼; ice chest: H 5 ⅝, W 4 ½, D 3; bureau: H 4 ¼, W 3 ½, D 1 ¹³⁄₁₆

Inscriptions: "W. B. LUCE, / HINGHAM MASS." in ink stamped inside drawers of work table and bureau; "HINGHAM—SOCIETY / ARTS—AND—CRAFTS" surrounding a graphic of a bucket printed on paper label on bottom of churn and back of bureau; "W.B.L 1907" in ink on bottom of churn; "Made June 1908 / WBL." in graphite on bottom of ice chest; "W.B.LUCE / HINGHAMASS" within graphic of Old Ship Church printed on paper label on bottom of ice chest

Provenance

Work table and churn: Courtesy of the Hersey family

Ice chest: Olive Loring, Hingham, Massachusetts; to John P. Richardson, Hingham; purchased by current owner at Willis Henry Auctions, November 19, 2011, lot 193. Courtesy of the Hersey family

Bureau: Collection of Marilee Boyd Meyer

Plate 19.4

Miniature looking glass, Luce & Ives Co., ca. 1910

Materials: probably tulip-poplar; glass

Dimensions: H 3 ⁵⁄₁₆, W 1 ⅝

Inscription: *"LUCE & IVES CO. / Hingham, Mass."* in ink stamped on back

Provenance: Courtesy of Boston Children's Museum

Plate 19.5

Germanic dollhouse, William B. Luce, ca. 1915

Provenance: Courtesy of Hingham Historical Society

Plate 19.6

Interior of Germanic dollhouse, William B. Luce, ca. 1915

Provenance: Courtesy of Hingham Historical Society

Plate 19.7

Set of miniature furniture, William B. Luce, 1904–34

Materials: Tulip-poplar

Dimensions: High chair: H 3 ¾, W (seat) 1 ¾, D (seat) 1 ½; rocking chair: H 4 ½, W (seat) 2, D (seat) 1 ½; bedstead: H 5 ¼, W 4 ¾, D 5 ½; Windsor rocking chair: H 4 ¼, W (seat) 2 ⅝, D (seat) 2 ¾; bureau: H 3 ½, W 3 ¾, D 1 ⅞; table: H 2 ¾, Diam 2 ½; standing looking glass: H 5 ¾, W 2 ½, D 2

Provenance: Courtesy of New-York Historical Society

20
RALPH THOMAS JONES (1886–1977)

IN 1915 RALPH BERGENGREN, A BOSTON WIT AND CHILDREN'S POET, REVIEWED THE work of several American toymakers for the Christmas issue of *House Beautiful* magazine. He wrote, "Mr. Jones has struck a happy medium between Toyland and reality, for although he finds his models in the colonial types, he follows them with a breadth and simplicity that gives his pigmy furniture a charm of its own and adapts it quite naturally to the world of playthings."[1]

Ralph T. Jones was one of the most prolific toymakers in Hingham's history. The son of a blacksmith, Jones grew up on High Street next to the workshop of Loring H. Cushing, a long-time family friend and icon in the local toymaking trade.[2] It seems probable that Jones learned the craft from Cushing, but little is known of his formative years. He had great talent, however; by 1910 he had begun manufacturing "Hand Made" wooden toys in the clapboarded barn behind his parents' house.[3] That same year he became an exhibitor at the prestigious Art Institute of Chicago, and he soon joined the Society of Arts and Crafts, Boston.[4] Business was brisk for the young craftsman; during especially busy periods he employed two or three women to help assemble and finish his miniature colonial furniture.[5] Patterns for several of these pieces still survive (pl. 20.1). For customers who preferred modern forms, he offered dining and library sets in the Mission style then popular (pl. 20.2). Jones also encouraged orders for bespoke "toys and novelties made from samples or drawings."[6] Between 1915 and 1918 his shop generated annual sales between $500 and $1,300.[7]

By the 1930s most of Hingham's toymakers had passed away, retired, or sought work in other industries. Jones, however, stayed the course—transforming his humble retail business into a thriving wholesale operation that employed upwards of thirty people. His goods became simpler in design and more crudely executed (pl. 20.3), but they found homes all over the country through agents in New York and beyond.[8] It was probably during this later period that he stopped marking his wares with printed paper labels.

As the century progressed, Cape Cod emerged as another outlet for his goods. Family members recall that Jones would load up his car in Hingham and drive south, selling his toys to tourist shops lining the commercial highway. In its 1948 catalogue, Cape Cod Products—a mail-order novelty retailer—marketed Jones's miniature furniture alongside the maritime-themed "heirloom bronzes" of the Provincetown sculptor William Boogar and the painted furniture of the renowned folk artist Peter Hunt.[9] Jones expanded his line of playthings to include items that evoked the historic character of the South Shore: miniature sea chests, cobbler's benches, whirligigs, and cranberry scoops (pl. 20.4).

In the 1960s Ralph Jones moved to his son's house in nearby Hanover, but he maintained a bench in the backroom of E. L. Margetts & Sons on Ward Street in Hingham, where he continued to craft toys and other specialty goods for friends and family. He displayed and sold many of these items at the Assinippi Diner, where he could be found every morning, in high spirits, enjoying his breakfast.[10] Jones stayed active in the trade until the day he passed away in 1977. The end of his remarkable career, spanning seven decades, brought a close to an extraordinary tradition of toymaking that had existed in Hingham since the eighteenth century.

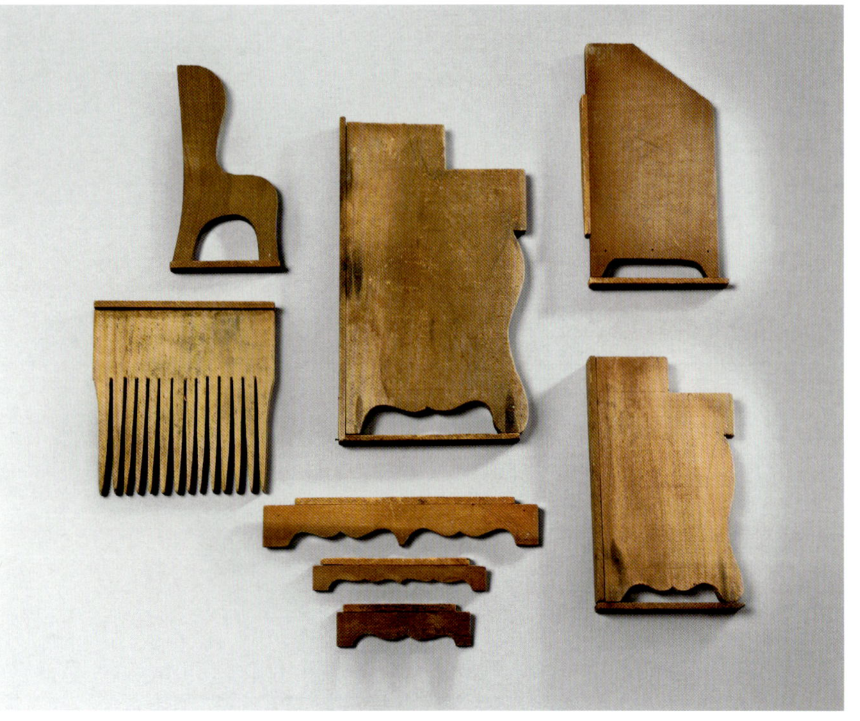

Plate 20.1

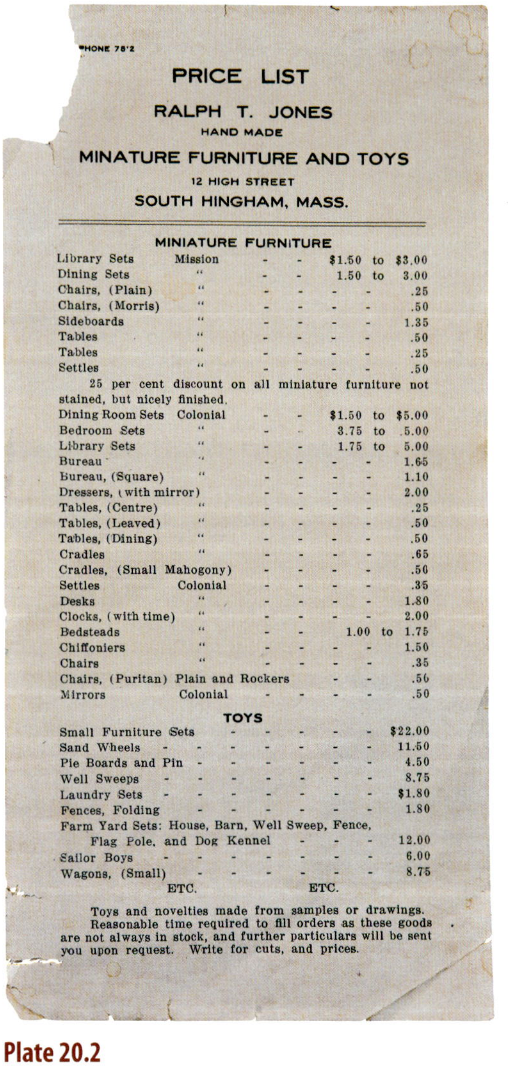

PHONE 78'2

PRICE LIST

RALPH T. JONES

HAND MADE

MINATURE FURNITURE AND TOYS

12 HIGH STREET

SOUTH HINGHAM, MASS.

MINIATURE FURNITURE

Library Sets	Mission	$1.50 to $3.00
Dining Sets	"	1.50 to 3.00
Chairs, (Plain)	"	.25
Chairs, (Morris)	"	.50
Sideboards	"	1.35
Tables	"	.50
Tables	"	.25
Settles	"	.50

25 per cent discount on all miniature furniture not stained, but nicely finished.

Dining Room Sets	Colonial	$1.50 to $5.00
Bedroom Sets	"	3.75 to 5.00
Library Sets	"	1.75 to 5.00
Bureau	"	1.65
Bureau, (Square)	"	1.10
Dressers, (with mirror)		2.00
Tables, (Centre)	"	.25
Tables, (Leaved)	"	.50
Tables, (Dining)	"	.50
Cradles	"	.65
Cradles, (Small Mahogony)		.50
Settles	Colonial	.35
Desks	"	1.80
Clocks, (with time)	"	2.00
Bedsteads	"	1.00 to 1.75
Chiffoniers	"	1.50
Chairs	"	.35
Chairs, (Puritan)	Plain and Rockers	.50
Mirrors	Colonial	.50

TOYS

Small Furniture Sets	$22.00
Sand Wheels	11.50
Pie Boards and Pin	4.50
Well Sweeps	8.75
Laundry Sets	$1.80
Fences, Folding	1.80
Farm Yard Sets: House, Barn, Well Sweep, Fence, Flag Pole, and Dog Kennel	12.00
Sailor Boys	6.00
Wagons, (Small)	8.75

ETC. ETC.

Toys and novelties made from samples or drawings.

Reasonable time required to fill orders as these goods are not always in stock, and further particulars will be sent you upon request. Write for cuts, and prices.

Plate 20.2

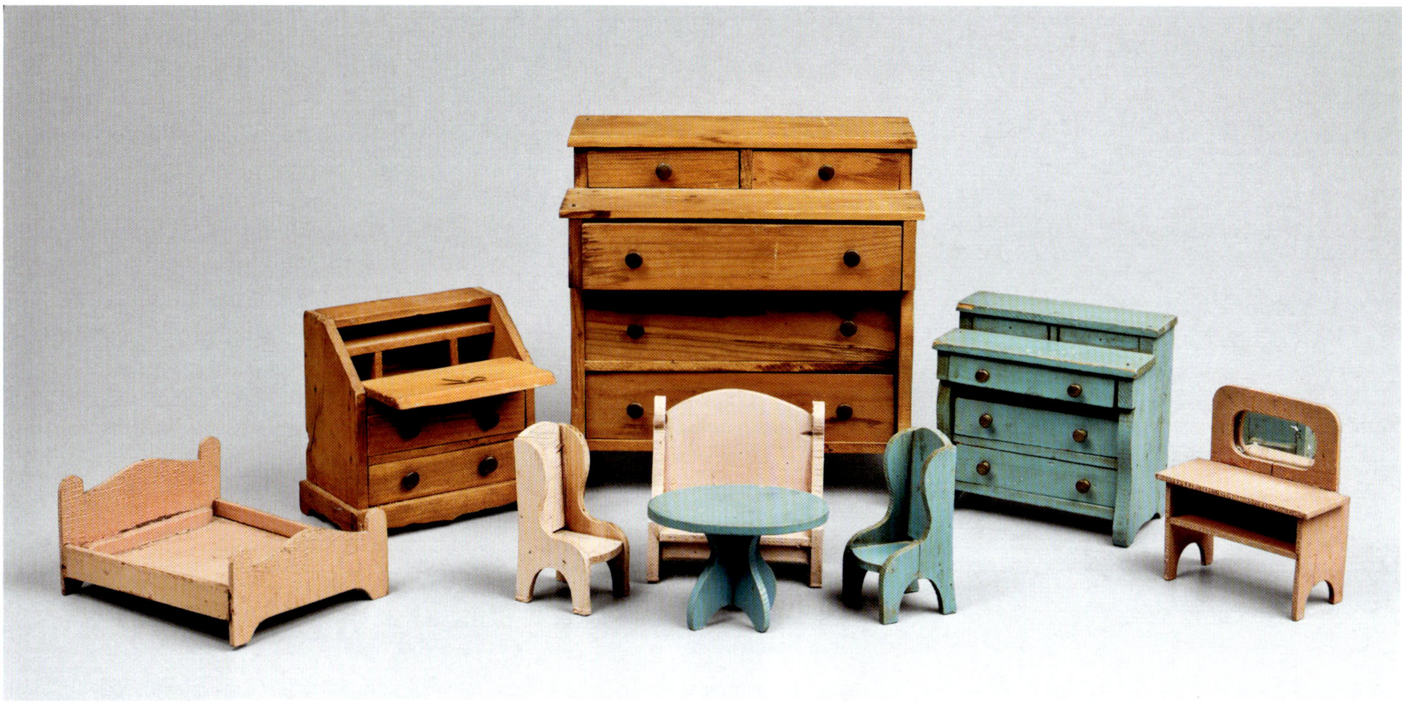

Plate 20.3

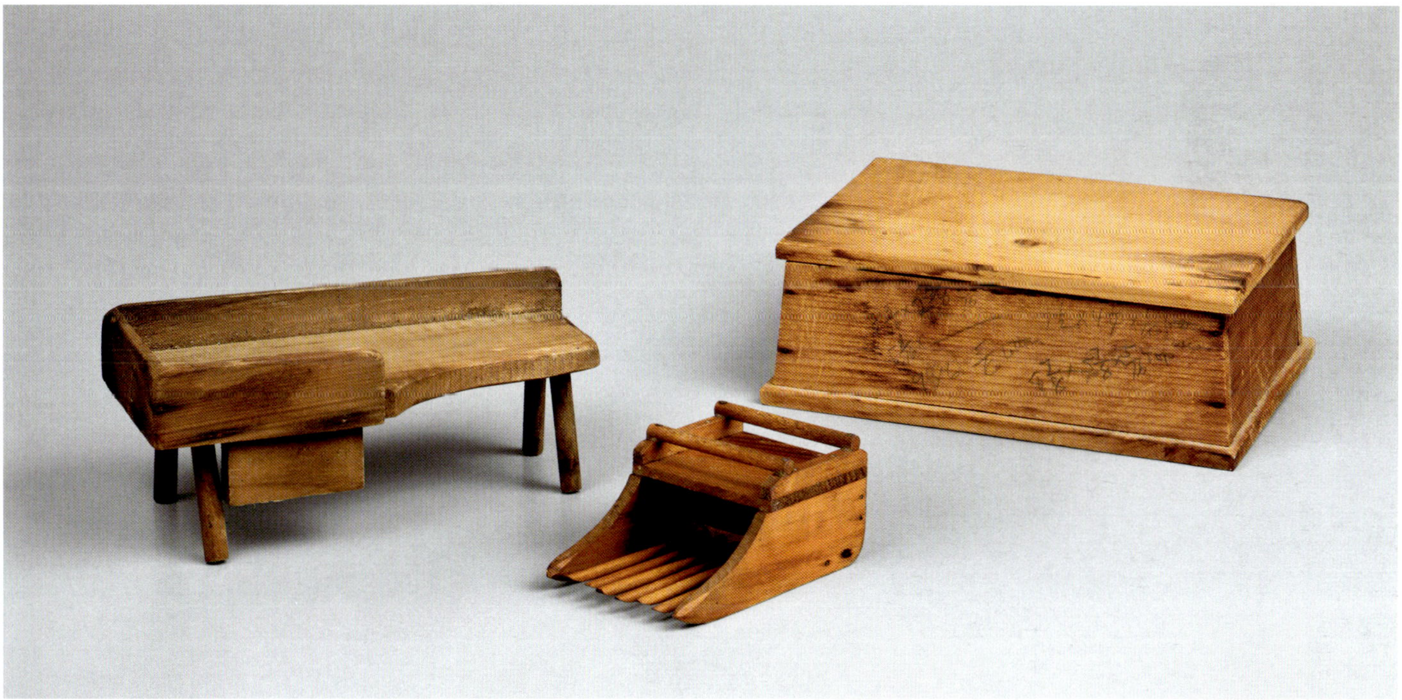

Plate 20.4

Plate 20.1

Patterns for miniature furniture, Ralph T. Jones, 1910–50
Materials: White pine and tulip-poplar; iron nails
Dimensions: Chair: H 5 5⁄8, W 3 1⁄8; cranberry scoop: H 5, W 5; bureau on left: H 8 5⁄8, W 5 1⁄4; molding: (top) H 1 1⁄4, W 7 1⁄2; (middle) H 3⁄4, W 4; (bottom) H 1, W 3 5⁄8; desk: H 6 5⁄8, W 3 7⁄8; bureau on right: H 6 3⁄8, W 3 7⁄8
Provenance: Private collection

Plate 20.2

Price list, Ralph T. Jones, ca. 1910
Dimensions: H 9 1⁄4, W 4
Provenance: Private collection

Plate 20.3

Miniature furniture, Ralph T. Jones, 1930–50
Materials: White pine; glass; brass knobs
Dimensions: Bed: H 2 1⁄2, W 4 3⁄16, D 5 1⁄2; desk: H 4 7⁄8, W 5, D 2 15⁄16; bureau at center: H 8 5⁄8, W 8 1⁄2, D 5 1⁄8; bureau on right: H 4 15⁄16, W 5 15⁄16, D 2 7⁄8; dressing table: H 4 1⁄8; W 3 3⁄4, D 1 7⁄8; chairs: H 3 3⁄4, W 1 5⁄8, D 1 3⁄4; bench: H 3 3⁄4, W 3 7⁄8, D 1 1⁄2; table: H 2 1⁄2, W 3 3⁄16, D 2 7⁄8
Provenance: Private collection

Plate 20.4

Miniature cobbler's bench, cranberry scoop, and sea chest, Ralph T. Jones, 1930–70
Materials: White pine; iron nails
Dimensions: Cobbler's bench: H 3 1⁄2, W 4 1⁄4, D 8; cranberry scoop: H 2 1⁄8, W 3, D 4 5⁄8; sea chest: H 3 1⁄8, W 9 1⁄4, D 6 1⁄4
Provenance: Private collection

21
N. D. CASS COMPANY (1896–1993)

IN THE EARLY 1940S YOUNG DICK KENNEY (B. 1934) AND HIS FRIENDS WOULD RAID the discard bin outside the Cass toy factory at Hingham Harbor, collecting odd parts to assemble playthings of their own creation.[1] Although the factory was new to town, its goods were not. Cass toys—available at the five-and-ten in the square—were well known to American children and had been for decades. In 1899 Nathan D. Cass (1875–1949) copurchased the Athol Paper Box Company in Athol, Massachusetts, manufacturers of paper and fiberboard boxes, travel cases, toy chests, and trunks for doll clothing.[2] Within a few years he became the sole proprietor of the firm and, in 1904, added a local toy factory to its holdings.[3] From that business emerged the N. D. Cass Company, one of the largest producers of wooden toys in the world, noted for their miniature furniture, games, rocking horses, and vehicles of all sorts, often decoratively painted in bright colors.[4] Expansion came quickly. In the 1920s Cass purchased a 5,400-acre site in Manchester, Vermont, that included a second factory and lumber mill to feed his growing operation.[5] Around the same time Cass opened a salesroom on Fifth Avenue in New York City to market its toys to a global audience.[6]

During World War II an estimated 70 percent of the Athol plant was converted to military manufacturing; Cass produced small wooden medical kits to hold hypodermic needles used by soldiers to inject morphine and penicillin on the battlefield.[7] When the demand for wooden toys skyrocketed because of the wartime rationing of rubber and metal—materials essential to most toy manufacturers—Cass seized the opportunity to snag a larger share of the market, acquiring additional factory space in places such as Hingham. In April 1943 the company took over an old automobile agency near the harbor, itself shut down because of the war. When it opened, the new factory employed fifty people, mostly women. Within six months the workforce had more than doubled to over 125 people.[8]

The factory adopted assembly-line production. Maple, beech, and pine boards were cut to size in the milling department and sent to "women operators who performed drilling, beveling and sanding operations." Many of the toys were dyed in large vats of red and blue stains. Others received silk screened decoration and were dried with large cooling fans. Finished toys traveled on a conveyor belt to yet another vat, where they received a power-spray coating of varnish. The Hingham factory turned out six thousand to seven thousand units a day, filling upwards of 60 percent of the firm's total orders.[9]

The "Pearl Harbor" military vehicles (pl. 21.1) were a bestseller for Cass and were produced in large numbers at the Hingham factory; an estimated 98,000 wheels a day were needed for these toys.[10] While local men assembled real destroyer escorts, tank landing ships, and other armored vessels at the Hingham Naval Shipyard, their counterparts at the Cass factory—only a few miles away—produced miniature wooden versions for children to play with on the home front (pl. 12.2). Other popular vehicles included a streamlined version of the General Lee tank, army transport trucks, 50-mm gun carriers, lumber carriers, invasion barges, and even airplanes.[11] These items were packaged in simple cardboard boxes bearing the name of the manufacturing plants and sold at Macy's, Marshall Field, and other leading department stores throughout the country. The Hingham factory closed after only a few years of operation, around 1945, but toy production resumed at the N. D. Cass Company's expanded facilities in Athol and continued until 1993.[12]

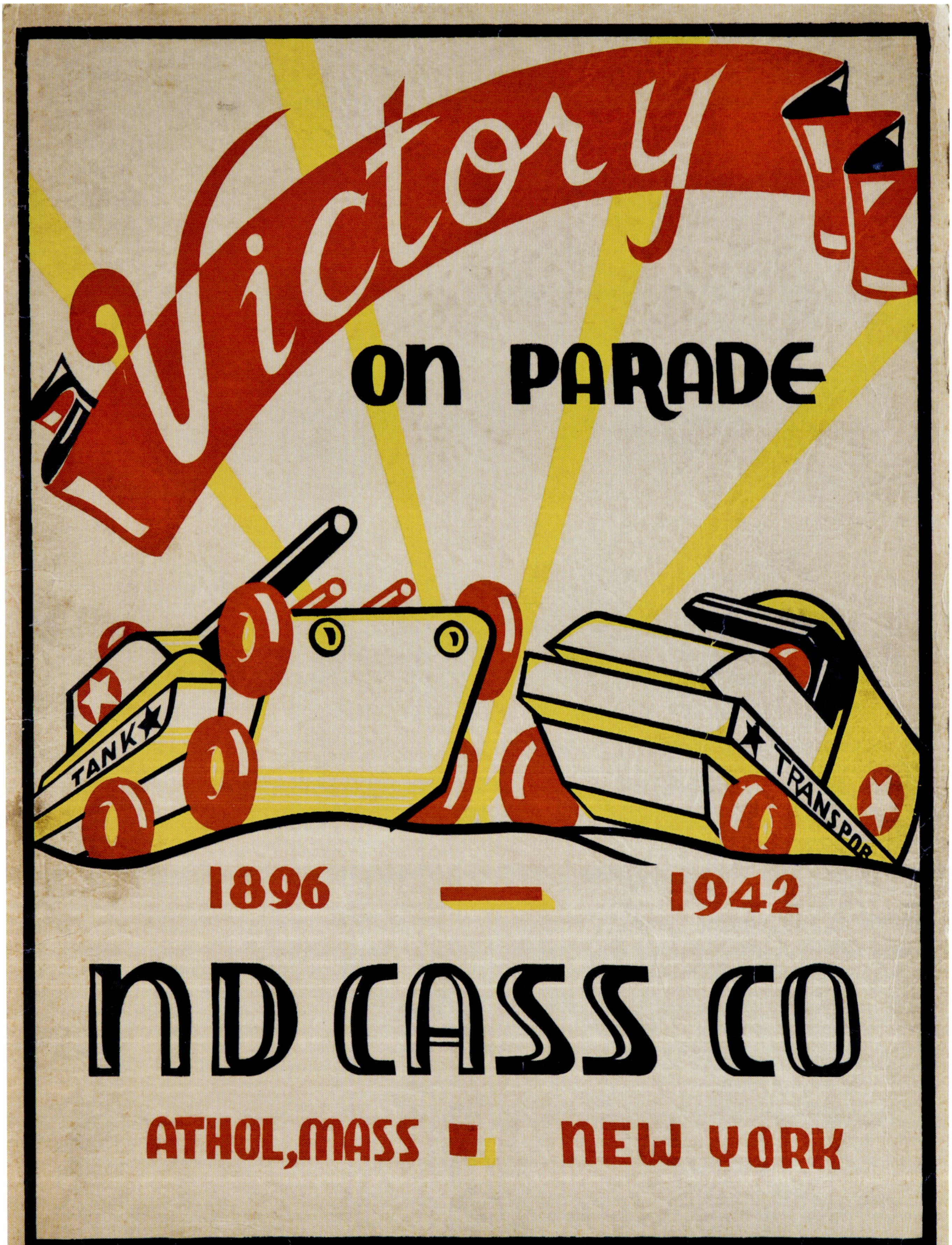
Victory
ON PARADE
TANK
TRANSPOR
1896 — 1942
ND CASS CO
ATHOL, MASS
NEW YORK

Plate 21.1

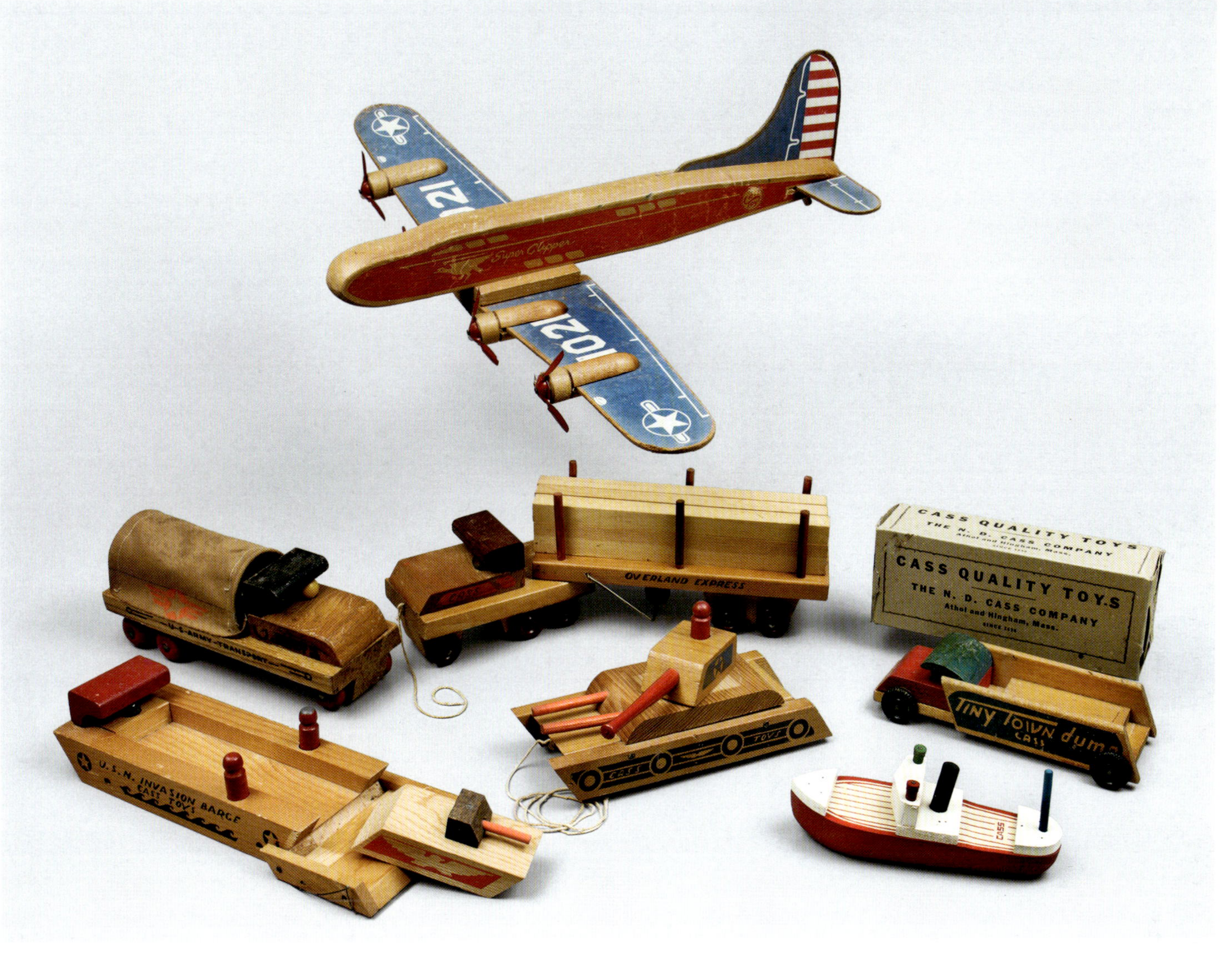

Plate 21.2

Plate 21.1

N. D. Cass Company "Victory on Parade" toy catalogue, 1942

Dimensions: H 11 ⁹⁄₁₆, W 9

Provenance: Courtesy of Ernie Britt

Plate 21.2

Miniature vehicles, N. D. Cass Company, 1942–46

Materials: White pine, ash, birch, and maple; particle board wings; canvas top; metal and rubber wheels; tin propellers

Dimensions: Airplane: H 5 ¾, W 25 ½, D 20 ⅝; U.S. Navy invasion barge: H 3, W 4 ⅞, D 12; U.S. Army transport truck: H 4 ½, W 3 ⅝, D 11; Overland Express lumber truck: H 5 ⅜, W 3 ½, D 16 ¾; box: H 2 ½, W 3 ¾, D 9 ⅜; Tiny Town dump truck: H 2 ¼, W 3 ¾, D 9; tank: H 4 ¾, W 4, D 10; boat: H 4, W 3 ½, D 8 ¾

Inscriptions: "Super Clipper" and "Cass Toys" printed on side of airplane; "U.S.N. INVASION BARGE / CASS TOYS" printed on side of barge; "CASS TOYS" and "U.S. ARMY TRANSPORT" printed on side of transport truck; "CASS" and "OVERLAND EXPRESS" printed on side of lumber truck; "TINY TOWN dump / CASS" printed on side of dump truck; "CASS TOYS" printed on side of tank; "CASS" printed on top of boat; "CASS QUALITY TOYS / THE N. D. CASS COMPANY / Athol and Hingham, Mass. / Since 1896" printed on sides of box

Provenance

Box: Courtesy of Bill Sarni

Miniature vehicles: Courtesy of Ernie Britt

NOTES

1 Peter Hersey (1757–1844)

1. For more information on the Derby School, known today as Derby Academy, see Betty Ring, *Girlhood Embroidery: American Samplers & Pictorial Needlework, 1650–1850,* 2 vols. (New York: Alfred A. Knopf, 1993), 1:152–53. See also Theodore Swan Roscoe's *History of Derby Academy, 1784–1984* (Hingham: Trustees of Derby Academy, 1984).

2. Homer Gregmore, "Historic Hingham," *New England Magazine* 23 (September 1905): 348–49.

3. The first known appearance of the nickname Bucket Town is in a satirical essay written by a group of students from Hingham's Derby School; see Salmagundi Society, "Salmagundi the Younger; or, The Whim Whams of ye Lads & Lasses of Buckettown," Hingham, Mass., 1816, 43–44, Ms. qAm. 2295, Special Collections, Rare Books and Manuscripts Department, Boston Public Library, Boston.

4. For a list of the various family dynasties in Hingham, see appendix C.

5. For information on Daniel Hersey, see John P. Richardson research notes, Hersey Family Collection. See also chap. 3, notes 7–8.

6. For information on northern white cedar, see Henry H. Gibson, *American Forest Trees* (Chicago: Hardwood Record, 1913), 97–100. In 1841 the estate of the cooper Jared Lane (1752–1839) advertised "Oak, cedar, & pinewood" of large growth and "cedar . . . valuable for coopers"; see *Hingham Journal,* January 2, 1841.

7. Of the more than forty Herseys who worked as coopers in Hingham, more than thirty lived on Hersey Street. The residences of others were nearby, on South and Lincoln streets, or are unknown; see appendix C.

8. A typed card pasted on the side of one of the buckets reads: "WOODEN FIRE BUCKET / Made in Hingham by P. Hersey, 1825. (name and date on bottom). / Used for many years as a fire bucket in the old Lincoln House, South Main Street near the Scituate town line. / Presented in 1934 by Henry N. Lincoln." For information on the Mordecai Lincoln House, see Edwin Victor Bigelow, *A Narrative History of the Town of Cohasset, Massachusetts* (Cohasset: Committee on Town History, 1898), 166–69, 188, 254.

9. Albert L. Pitcher, *Origin and Growth of the Fire Department in Hingham, Massachusetts, 1635–1942* (Weymouth, Mass.: Smith Print, 1946), 11, 18.

2 Jared Lane (1752–1839)

1. Four companies of Hingham militia marched to Lexington and Concord, but they arrived too late to participate in the battles. See George Lincoln et al., *History of the Town of Hingham, Massachusetts, Historical* (Hingham: By the town, 1893), 1:278; see also Lorena Laing Hart and Francis Russell Hart, *Not All Is Changed: A Life History of Hingham* (Hingham: Hingham Historical Commission, 1993), 59.

2. Hawkes Fearing, account book, 1775–78, Box FP56:1, Hingham Historical Society, Hingham.

3. Ibid.

4. For more information on Lane's military service, see Office of the Secretary of State, *Massachusetts Soldiers and Sailors of the Revolutionary War,* vol. 9 (Boston: Wright & Potter, 1902), 477.

5. Lane's property is described in auction notices that appeared in the *Hingham Journal* following his death in 1839; see *Hingham Journal,* January 2, 1841, and January 8, 1842.

6. Isaiah Hersey, account book, 1824–42, Hersey Farm Collection.

7. Hingham woodenware was commonly painted shades of blue, green, and gray. The coloring may simply have been decorative, though it probably also served as a brand identifier. The 1749 estate inventory for the cooper Obadiah Groce included a large quantity of indigo, possibly for use as a paint pigment; Obadiah Groce, inventory, filed September 15, 1753, Suffolk County Probate Records, 49:52.

8. Lane's family members served as Hingham's leading sign and fancy painters. His brother Rufus identified himself as a painter and glazier; see George Lincoln et al., *History of the Town of Hingham, Massachusetts, Genealogical* (Hingham: By the town, 1893), 2:417. Rufus's sons, Rufus and Charles, carried on an extensive painting business through the late 1820s; see Hingham merchant's account book, 1828, box FP62:2, Hingham Historical Society, Hingham. For more information on the Lanes see cat. 5, note 8.

3 Elijah Lewis Jr. (1762–1819)

1. A handwritten note tied to the pail reads "One of the oldest examples of Hingham woodenware I have owned. Acquired from Mrs. Foley after Mason A. Foley's death in 1968. Acquired in the 1930s by Mr. Foley when buildings were torn down on Summer St. between Ship St. and North. Bldg. backed up to hill below Cottage St. This pail was made by Elijah Lewis (1762–1819). He resided on Pleasant Street but had a cooper's shop at the harbor where he was also a packetman. Mr. Foley acquired Lewis's tools & other woodenware which he sold but kept this pail & a pantry type box. That also bears mark of William Davis (b. 1805), Hingham shipmast, spar, etc. maker. An early strap sided Hingham Pail. J. P. Richardson, 192 So. Pleasant St."

2. An inventory of Lewis's estate also included coopering tools, timber, staves, twenty-one dozen barrel covers, and one dozen tubs; Elijah Lewis, inventory, filed May 17, 1819, Plymouth County Probate Records, 50:235.

3. For a contemporary description of different types of Hingham woodenware, see François André Michaux, *The North American Sylva; or, A Description of the Forest Trees of the United States, Canada and Nova Scotia,* 3 vols. (Philadelphia: Thomas Dobson, 1818–19), 3:254–56.

4. George Lincoln et al., *History of the Town of Hingham, Massachusetts, Genealogical* (Hingham: By the town, 1893), 2:190.

4 R. Bates (act. 1780–1830)

1. Charlemagne Tower, *Tower Genealogy: An Account of the Descendants of John Tower of Hingham, Mass.* (Cambridge, England: John Wilson and Son, 1891): 91-2.

2. Only a few Hingham dumb bettys are known to exist. There is an example with iron hoops in the Hersey Family Collection that bears the mark of the cooper Thomas Hobart (1817–55). For a miniature toy example by Edward Lewis Hersey (1822–54), see pl. 13.2. The identity of R. Bates is unknown, which suggests that he worked for only a brief time. A butter churn at the Braintree Historical Society, a butter churn at the Quincy Historical Society, and a washtub strainer at the Hingham Historical Society also bear his mark, which is typically burned rather than impressed into the wood.

3. For a description of laundering devices used in early America, see Kenneth L. Smith, *Estate Inventories: How to Use Them* (Morgantown, Pa.: Masthof Press, 2000), 83, 111.

5 Elisha Burr (1776–1863)

1. Orderly Book of the Hingham Rifle Company, 1812–43, Box MR5:2, Hingham Historical Society, Hingham.

2. In 1775 several members of the Burr family wholesaled canteens to the Hingham merchant Hawkes Fearing. He, in turn, sent them to Roxbury for the Massachusetts militia; see Hawkes Fearing, account book, 1775–78, Box FP65, Hingham Historical Society, Hingham.

3. No fewer than seventeen members of the Burr family in Hingham worked as coopers from the seventeenth through the nineteenth century. Elisha Burr's father, David (1745–1814), grandfather Elisha (1703–1778), and great-great-uncle Simon Jr. (b. 1655/56) were all active in the trade; see George Lincoln et al., *History of the Town of Hingham, Massachusetts, Genealogical* (Hingham: By the town, 1893), 2:101–6; see also appendix C.

4. An early handwritten note pasted to the back of a canteen at the Hingham Historical Society reads, "These canteens were made for and carried by The Soldiers in the American Revolution. They were afterward used by The Soldiers in the War of 1812. M. + F. Burr took them in their store and sold them in Boston. A bill of the same of over 1,000 made by Elisha Burr for M. &. F. Burr is preserved. The price paid was fifteen cents each. The hoops are of native white oak. The heads are of eastern white Pine bought from Maine." A second, smaller note reads, "Canteen belonging to David Sprague, Born in Hingham May 14- 1754. Died [in Hingham] August 7- 1832. A soldier of The Revolution. See obituary notice in *Hingham Gazette* Aug. 10, 1832. Resided at Hingham Centre. Made by Elisha Burr. Born in Hingham Dec. 3- 1703. Died [in Hingham] May 20- 1778. Cooper, Constable 1774. Resided on Leavitt St. Hingham Centre."

5. The canteen was donated to the Hingham Historical Society in 1917 with a history of its belonging to the Washington Light Infantry, a militia unit from Boston. This history has not been substantiated, though a second, identical canteen with the painted initials PM also has a long history of ownership in Hingham and is marked by a local cooper, which suggests that both canteens were used by a local militia. For information on Sprague, see Lincoln et al., *History of Hingham, Genealogical*, 3:171.

6. For a discussion of early militias and their decorated equipment, see John Obed Curtis and William H. Guthman, *New England Militia Uniforms and Accoutrements* (Sturbridge, Mass.: Old Sturbridge Village, 1971).

7. Lincoln et al., *History of Hingham, Genealogical*, 2:420–21.

8. Although R. & C. Lane became traders of fishing equipment and manufacturers of salt, they carried on the business of painting at least through the late 1820s; see Hingham merchant's account book, 1828, box FP62:2, Hingham Historical Society, Hingham. The Lane brothers probably learned painting and glazing from their father, Rufus (1758–1801), who was also active in the trade; see Lincoln et al., *History of Hingham, Genealogical*, 2:419. Henry Nye apprenticed with Rufus and later formed a partnership with Isaac Lane; see Solomon Lincoln's "Biographical Sketches of Hingham Residents," Box FP67:1, Hingham Historical Society, Hingham. The account book of the wealthy merchant Samuel Norton shows extensive work performed by Lane and Nye in 1798, including painting floors, furniture, fire buckets, and a coat of arms; see Samuel Norton, account book, 1798–1802, Wigglesworth Family Papers, 1682–1966, box 19, Massachusetts Historical Society, Boston. In addition to the Lanes, several members of the Sprague family were skilled with a paintbrush, notably Blossom Sprague (1784–1860), a carriage painter. According to tradition, his nephew Isaac Sprague Jr. apprenticed to him before becoming a famed horticultural and wildlife painter; see Emanuel D. Rudolph, "Isaac Sprague: 'Delineator and Naturalist,'" *Journal of the History of Biology* 23, no. 1 (Spring 1990): 93–94.

9. Account Book of the Hingham Rifle Company, 1813–41, Box MR5:3, Hingham Historical Society, Hingham; Lincoln et al., *History of Hingham, Genealogical*, 2:431.

10. A second oval militia canteen marked by Burr and without decoration is in the Hersey Family Collection.

11. Several members of the Sprague family marked their goods with initials; see appendix C. The initials PS, attributed to Peter Sprague, are stamped on two circular boxes at the Hingham Historical Society and on three circular decorated canteens formerly in the Guthman Collection; see Guthman, "Decorated American Militia Equipment," 129. The initials MS, attributed to Moses Sprague, are stamped on a circular canteen that descended through the Burr-Studley family, and on another in a private collection; the latter is decorated with the initials ELI, possibly for Easton Light Infantry; see cat. 6. The initials AS, attributed to Adna or Amos Sprague, are found on round boxes. The mark "J. BURR" was used by John Burr (1791–1857), Elisha's cousin, and is found on a number of circular canteens; for an example, see Curtis and Guth-

man, *New England Militia Uniforms and Accoutrements,* 81. Other canteens with local histories are stamped with the initials JR and NR. These marks probably belong to members of the Ripley family, of which the brothers and coopers John (1787–1852) and Nehemiah (1780–1863) are good candidates; see Lincoln et al., *History of Hingham, Genealogical,* 3:135–36. The initials JR appear on a canteen that descended through a Hingham family and is now in the Hersey Family Collection. The initials NR appear on a decorated canteen that descended through the Burr-Studley family of Hingham. They are found on a second canteen that bears an early paper label inscribed, "This canteen was made for the War of 1812 by John Burr, born 1791, died 1857"; see Mike O'Donnell, *U.S. Army & Militia Canteens, 1775–1910* (Alexandria, Va.: O'Donnell Publications, 2008), 49.

12. Sometime in the twentieth century Carl Burr (1884–1961), son of Elisha Burr (1839–1909) altered the brand to "C. BURR" by removing the central bar of the letter E. The brand was later restored by Hanover blacksmith Andy Addoms.

13. For a discussion of the Burr family of bird carvers, see Bob Mosher, "The Burr Family of Hingham," *Decoy Magazine* (May–June 2004): 8–14.

6 Moses Sprague (1776–1850)

1. Benjamin Drew, *Burial Hill, Plymouth, Massachusetts: Its Monuments and Gravestones Numbered and Briefly Described* (Plymouth: D. W. Andrews, 1894), 69.

2. Logbook for the brig *Elder Brewster,* 1829–32, location unknown. For more information on the *Elder Brewster,* see *Ship Registers of Plymouth, Massachusetts, 1789–1909,* part I (Boston: National Archives Project, WPA, 1939), 35–36.

3. For more information on sailors' arts and crafts, see J. Welles Henderson and Rodney P. Carlisle, *Marine Art & Antiques: Jack Tar, a Sailor's Life, 1750–1910* (Woodbridge, Suffolk, U.K.: Antique Collectors' Club, 1999), 107–30.

4. Martin Sprague (1800–67), a distant cousin, was also active at boxmaking in the nineteenth century; however, the appearance of the initials M.S on canteens made around or before the early 1820s suggests that the marking was used by the elder Sprague.

5. At least twenty-two members of the Sprague family worked as coopers in the eighteenth and nineteenth centuries; see appendix C. Several of these craftsmen marked woodenware with the initials A.S, B.S, P.S, and M.S

6. Boxes with virtually identical construction often bear the marks used by the Murdock family of coopers in Winchendon and Boston.

7 Bela Sprague (1804–1878)

1. François André Michaux, *The North American Sylva; or, A Description of the Forest Trees of the United States, Canada and Nova Scotia,* 3 vols. (Philadelphia: Thomas Dobson, 1818–19), 3:254–56.

2. In 1761 the Hingham trader and ship captain Seth Stowers (1735–88) sailed to Louisbourg and Quebec, where he sold dozens of buckets and boxes; see Seth Stowers, account book, 1761–88, Hersey Family Collection. In 1817 John G. Ladd & Co. of Alexandria, Va., advertised nests of Hingham boxes and other woodenware recently received from the brig *Economy* from Boston; Alexandria Gazette, April 11, 1817.

3. *Philadelphia Federal Gazette,* May 15, 1791, 3.

4. No fewer than twenty-two members of the Sprague family worked as coopers in Hingham in the eighteenth and nineteenth centuries; see appendix C.

5. Bela Sprague's only residence appears to have been his parents' house on Union Street. The family's one-acre house lot included a small shop; see Bela Sprague, inventory, filed November 30, 1847, Plymouth County Probate Records, 89:512.

6. Although Hingham coopers are known to have moved to different parts of New England, where they undoubtedly worked in the same tradition, most boxes with opposing tails of this type were probably made in Hingham; the town held a tight grip on the wholesale market for these items and other woodenware until after the mid-nineteenth century. Boxes of this type are often misattributed to Shaker craftsmen; the Harvard and Shirley communities in Massachusetts owned Hingham buckets and probably other woodenware purchased during frequent trips to Boston, where the Shakers sold large quantities of seeds and sieves.

7. "Price Current of Woodenware," broadside, December 8, 1840, Hersey Family Collection.

8. For more on the different markings used by Sprague coopers, see cat. 5, note 11; cat. 6; and appendix C.

8 Wilder Bucket Factory (1830–1891)

1. The invention of the stave-jointer allowed coopers to dress the vertical edges of staves with tongue-and-groove joints. The first patent for this machine appears to have been issued in 1811 to Barnabas Langdon and William Mowry of Washington County, N.Y. (U.S. Patent 1,539X). Cylinder saws allowed coopers to cut the convex and concave sides of a stave; the first patent for this machine appears to have been issued in 1828 to Lewis R. Bump of Wareham, Mass. (U.S. Patent 5,212X).

2. George Lincoln et al., *History of the Town of Hingham, Massachusetts, Genealogical* (Hingham: By the town, 1893), 3:326.

3. "History of the Bucket Mill," *Hingham Journal,* April 4, 1902, 1.

4. For a description of Wilder's factory, see U.S. Department of the Treasury, *Steam Engines: Letter from the Secretary of the Treasury, Transmitting, in Obedience to a Resolution of the House, of the 29th of June Last, Information in Relation to Steam Engines, & c., Doc. No. 21* (Washington D.C.: United States Department of the Treasury, 1839), 43. For information on Holmes Hinkley, see Walter Swain Hinchman, *Holmes Hinkley: An Industrial Pioneer, 1793–1866* (Cambridge, Mass.: Riverside Press, 1913).

5. In 1836 M. Humphrey & Co., run by Moses Humphrey, opened the second bucket factory in Hingham to rely on steam-powered machinery. Outside the town there were several woodenware

factories larger than C. & A. Wilder's; manufacturers in Winchendon, Massachusetts, had begun using specialized coopering machinery in the 1820s. By the 1850s, these men were the chief producers of woodenware in New England and probably the country; see U.S. Federal Census, Non-Population Schedules, 1850–60, Winchendon, Worcester County, Massachusetts. See also Abijah Perkins Marvin, *History of the Town of Winchendon* (Winchendon, Mass.: Published by the author, 1868), 369, 371–72.

6. Although the firm was named C. Wilder & Sons, Crocker Wilder's only son, William Crocker Wilder (1829–91), did not begin regular work at the factory until 1846; see "History of the Bucket Mill."

7. Ibid.

8. "C. & A. Wilder and the Elijah Whiton Factories," *Hingham Journal,* May 6, 1853, 2.

9. Lincoln et al., *History of Hingham, Genealogical,* 3:322, 375. In 1847 the factory is described in a newspaper advertisement as C. Wilder & Son; *Hingham Patriot,* November 12, 1847, 3. By 1850 the "Hingham Business Directory" lists the factory as C. & A. Wilder; "Hingham Business Directory," *Hingham Journal,* February 1, 1850, 1.

10. "C. & A. Wilder and the Elijah Whiton Factories." See also "Hingham Buckets," *Hingham Journal,* April 23, 1858, 2.

11. Lincoln, *History of Hingham, Genealogical,* 3:323. For William C. Wilder's obituary, see "William Crocker Wilder," *Hingham Journal,* February 27, 1891, 1. The company was known as C. & A. Wilder from at least 1850 until about 1870–71. The 1871 report of the Hingham Agricultural and Horticultural Society lists the company as C. Wilder & Son, though it is described, probably mistakenly, as C. & A. Wilder the following year. After 1871 the company appears only as C. Wilder & Son in directories, advertisements, and other public records.

12. Goods made between 1870 and 1891 are stamped "C. WILDER & SON / SO. HINGHAM, MASS."; this mark is in the shape of an oval with a decorative flourish in the center. Goods made between 1830 and 1850 are stamped "C. WILDER & SONS / HINGHAM"; this mark is horizontal.

13. *Transactions of the Hingham Agricultural and Horticultural Society, from October, 1858 to March, 1861* (Hingham: Hingham Agricultural and Horticultural Society, 1861), 69.

14. George Lincoln et al., *History of the Town of Hingham, Massachusetts, Historical* (Hingham: By the town, 1893), 1:164.

15. Appraisal of Partnership Property, C. Wilder & Son, May 5, 1891, in John P. Richardson Papers, "Hingham Woodenware Chronology, Individual Coopers, Toy, and Box Makers," Hersey Family Collection. Although A. J. Sprague took control of the factory, it was his father, Seth, who purchased it from the Wilder estate. Details of the transaction are recorded in the pages of a small handwritten journal dated May 28, 1891; see John P. Richardson Papers, "Hingham Woodenware Chronology."

16. A. J. Sprague ran the factory from 1891 to about 1895. It was during this period that he produced woodenware stamped "A. J. SPRAGUE / SO. HINGHAM, MASS." For a history of Sprague's involvement in the factory, see "History of the Bucket Mill." Curiously, a 1902 town directory lists Sprague as a woodenware manufacturer off Main Street in South Hingham, which suggests that he may have run the factory after Streeter took over; *The Weymouth, Mass., Directory, Including Hingham and Braintree, 1902–3* (Boston: W. E. Shaw, 1902), 112.

17. Apparently Streeter represented a number of investors from Winchendon; see "Important Transfer," *Hingham Journal,* March 30, 1900, 8; and "South Hingham," *Hingham Journal,* October 5, 1900, 4.

18. "History of the Bucket Mill." See also "South Hingham," *Hingham Journal,* May 30, 1902, 4; and "Important Transfer." In 1902 Charles H. Morse, executor of the Wilder estate, purchased the bucket mill from Streeter and reopened the factory, though it closed shortly thereafter; "Legal Notice," *Hingham Journal,* June 27, 1902, 5; "South Hingham," *Hingham Journal,* September 19, 1902, 8; and "South Hingham," *Hingham Journal,* October 31, 1902, 8.

9 Leonard Gardner (1809–1890)

1. The *Boston Globe* article was reprinted in the *Spirit Lake (Iowa) Beacon,* "Stubbed His Toe. How Leonard Gardner's Misfortune Became a Boon," November 13, 1885, 13. For further descriptions of Gardner's mathematical accomplishments, see *Boston Globe,* "He Juggled Figures. Mathematician Gardner Dead," November 16, 1890, 3; *Boston Daily Globe,* "Gathering Valuable Works," January 29, 1891, 14; *Decatur (Ill.) Daily Review,* "A Genius at Figures," November 8, 1879, 3; and Bates Torrey, "Weymouth Folklore," *History of Weymouth, Massachusetts, Historical* (Weymouth, Mass.: Weymouth Historical Society, 1923), 2:792-3

2. *Spirit Lake Beacon,* "Stubbed His Toe," November 13, 1885, 13.

3. Charles Gardner, account book, 1796–1801, box FP73, Hingham Historical Society, Hingham.

4. *Spirit Lake Beacon,* "Stubbed His Toe," November 13, 1885, 13.

5. George Lincoln et al., *History of the Town of Hingham, Massachusetts, Genealogical* (Hingham: By the town, 1893), 3:322. For information on Wilder, see chap. 1 and cat. 8.

6. Joshua Tower took over Wilder's first factory around 1846; see appendix C. Elijah Whiton, a self-taught engineer, operated a factory at the entrance of Hersey Street near Hobart Street; see chap. 1 and George Lincoln et al., *History of the Town of Hingham, Massachusetts, Historical* (Hingham: By the town, 1893), 2:164.

7. For information on Gardner's move to Weymouth, see *Boston Globe,* "He Juggled Figures," November 16, 1890, 3. For information on Loud's box mill, see *Biographical Review: Containing Life Sketches of Leading Citizens of Norfolk County, Massachusetts* (Boston: Biographical Review Publishing Co., 1898), 216. Gardner's move to South Weymouth may have been

prompted by the opening of Wilder's new bucket factory in 1845. Around the same time the first Wilder factory was taken over by his brother-in-law, Joshua Tower. Other Hingham coopers who made the move to Weymouth include Perez L. Fearing and Quincy Hersey.

8. U.S. Federal Census, Non-Population Schedules, 1850–60, Weymouth, Norfolk County, Massachusetts.

9. Buckets stamped "L. GARDNER/S. WEYMOUTH" are constructed identically to the example in pl. 9.

10. "Letter from Liberty Plain," *Hingham Journal,* March 2, 1866. Gardner first advertises in Hingham in 1867; *The Plymouth County Directory and Historical Register of the Old Colony* (Middleboro, Mass.: Stillman B. Pratt & Co., 1867), 53.

11. For information on Shaker woodenware, see June Sprigg and Jim Johnson, *Shaker Woodenware: A Field Guide,* 2 vols. (Great Barrington, Mass.: Berkshire House, 1991).

12. *The Weymouth, Quincy, and Braintree Directory for 1870–71* (Boston: D. Dudley & Co., 1870), 57. Town directories for subsequent years list Gardner's occupation as carpenter, painter, and sealer of weights and measures.

13. *Spirit Lake Beacon,* "Stubbed His Toe," November 13, 1885, 13.

14. Lincoln et al., *History of Hingham, Historical,* 2:77.

10 Nelson Corthell (1815–1900)

1. Of the ten traditional coopers working in Hingham, only one was under the age of sixty, and half of them worked primarily as toymakers; see U.S. Federal Census, Population Schedule, 1880, Hingham, Plymouth County, Mass. See also appendix D.

2. George Lincoln et al., *History of the Town of Hingham, Massachusetts, Genealogical* (Hingham: By the town, 1893), 2:143.

3. Bela Sprague was the other cooper who exhibited woodenware; see *Transactions of the Hingham Agricultural and Horticultural Society for the Year 1877* (Hingham: Hingham Agricultural and Horticultural Society, 1878), 49.

4. For information on Hingham as a resort town, see Lorena Laing Hart and Francis Russell Hart, *Not All Is Changed: A Life History of Hingham* (Hingham: Hingham Historical Commission, 1993), 112, 161–68.

5. *Directory of Hingham and Hull, Mass., 1894* (Quincy, Mass.: J. H. Hogan, 1894), 119.

6. Mary Earle Gould, *Early American Wooden Ware & Other Kitchen Utensils* (Springfield, Mass.: Pond-Ekberg, 1942), 170–71.

7. A handwritten note stored inside the bucket reads, "Acquired from the Burns estate collapsing shed. Emerald St., Hingham. Small Hingham Bucket made by Nelson Corthell (1815–), box cooper Resided on Hersey St. His son Nelson Francis Corthell was killed at the Battle of Bull Run, Va. Aug. 1862. N. Corthell had the last of the small coopers shops. Rare, JPR."; Lincoln et al., *History of Hingham, Genealogical,* 2:100.

8. Small buckets with the same construction features occasionally bear the mark of the Alvin Streeter Company, a bucket manufacturer in Hingham.

11 Cotton Hersey (1792–1881)

1. For information on the origin and development of Hersey Street, see chap. 3.

2. George Lincoln et al., *History of the Town of Hingham, Massachusetts, Genealogical* (Hingham: By the town, 1893), 2:309.

3. Reuben Hersey is Hingham's earliest documented toymaker; see chap. 3. See also "The Box and Toy Business," *Hingham Journal,* August 11, 1876, 1.

4. For a box by Samuel Hersey, see pl. 12.1.

5. For a history of Cotton Hersey's house at 168–170 Hersey Street, see Julian Loring's notes on Hingham houses, Hingham Historical Society, Hingham.

6. By the early 1850s Samuel Hersey, Caleb Hersey, Joseph Hersey Jr., and Edward Lewis Hersey had begun advertising themselves as toymakers on Hersey Street; see appendix D.

7. A handwritten note by John P. Richardson is kept inside the box. It reads "Very rare or unique box with original yellow paint by Cotton Hersey of Hingham, Mass. Bears mark of 'C. Hersey' top and bottom. Inside cover inscribed in pencil, 'Made by Cotton Hersey for Lydia Hersey, Sept. 4, 1829.' Lydia, his daughter, was born in 1824 & died in 1843. The initials 'B.H.' probably stand for a son Bela b. 1817. Cotton Hersey was born in 1792 & died in 1881. He was a 'toy manufacturer.' Resided on Hersey St."

8. For information on Hingham canteens, see chap. 1 and cat. 5. Cotton Hersey was among the small band of soldiers who rescued an American gunboat that was driven into Scituate Harbor by a British frigate during the War of 1812; *Boston Journal,* January 11, 1878.

12 Samuel Hersey (1820–1909)

1. Hersey was awarded a gratuity of two dollars for his goods; see *Transactions of the Hingham Agricultural and Horticultural Society for the Year 1873* (Hingham: Hingham Agricultural and Horticultural Society, 1873), 49.

2. Reuben Hersey is Hingham's earliest documented toymaker; see chap. 3 and "The Box and Toy Business," *Hingham Journal,* August 11, 1876, 1.

3. For information on Cotton Hersey, see cat. 11. For information on the many Hersey coopers and toymakers on Hersey Street, see chaps. 2 and 3. See also appendices A, C, and D.

4. Samuel and his brother Edmund probably apprenticed to their father, Samuel French Hersey. It is unknown whether Samuel French Hersey also made toys.

5. Samuel Hersey and his cousin Edward Lewis Hersey were the first Hingham coopers to advertise themselves as toyware and box manufacturers; see "Hingham Business Directory," *Hingham Journal,* February 1, 1850, 1. For information on Edward Lewis Hersey, see cat. 13.

6. Hersey last advertised himself as a toymaker in the *Directory of Hingham and Hull, Mass., 1894* (Quincy, Mass: J. H. Hogan, 1894), 83. The 1900 U.S. Census lists him once again as a cooper, though at age eighty he had probably retired from the trade.

7. A brief description of Hersey's summer business was recounted by a fellow toymaker, Loring H. Cushing, in "Artist Toy Makers of America," *Boston Daily Globe*, December 24, 1911, SM3. See also chap. 2. For information on Hingham as a resort town, see Lorena Laing Hart and Francis Russell Hart, *Not All Is Changed: A Life History of Hingham* (Hingham: Hingham Historical Commission, 1993), 112, 161–68. Hersey's goods may have been wholesaled outside the region. In 1863 Lophorkash Emporium, a fancy and dry goods store in New London, Connecticut, advertised "a set of beautiful Hingham toyware"; *New London Chronicle*, June 12, 1863, 3.

8. The rocking chair was discovered among a group of mid-nineteenth-century American and German dollhouse furniture that belonged to Clara and Elizabeth Hoar. For information on their father, see Moorfield Storey and Edward Emerson, *Ebenezer Rockwood Hoar: A Memoir* (Boston: Houghton Mifflin, 1911).

9. U.S. Federal Census, Non-Population Schedule, 1860, Hingham, Plymouth County, Mass.

13 Edward Lewis Hersey (1822–1853)

1. Samuel Hersey also advertised himself as a "Toy Ware & Box Manufacturer": see "Hingham Business Directory," *Hingham Journal*, February 1, 1850, 1. Edward and Samuel were also the only toymakers to list themselves in the *Massachusetts Register* for 1852. It is notable that no such craftsmen were listed in nearby Boston, though a few dealers in general merchandise and fancy goods included toys, probably imported from Europe, among their inventory. For information on Samuel Hersey, see cat. 12 and chap. 2.

2. Vital Records identify the cause of Hersey's infant daughter's death as "lung fever," but she probably died of the same disease that took her parents; Massachusetts Vital Records, 1841–1910, 67:253.

3. Hersey died of consumption, better known today as tuberculosis; ibid., 76:204.

4. Samuel Hersey may have acquired the finished and unfinished toys from Edward's estate. A miniature box in the collection of Historic New England bears the mark of Edward and the label of Samuel. Edward L. Hersey, inventory, filed August 31, 1852, Plymouth County Probate Records, 96:348.

5. Hersey probably learned the trade of cooper from his father, Joseph Hersey (1781–1863), who married Reuben Hersey's daughter. Edward's brother, Joseph Jr. (1806–65), also worked briefly as a toymaker. For information on the coopers and toymakers who worked on Hersey Street, see chap. 3 and appendices A, C, and D.

6. Census records show Edward Lewis Hersey living in the house of his deceased grandfather in 1850. Reuben's daughter, Eliza Lane Hersey, also lived in the house; U.S. Federal Census, Population Schedule, 1850, Hingham, Plymouth County, Mass. By the time of his death in 1853, he had also acquired a quarter-acre plot with a shop.

7. These pieces descended in the same collection as a small bucket, dipper, stool, and drying rack that sold at Willis Henry Auctions, Marshfield, Mass., November 21, 2009, lots 91–94 and 96–97. The bucket and piggin both contained white stickers inscribed in ink: "Loaned by Aurelia Fearing." The piggin, drying rack, stool, and dumb betty are illustrated in William Blake Luce's "The Making of Toys in Hingham, Mass.," *Palette and Bench* 1, no. 1 (December 1910): 10–13.

8. A similar yellow stain is applied to a miniature bucket made by Cotton Hersey and owned by a member of the Canterbury Shaker community; see pl. 11.6.

14 William Sewall Tower (1826–1898)

1. Marshall McClintock and Inez McClintock, *Toys in America: A Profusely Illustrated History* (Washington, D.C.: Public Affairs Press, 1961), 83–84.

2. Tower was employed at the Wilder bucket factory by 1858; see *Transactions of the Hingham Agricultural and Horticultural Society, from October, 1858 to March, 1861* (Hingham: Hingham Agricultural and Horticultural Society, 1861), 69. He is identified as a turner in the 1855 Massachusetts State Census. For additional information on Tower, see chap. 2.

3. *The Tenth Exhibition of the Massachusetts Charitable Mechanic Association at Faneuil and Quincy Halls, in the City of Boston, September and October 1865* (Boston: Wright & Potter, 1865), 118. In 1869 he made and exhibited a matching calling card tray, described as a "fancy top card table"; *Eleventh Exhibition of the Massachusetts Charitable Mechanic Association at Faneuil and Quincy Halls, in the City of Boston, September and October 1869* (Boston: Wright & Potter, 1869), 126.

4. *Hingham Journal*, August 20, 1869, 2.

5. "Letter from South Hingham," *Hingham Journal*, July 23, 1869, 2.

6. McClintock and McClintock, *Toys in America*, 85–93.

7. Charles Tower is listed in early advertisements as a principal in the firm; *Hingham Journal*, August 27, 1869, 2. George Tower is also known to have worked at the factory during its early beginnings; U.S. Federal Census, Population Schedule, 1870, Hingham, Plymouth County, Mass. For information on the Society of Arts and Crafts, Hingham, see Susan B. Willard, "The Hingham Society of Arts and Crafts," in *Hingham: A Story of Its Early Settlement and Life, Its Ancient Landmarks, Its Historic Sites and Buildings* (Hingham: Daughters of the American Revolution, Old Colony Chapter, 1911), 120–23. See also C. Chester Lane, "Hingham Arts and Crafts, Their Aims and Objects," *Craftsman* 3 (December 1903): 276–81.

8. The McClintocks were the first to write about the "Tower Toy Guild." In *Toys in America* they relied heavily on information from Ella Cushing, the aged niece of the Hingham toymaker Loring H. Cushing; he worked for William Tower for more than two decades, but he retired before the factory closed in 1900. Following the founding of the Society of Arts and Crafts, Hingham, in 1901, Cushing began making toys again, but

at his own shop and under his own name. Other craftsmen, including Augustus Hudson, a former cabinetmaker, also consigned goods to be exhibited and sold by the society. The involvement of Jacobs's Hatchet Factory in the toy industry remains a mystery. The factory's records survive at the Hingham Historical Society and do not include any accounts for the manufacture of toys or a relationship with Tower. There are, however, two miniature hatchets that survive with local histories. One is cast entirely in lead and is stamped J on the axe head; the other has a wood handle and was purportedly one of two owned by S. Randall Lincoln (b. 1879), who claimed that both were made at Jacobs's factory; both axes are in the Hersey Family Collection.

9. U.S. Federal Census, Non-Population Schedule, 1870, Hingham, Plymouth County, Mass.; *Hingham Journal,* March 31, 1893, 1.

10. *Hingham Journal,* June 28, 1878: 3; *Hingham Journal,* June 9, 1882, 3; and *Directory of Hingham and Hull, Mass., 1894* (Quincy, Mass.: J. H. Hogan, 1894), 25.

11. "Manufacturers at South Hingham," *Hingham Journal,* June 2, 1876, 2.

12. U.S. Congress, House, *Reports of the United States Commissioners to the Paris Universal Exposition, 1878,* 46th Congress, 3rd sess., 1880, H. Ex. Doc. 42, part 1:347, 369, and 437.

13. For a description of Tower's display of toys at the Boston Mechanic Association's Fair, see "Hingham Toy Ware," *Hingham Journal,* October 21, 1881, 1. A miniature wagon made by Tower was owned by John P. Richardson and includes his handwritten note, "Tower Toy Co. toy cart/Two wheels, marked with Tower Toy Company stencils. Acquired from Randall Lincoln (b. 1879) who purchased it at a sale near the north corner of High + Main Sts. Randall said that Wm. Tower had a shop at the edge of Main Street. At some point it was cleared out and the contents sold. Randall purchased two or three boxes of small toys, tools, etc. Randall remembered the Tower shop with its sign as a young man. Randall thought this car may have been a reject because of improper stamped impression on bottom, and was used as a display piece. Randall also purchased two small toy wheelbarrows made by Tower. He thought that he had given one to Julian + Olive Loring who lived near his So. H. Main St. home. Neither of the Lorings recalled a wheelbarrow, however Olive showed me a set of toy tools about 6″ high (rake, shovel, hoe) with a Tower Co. label. After Olive's death I asked their daughter from Cape Cod about them. She did not [know]."

14. *Boston Daily Globe,* "Toy Makers of Hingham," March 13, 1910, SM11.

15 Elijah Loring Corthell (1836–1917)

1. Tower's miniature pails are occasionally stamped "PATENT/APPLIED FOR" on the bottom, though he appears to have never received a patent for his goods. Both Tower and Corthell exhibited their toys at the Hingham Agricultural and Horticultural Society Fair in 1877; see *Transactions of the Hingham Agricultural and Horticultural Society for the Year 1877* (Hingham: Hingham Agricultural and Horticultural Society, 1878), 50–51. An octagonal butter churn is illustrated in an advertisement for Tower's Toy Bazaar that appeared in the *Hingham Journal,* June 28, 1878. The same advertisement includes images of a miniature wheelbarrow, rake, shovel, washtub, and covered pail—all forms described in Corthell's price list.

2. In 1835 Elijah C. Corthell married Mary French Hersey, daughter of Samuel French Hersey, a cooper and farmer. The newlyweds lived on Hersey Street beside Mary's brother Samuel, a prominent toymaker; see George Lincoln et al., *History of the Town of Hingham, Massachusetts, Genealogical* (Hingham: By the town, 1893), 2:142.

3. In 1861 E. C. Corthell & Sons exhibited at the Hingham Agricultural and Horticultural Society Fair; see *Transactions of the Hingham Agricultural and Horticultural Society from October 1858 to March 1861* (Hingham: Hingham Agricultural and Horticultural Society, 1861), 69. By 1867 Reuben Hersey Corthell, the second son, was working as foreman at the Hingham Woodenware Co.; see *The Plymouth County Directory and Historical Register of the Old Colony* (Middleboro, Mass.: Stillman B. Pratt, 1867), 52. By 1875 Elijah L. Corthell identified himself as a toy manufacturer; see Hingham Militia Enrollments, 1868–70, box MR3:16, Hingham Historical Society, Hingham.

4. U.S. Federal Census, Non-Population Schedule, 1880, Hingham, Plymouth County, Mass.

5. Unskilled laborers earned $1.25/day. The men worked eight-hour shifts between November and May.

6. *Resident and Business Directory of Hingham, Mass.* (Needham, Mass.: Local Directory Publishing Co., 1885), 82. Corthell's trade card for his picture frame manufacturing business was also discovered in his house.

7. *The Weymouth, Mass., Directory, Including Hingham and Braintree, 1902–3* (Boston: W. E. Shaw, 1902), 101. *Directory of Hingham and Cohasset, Massachusetts, 1915* (Boston: Union Publishing Co., 1915), 30.

16 George William Fearing (1838–1918)

1. The society began to promote itself in earnest in 1903. Fearing was mentioned by name in most of this early material, including an announcement in the *New York Times,* August 23, 1903, 9.

2. Fearing's shop was apparently in the basement of his house; see "A Slender Santa Claus," *Boston Journal,* November 8, 1903, 3. For information on Luce, see cat. 19.

3. Images of Fearing were reproduced in the following publications: J.S.F., "A Christmas Scrapbook," *Good Housekeeping,* July–December 1904, 740–41; "Hingham Village Industries," *Southern Workman* 32, no. 9 (September 1903): 537–41; and "Toy Makers of Hingham," *Boston Daily Globe,* March 13, 1910, SM11.

4. *Southern Workman* 32, no. 9; "The Society of Arts & Crafts, Hingham, Massachusetts," *Handicraft* 4, no. 3 (June 1911): 99–102; "The Hingham Society of Arts and Crafts," in *Hingham: A Story of Its Early Settlement and Life, Its Ancient Landmarks, Its Historic Sites and Buildings* (Hingham: Daughters of the American

Revolution, Old Colony Chapter, 1911), 120–23; and "Arts and Crafts Exhibit at Hingham, Mass.," *Drawing and Manual Training Journal* 2, no. 3 (December 1903): 123–24.

5. For information on Perez L. Fearing, see George Lincoln et al., *History of the Town of Hingham, Massachusetts, Genealogical* (Hingham: By the town, 1893), 2:225–26. Fearing's factory employed eight men and produced 8,600 buckets annually; U.S. Federal Census, Non-Population Schedule, 1850, Weymouth, Norfolk County, Mass.

6. For a general chronology of Fearing's work history see "Toy Makers of Hingham." In 1870 Fearing worked in a bucket factory, probably the one run by Crocker and Alden Wilder; U.S. Federal Census, Population Schedule, 1870, Hingham, Plymouth County, Mass.

7. Fearing worked for Tower some time between 1870 and 1877, when local military enlistments record his occupation as peddler; see Hingham Militia Enrollments, 1868–70, box MR3:16, Hingham Historical Society, Hingham. A billhead from 1878 identifies him as a lumber dealer; see Hingham Records, 1637–1931, Ms. N-1404, box 11, Massachusetts Historical Society, Boston.

8. Fearing's toy and cooper's shop is first mentioned in an announcement in the *Hingham Journal*, April 14, 1888, 1.

9. Fearing's carved hickory brand is mentioned in "Arts and Crafts Exhibit at Hingham, Mass.," 123–24.

10. J.S.F., "A Christmas Scrapbook."

11. Fearing's miniature reproduction of the Governor Carver chair is mentioned in "Toy Makers of Hingham." For the original chair at Pilgrim Hall Museum in Plymouth, see Robert F. Trent and Karin Goldstein, "Notes about New 'Tinkham' Chairs," in *American Furniture* annual, ed. Luke Beckerdite (Milwaukee, Wis.: Chipstone Foundation, 1998), 224.

17 Loring Hersey Cushing (1840–1915)

1. Cushing's assistant, Alonzo F. Cushing, was also aboard the carriage when it crashed; see "Nearly an Accident," *Boston Evening Journal*, August 17, 1881, 3.

2. For information on Tower, see cat. 14. Two early interviews with Cushing provide the most details about his career; see "Toy Makers of Hingham," *Boston Daily Globe*, March 13, 1910, SM11; see also "Artist Toy Makers of America: Reappearance of Handmade Playthings in Nursery," *Boston Daily Globe*, December 24, 1911, SM3.

3. For information about his involvement in this aspect of the firm, see "Toy Makers of Hingham."

4. Cushing told a reporter that he left the firm following Tower's death in 1898. This chronology, however, does not conform to listings in town directories, which identify him as a stitcher in 1894. It seems more likely that he left the factory after Tower relinquished control to his son-in-law some time in the early 1890s; see chap. 2 for more information about the change in ownership. For an excellent overview of Cushing's career and a line drawing of the toymaker at work in his shop, see "Hingham's Toymakers Traveling the Path Which Was Laid Out 75 Year Ago," *Boston Herald Magazine*, October 1, 1911, 29.

5. For an overview of the Society of Arts and Crafts, Hingham, see Susan B. Willard, "The Hingham Society of Arts and Crafts," in *Hingham: A Story of Its Early Settlement and Life, Its Ancient Landmarks, Its Historic Sites and Buildings* (Hingham: Daughters of the American Revolution, Old Colony Chapter, 1911), 120–23. See also C. Chester Lane, "Hingham Arts and Crafts, Their Aims and Objects," *Craftsman* 3 (December 1903): 276–81.

6. Cushing's shop is described in "Toy Makers of Hingham"; see also "Artist Toy Makers of America."

7. Fearing Burr and George Lincoln, *The Town of Hingham in the Late Civil War, with Sketches of Its Soldiers and Sailors* (Hingham: By the town, 1876), 123.

8. A later photograph of Cushing in his Grand Army of the Republic uniform is in the Hersey Family Collection.

9. Cushing describes the construction and finishing of his toys in "Artist Toy Makers of America."

18 Augustus Lincoln Hudson (1830–1922)

1. For a discussion of the bird carvers of Hingham, see Bob Mosher, "The Burr Family of Hingham," *Decoy Magazine* (May–June 2004): 8–14; and Cap Vinal, *Joseph W. Lincoln: Accord, Massachusetts, 1/26/1859–2/16/1938* ([Rockland, Mass.: Cap Vinal], 2002).

2. George Lincoln et al., *History of the Town of Hingham, Massachusetts, Genealogical* (Hingham: By the town, 1893), 2:358.

3. In 1859 Hudson exhibited a beautifully carved black walnut sideboard described in the *Transactions of the Hingham Agricultural & Horticultural Society, from October, 1858, to March, 1861* (Boston: Hingham Agricultural and Horticultural Society, 1861), 67, 69. Hudson is also listed as a carver in the enrollment of the Lincoln Light Infantry, 1854, box MR1, Hingham Historical Society, Hingham. For a history of the sign, see Torrent Engine Co., records, 1826–91, June 2, 1860, Bare Cove Fire Museum, Hingham.

4. *Transactions of the Hingham Agricultural & Horticultural Society from October, 1858, to March, 1861*, 159, 161.

5. According to tradition, the original sign deteriorated. A reproduction made in 1942 is at the Bare Cove Fire Museum, Hingham. A second reproduction made in 2006 by Hassan Wood Carvers and Sign Co. of Cohasset is hanging on the Torrent Engine House in West Hingham.

6. A history of the sign and its repairs is recorded in the official records of the Torrent Engine Co. Entries dated May 4 and June 4, 1874, indicate that Augustus Hudson, not his brother, may have been responsible for the repainting that occurred that year, but there is no mention of who executed the original scene: a depiction of the men of the Torrent Engine Co. demonstrating their pump to a crowd at the center of town. In 1883 a Mr. Spring was commissioned to paint a new scene of his own design, probably the natural waterfall found on many photo-

graphs from the turn of the century. For the most information available on William Hudson, see Julian Loring's "Notes on the Artist William Hudson, Jr.," unpublished manuscript (June 1976), John P. Richardson Papers, Hingham Historical Society, Hingham.

7. William and Augustus Hudson, 1866, Photographic Sensitizing-Box, U.S. Patent 51,834, filed June 16, 1865, issued January 2, 1866. In 1878 Augustus Hudson was issued a second patent, his brother serving as witness; see Augustus Hudson, 1878, Improvement in Photographic Cameras, U.S. Patent 201,248, filed January 12, 1878, issued March 12, 1878.

8. In 1869 Hudson exhibited twenty-four stereoscopes at the Hingham Agricultural and Horticultural Society Fair; see *Transactions of the Hingham Agricultural & Horticultural Society for the Year 1869* (Hingham: Hingham Agricultural and Horticultural Society, 1870), 57. He exhibited a collection of stereoscopes again in 1876; see *Transactions of the Hingham Agricultural & Horticultural Society for the Year 1877* (Hingham: Hingham Agricultural and Horticultural Society, 1878, 33. He is identified primarily has a cabinetmaker in town directories and records, though the 1880 Federal Census lists his occupation as stereoscope maker. He is listed as a manufacturer of photo-gelatin dry plates in the 1885 *Resident and Business Directory of Hingham, Mass* (Needham, Mass.: Local Directory Publishing Company, 1885). By 1894 and until his death in 1922, town directories list Hudson's occupation only as cabinetmaker.

9. An illustration of Hudson's work table and toys appears in *Handicraft* 4, no. 6 (September 1911): 227.

10. For a virtually identical bedstead, but with a triangular headboard, see Decorative Arts Photographic Collection, 66.2906, Winterthur Museum, Library, & Gardens. For an excellent overview of Hudson's involvement with the Society of Arts and Crafts, including a line drawing of the toymaker working at his bench, see "Hingham's Toymakers Traveling the Path Which Was Laid Out 75 Year Ago," *Boston Herald Magazine,* October 1, 1911, 29.

19 William Blake Luce (1860–1924)

1. For Richardson's obituary, see *Hingham Journal,* February 11, 2011.

2. A description of this event is recorded in John P. Richardson's research notes; see William Blake Luce, John P. Richardson Papers, Hingham Historical Society, Hingham.

3. For Luce's glass-plate negatives, see the John P. Richardson Photograph Collection, Hingham Historical Society, Hingham.

4. For a brief history of Luce's career, see his obituary, "William Blake Luce," *Hingham Journal,* October 10, 1924, 4.

5. In 1883 Luce won an amateur photography contest at the Institute of Technology in Boston; see *Photographic Times and American Photographer* 13, no. 156 (December 1883): 672–73. Luce received the following patents: Reversible Car Seat, January 31, 1888 (no. 377,205); Photographic Shutter, June 10, 1890 (no. 429,929); Photographic Camera, April 1, 1890 (no. 424,564); Electric Hand Fan, August 4, 1891 (no. 457,127); Electric Indicator, June 27, 1893 (no. 500,404); Electric Arc Lamp, September 12, 1893 (no. 505,051); Electric Arc Lamp, October 31, 1893 (no. 507,916); Electric Arc Lamp, February 6, 1894 (no. 514,425); Electric Arc Lamp Clutch, May 14, 1895 (no. 539,365); Equalizing Spring Connections, June 22, 1897 (no. 585,152); Toy Flying Machine, February 21, 1911 (no. 984,683). Perry Mason & Co. owned the *Youth's Companion.* In 1890 the firm advertised in the magazine an "Argus 12-Shot Repeating Camera," which Luce had patented. The camera was not widely popular and only a few extant examples are known: see Rob Niederman, "Argus 12-Shot Repeating Camera, c. 1890, Perry Mason & Company, Boston, Mass," www.antiquewoodcameras.com/argusrpt.html (accessed November 29, 2012).

6. William B. Luce, *Kites and Experiments in Aerial Photography, with Illustrations* (Hingham Centre; published by the author, 1898). See also *Photo Miniature: A Magazine of Photographic Information* 5, no. 57 (December 1903): 439–40.

7. For an overview of the Hingham Society of Arts and Crafts, see Susan B. Willard, "The Hingham Society of Arts and Crafts," in *Hingham: A Story of Its Early Settlement and Life, Its Ancient Landmarks, Its Historic Sites and Buildings* (Hingham: Daughters of the American Revolution, Old Colony Chapter, 1911), 120–23. See also C. Chester Lane, "Hingham Arts and Crafts, Their Aims and Objects," *Craftsman* 3 (December 1903): 276–81.

8. Luce also published the first illustrated history of toymaking in Hingham; see William Blake Luce, "The Making of Toys in Hingham, Mass.," *Palette and Bench* 1, no. 1 (December 1910): 10–13. For another looking glass by Luce, see Judith Armitstead, "Doll Furniture and Toys of the Early Hingham Craftsmen, Part II," *Antique Doll Collector* 13, no. 5 (June 2010): 46.

9. Luce & Ives exhibited fourteen pieces of toy furniture at the 1910 Hingham Agricultural and Horticultural Society Fair. Augustus P. Ives is the only person with the surname Ives who is listed in local directories around 1910.

10. F. W. Coburn, "Fifteen Years of Arts and Crafts at Hingham," *Boston Herald,* July 23, 1916, D3. In 1900 English humorist Jerome K. Jerome published *Three Men on the Bummel,* the comical journey of three men touring the German Black Forest by bicycle.

11. An original piece of Luce's letterhead is stored in a box of miniature furniture parts at the Old Ordinary, Hingham Historical Society.

12. No examples of this miniature furniture are known to have a maker's mark, but they can be firmly tied to Luce on the basis of the survival of a large collection of finished and unfinished toys made from the same patterns, all of which were salvaged from Luce's studio.

13. Town directories indicate that Stuart took charge of the shop in 1924 and continued to operate it until about 1934.

20 Ralph Thomas Jones (1886–1977)

1. Ralph Bergengren, "Unusual American Toys," *House Beautiful*, December 1915, 17.

2. Jones lived and worked at 12 High Street. Cushing lived and worked at 14 High Street. Cushing's last will and testament indicates that he had a close relationship with the Jones family; Loring H. Cushing, will, recorded July 6, 1912, Plymouth County Probate Records, 20:349. For more information on Cushing, see cat. 17.

3. U.S. Federal Census, Population Schedule, 1910, Hingham, Plymouth County, Mass. An early photograph of the Jones family house and barn on High Street is in the possession of the descendants of Ralph T. Jones.

4. *Catalogue of the Ninth Annual Exhibition of Original Designs for Decorations and Examples of Art Crafts Having Distinct Artistic Merit* (Chicago: Art Institute of Chicago, December 6–23, 1910), 676–88. *The Annual Report of the Society of Arts & Crafts, Boston, Massachusetts* (Boston: Society of Arts and Crafts, Boston, 1911), 47. Jones was also member of the National Society of Craftsmen, New York; see Florence N. Levy, ed., *American Annual Art*, vol. 8 (Washington, D.C.: American Federation of Arts, 1916), 429–30.

5. Ralph T. Jones's Business Reports for the Annual Reports on the Statistics of Manufacturers, 1915–18, are in a private collection.

6. See pl. 20.2.

7. Ralph T. Jones's Business Reports, 1915–18.

8. "Cohasset Man Has Help Celebrating 90th Birthday," *Quincy (Mass.) Patriot Ledger*, October 15, 1976.

9. "Merry Christmas from Cape Cod," sales catalogue, Cape Cod Productions, ca. 1948, in a private collection.

10. "Cohasset Man Has Help Celebrating 90th Birthday." Jones original workshop on High Street was acquired by the Margett family and moved to Ward Street in the 1950s.

21 N. D. Cass Company (1904–1993)

1. Richard Kenney, telephone interview, with the author, September 19, 2013.

2. According to the packaging and promotional material for the N. D. Cass Company, the firm was established in 1896. This was likely the year that Nathan Cass began working at the box factory of Horace Hager, where he may have produced toys. Cass and David E. Gilkey did not purchase Hager's factory until 1899; the 1954 *History of Athol* incorrectly lists this date as 1896; see William G. Lord's, *History of Athol, Massachusetts* (Athol: William G. Lord, 1954): 306–7. Multiple newspaper announcements confirm that Cass and Gilkey purchased the factory in 1899 and continued to operate under the same name. See *Athol Chronicle*, November 30, 1899; *Athol Almanac*, November 27, 1899; and "Old-Time Notes: Thirty Six Years Ago," *Athol Transcript*, January 30, 1935. The author is grateful to Richard Chaisson for sharing his extensive files on Cass toys.

3. The *History of Athol* incorrectly states that Gilkey sold his share to Cass in 1902. According to multiple newspaper announcements, the sale took place in 1901; see *Athol Daily News*, October 17, 1901; *Athol Almanac*, October 21, 1901; and *Athol Transcript*, October 22, 1901. In 1904 Cass purchased the Athol planing mill of Charles H. Porter and converted it into a toy factory; see *Athol Transcript*, January 26, 1904; *The Athol Chronicle*, January 28, 1904; and *Athol Church Record*, January 23, 1904.

4. Cass operated his factory—the Athol Paper Box Company—and his toy factory as separate entities. Although he was producing toys in earnest by 1904, N. D. Cass Company was not certified by the state until 1908; see *Report of the Tax Commission for the State of Massachusetts for the Year ending November 30, 1908* (Boston: Wright & Potter Printing Co., 1909), 302.

5. For a detailed description of the Vermont operation, see "Making Wooden Toys: The Processes From Tree to Crate; Interesting Facts about the N. D. Cass Industry, The New Plant at Manchester, Vt.," *Athol Transcript*, September 27, 1921.

6. The New York office opened in 1923; see "Main Cass Company Building Sold," *Athol Daily News*, September 20, 1997.

7. For a description of items produced by the N. D. Cass Company for the wartime effort, see "Demand for New Wooden Toys Strains Cass Co.'s Capacity," *Athol Daily News*, November 25, 1942. See also Athol War Historical Committee, *Athol in World War II: A History* (Athol: Transcript Press, 1950), 75.

8. Albert D. Hughes, "Toys by the Thousand Pour from Old Garage in Hingham," *Christian Science Monitor*, December 9, 1943: 5; "Demand for New Wooden Toys Strains Cass Co.'s Capacity," *Athol Daily News*, November 25, 1942; and "Millions of Christmas Toys Come from Athol This Year," *Athol Daily News*, November 27, 1942.

9. Hughes, "Toys by the Thousand . . . ," p. 5.

10. "Millions of Christmas Toys Come from Athol this Year."

11. Cass produced a wide range of toys, including wooden games, rocking horses, sleds, and doll furniture. For a description of some of the toys produced at the Hingham factory, see Hughes, "Toys by the Thousand . . . ," p. 5

12. The Hingham factory probably closed in 1945 to accommodate production in a new facility in Athol; see "New Cass Toy Plant to Employ 150 Soon," *Athol Daily News*, March 5, 1945. Cass closed its main plant in Athol in 1993 but continued manufacturing other wooden items until 1997; see "Main Cass Company Building Sold," *Athol Daily News*, September 20, 1997. See also "Production of Toys Ended Run in 1993," *Athol Daily News*, January 20, 2012.

APPENDIX A
HISTORY OF A BUCKET

In August 1920, Herbert W. Hersey (1867–1959) loaned the coopering tools that had belonged to his grandfather, Caleb Strong Hersey (1813–94) (see fig. 1.5), to the Hingham Historical Society. Accompanying the loan was an extraordinary essay and two pencil sketches by Herbert that document in rich detail the process of bucketmaking in his grandfather's era, the early nineteenth century. From the harvesting of wood to the layout of a shop, he captured what the anthropologist hopes to record: a highly specialized activity as it takes place within the context of its community and its time.

Hersey wrote as a well-qualified participant-observer, and he made use of available informants. Though a grocer by trade, he had grown up in his grandfather's house, behind which the old man's busy coopering and toymaking shop stood. A fifth-generation Hingham craftsman, Caleb Strong Hersey was grandson of Peter Hersey (1757–1844), a prominent early cooper (see cat. 1). Herbert's father, Francis Hersey (b. 1835), was also a cooper, who probably was trained by his own father, Caleb. The young Herbert likely spent hours in the fragrant, bustling workshops of his kinsmen, helping, watching, and listening. In writing his essay many years later, Herbert recalls how he "ha[d] his mind refreshed at times by his mother [Susan Greeley Hersey, b. 1848] who remember[ed] clearly the doings of the past 73 years."[1]

Herbert's essay is reproduced here in order to give a sense of the great care he devoted to producing these documents and the wealth of generational knowledge he possessed. His manuscript describes an intensive, exacting enterprise dependent upon cooperative labor, reliable business networks, and transportation, wood lore, elaborate planning, workspace efficiency, time management, brute strength, superb hand skills, and a craftsman's eye. Implicit in Herbert's accounts as well are the communities, technologies, and landscapes in which this undertaking took place that together provide a marvelous glimpse of a past world in action. Above all, however, Herbert Hersey's words illuminate the historical link between kinship and craft, in which generational knowledge, strengthened by rich personal associations, blended to become a valued and honorable legacy.

1. Herbert W. Hersey to William W. Lunt, August 28, 1920, Hingham Historical Society, Hingham, Massachusetts.

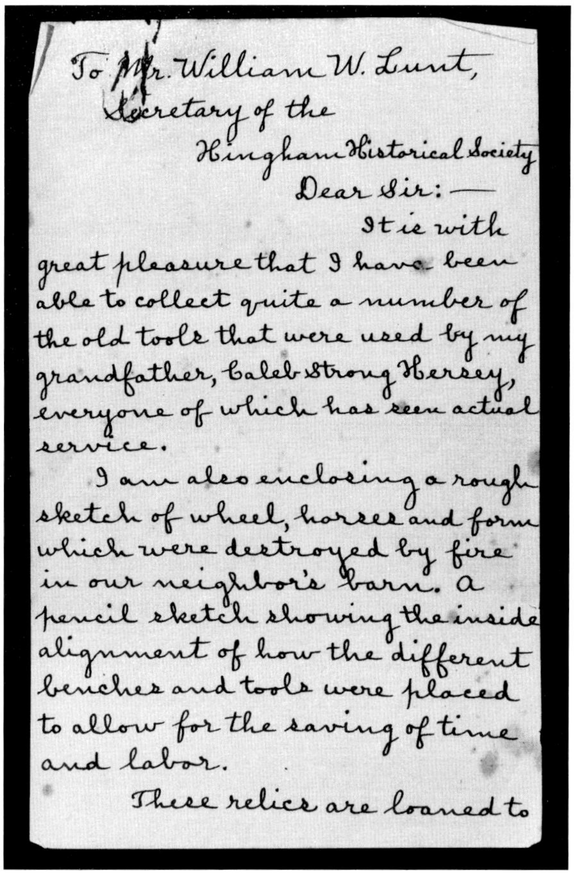

To Mr. William W. Lunt,
Secretary of the
Hingham Historical Society
Dear Sir:—

It is with great pleasure that I have been able to collect quite a number of the old tools that were used by my grandfather, Caleb Strong Hersey, everyone of which has seen actual service.

I am also enclosing a rough sketch of wheel, horses and form which were destroyed by fire in our neighbor's barn. A pencil sketch showing the inside alignment of how the different benches and tools were placed to allow for the saving of time and labor.

These relics are loaned to

1

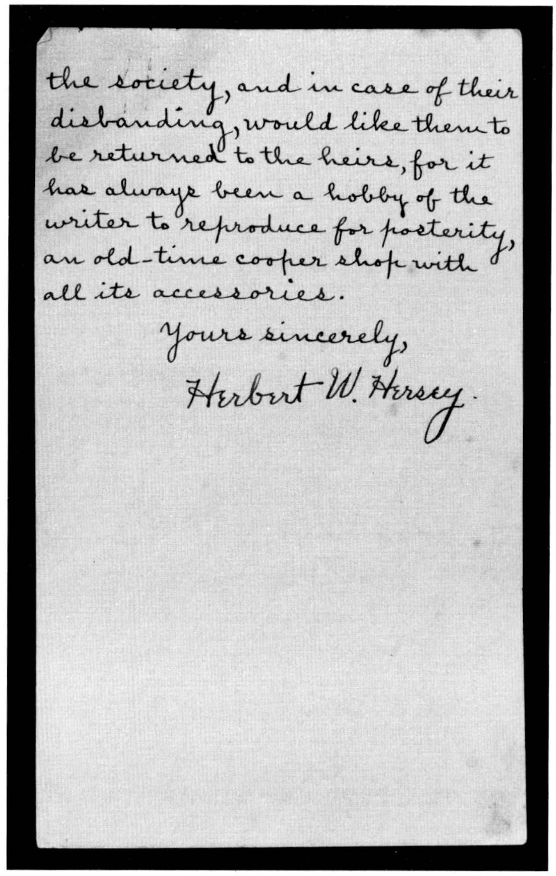

the society, and in case of their disbanding, would like them to be returned to the heirs, for it has always been a hobby of the writer to reproduce for posterity, an old-time cooper shop with all its accessories.

Yours sincerely,
Herbert W. Hersey.

2

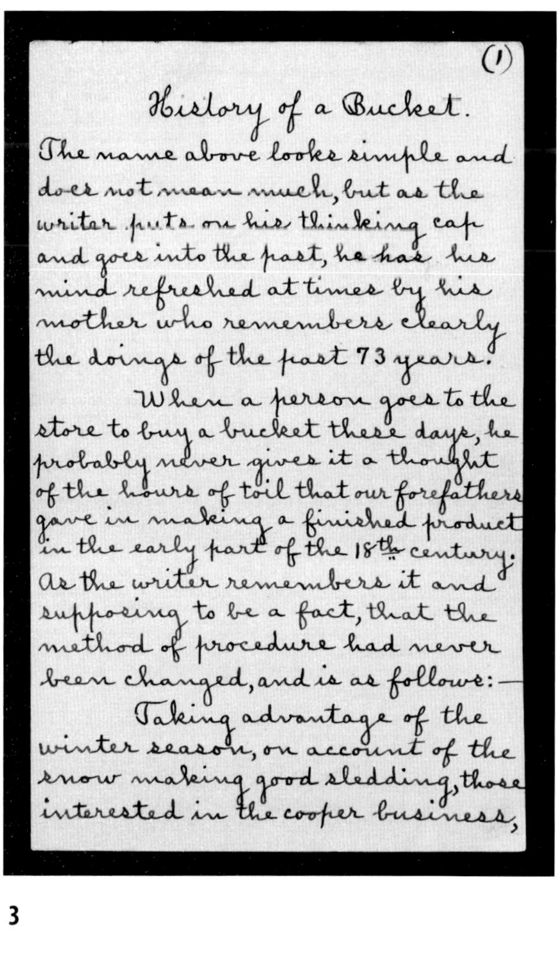

(1)

History of a Bucket.

The name above looks simple and does not mean much, but as the writer puts on his thinking cap and goes into the past, he has his mind refreshed at times by his mother who remembers clearly the doings of the past 73 years.

When a person goes to the store to buy a bucket these days, he probably never gives it a thought of the hours of toil that our forefathers gave in making a finished product in the early part of the 18th century. As the writer remembers it and supposing to be a fact, that the method of procedure had never been changed, and is as follows:—

Taking advantage of the winter season, on account of the snow making good sledding, those interested in the cooper business,

3

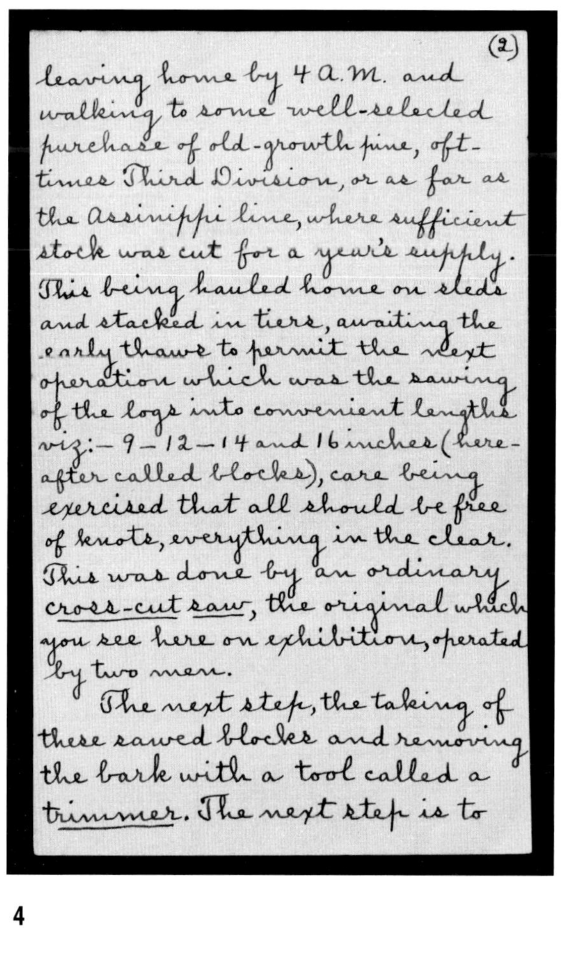

(2)

leaving home by 4 A.M. and walking to some well-selected purchase of old-growth pine, oft-times Third Division, or as far as the Assinippi line, where sufficient stock was cut for a year's supply. This being hauled home on sleds and stacked in tiers, awaiting the early thaws to permit the next operation which was the sawing of the logs into convenient lengths viz:— 9—12—14 and 16 inches (here-after called blocks), care being exercised that all should be free of knots, everything in the clear. This was done by an ordinary cross-cut saw, the original which you see here on exhibition, operated by two men.

The next step, the taking of these sawed blocks and removing the bark with a tool called a trimmer. The next step is to

4

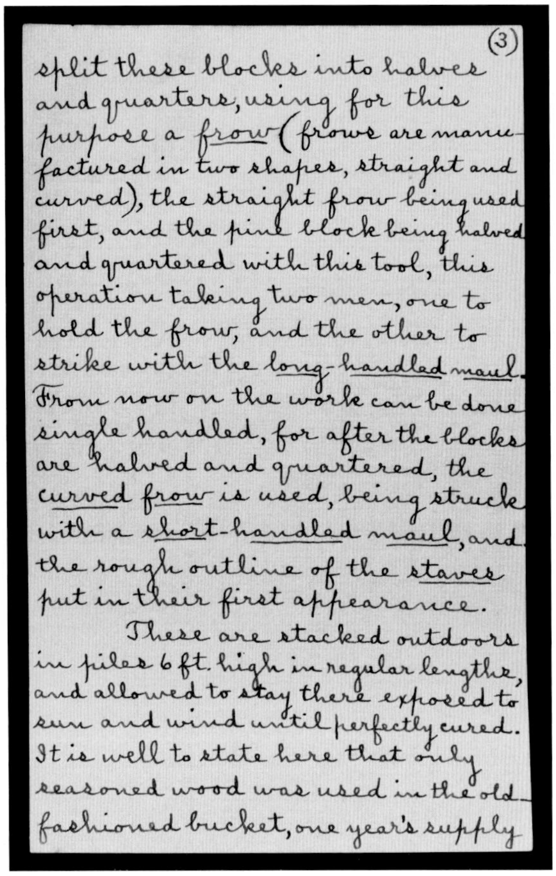

(3)

split these blocks into halves and quarters, using for this purpose a frow (frows are manufactured in two shapes, straight and curved), the straight frow being used first, and the pine block being halved and quartered with this tool, this operation taking two men, one to hold the frow, and the other to strike with the long-handled maul. From now on the work can be done single handled, for after the blocks are halved and quartered, the curved frow is used, being struck with a short-handled maul, and the rough outline of the staves put in their first appearance.

These are stacked outdoors in piles 6 ft. high in regular lengths, and allowed to stay there exposed to sun and wind until perfectly cured. It is well to state here that only seasoned wood was used in the old-fashioned bucket, one year's supply

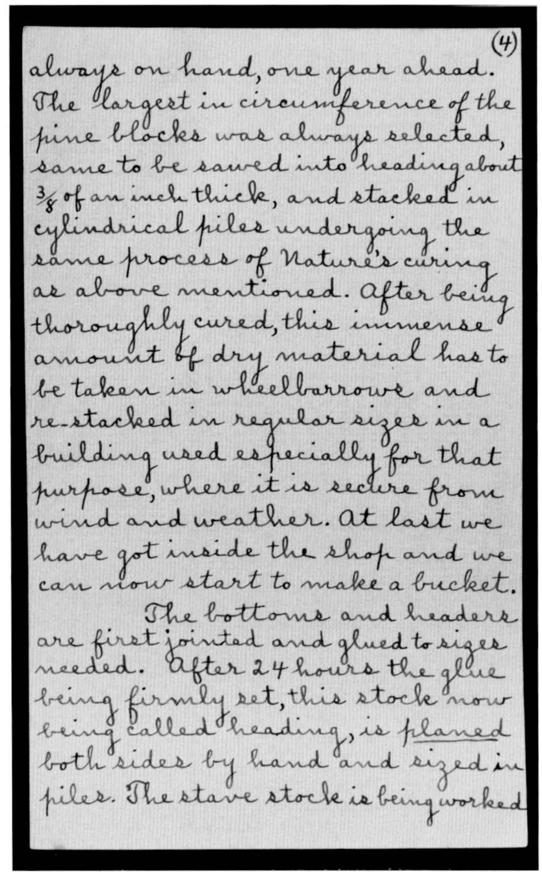

(4)

always on hand, one year ahead. The largest in circumference of the pine blocks was always selected, same to be sawed into heading about 3/8 of an inch thick, and stacked in cylindrical piles undergoing the same process of Nature's curing as above mentioned. After being thoroughly cured, this immense amount of dry material has to be taken in wheelbarrows and re-stacked in regular sizes in a building used especially for that purpose, where it is secure from wind and weather. At last we have got inside the shop and we can now start to make a bucket.

The bottoms and headers are first jointed and glued to sizes needed. After 24 hours the glue being firmly set, this stock now being called heading, is planed both sides by hand and sized in piles. The stave stock is being worked

5

6

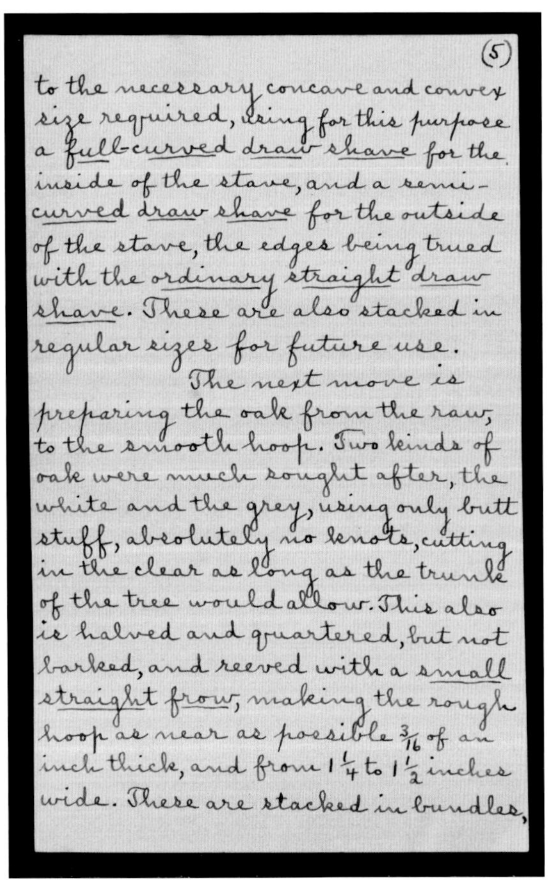

(5)

to the necessary concave and convex size required, using for this purpose a full-curved draw shave for the inside of the stave, and a semi-curved draw shave for the outside of the stave, the edges being trued with the ordinary straight draw shave. These are also stacked in regular sizes for future use.

The next move is preparing the oak from the raw, to the smooth hoop. Two kinds of oak were much sought after, the white and the grey, using only butt stuff, absolutely no knots, cutting in the clear as long as the trunk of the tree would allow. This also is halved and quartered, but not barked, and reeved with a small straight frow, making the rough hoop as near as possible 3/16 of an inch thick, and from 1 1/4 to 1 1/2 inches wide. These are stacked in bundles,

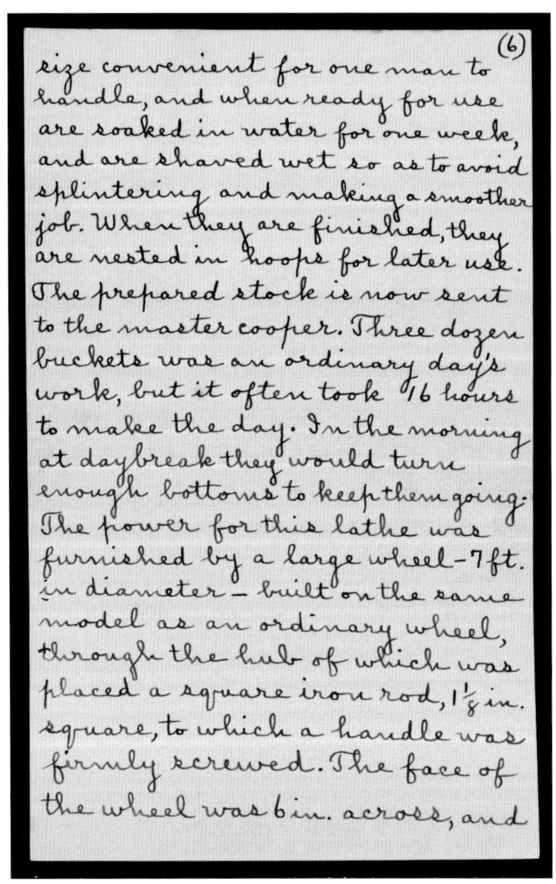

(6)

size convenient for one man to handle, and when ready for use are soaked in water for one week, and are shaved wet so as to avoid splintering and making a smoother job. When they are finished, they are nested in hoops for later use. The prepared stock is now sent to the master cooper. Three dozen buckets was an ordinary day's work, but it often took 16 hours to make the day. In the morning at daybreak they would turn enough bottoms to keep them going. The power for this lathe was furnished by a large wheel – 7 ft. in diameter – built on the same model as an ordinary wheel, through the hub of which was placed a square iron rod, 1 1/8 in. square, to which a handle was firmly screwed. The face of the wheel was 6 in. across, and

7

8

(7)

underneath the rim of the wheel, firmly bolted to the rim, were cast iron weights, cast in the form of the wheel 1½ in. square, making the wheel weigh when complete nearly 300 pounds. Over this wheel and running to a push pulley, was a 4 in. leather belt. This belt could be moved by a slide throwing it into gear, and the lathe was in motion. The size of the bottoms to be cut was regulated by the iron gauge, which worked on an endless screw, as seen by the exhibit, and was the only piece of lathe saved from the fire when Deacon Thayer's barn burnt. When the heading leaves this machine, it is in a perfect round,—sized, beveled (or chamferred) and sand-papered.

The next step is the setting

9

(8)

up of the bucket. During this process the workman had everything so arranged that he never had to move far from one spot, thereby saving labor and time. Working at a bench 20 in. wide, a little above waist high, the shaved staves stacked at his left, the bottoms at his right, and the long jointer placed at his immediate left, at height of bench, firmly wedged, knife edge uppermost, which allowed him to joint edges of the staves, by pushing the stave to the knife, repeating the process as many times necessary to complete a perfect joint. Then taking the shaved and jointed stave, he places upon a fixed oaken form, this form concaved to fit the back of the stave, the stave being held in place by a 1½ in. strap, which was fastened at the back of the form, and continued over the form,

10

(9)

connecting with a bucket filled with scrap iron, to the exact amount of weight needed to hold the stave firmly in place while the chime was being cut. This stave now being shaved, jointed and chimed, was firmly driven to the bottom, chamfer side down, driven with a small hand maul until firmly seated. Then about 2½ in. from top of stave a wooden peg is inserted which is put there to hold the top part of staves in position. The process is then repeated until the entire round of the bucket is completed. It then goes to the boss cooper who works upon the bench as here seen on exhibit (Notice the three holes cut in bench to receive the ears of the buckets or tubs and hold them in position during the process of coopering).

N.B.,—(We here see the advantage of having our hoops nested and

11

(10)

sized, the nesting and sizing being done while wet, they hold their shape after becoming thoroughly dry). Taking the hoop and placing it over the bucket until it shows a practical fit, then removing same, careful not to change the size already obtained, placing the hoop after boring the toe of the hoop (which is the sharpened end of the hoop), and inserting a large hand-made copper nail, called toe nail, carefully placing the hoop bottom side up, on a large flat iron made for the purpose (See exhibit), clinching same until perfectly flat with the inside of hoop, three other smaller copper tacks are used to make the hoop rigid and avoid twisting or splitting of hoop while being driven on. The driving on process being accomplished on the bench above referred to, a moderate heavy hammer being used, and a wooden swage being

12

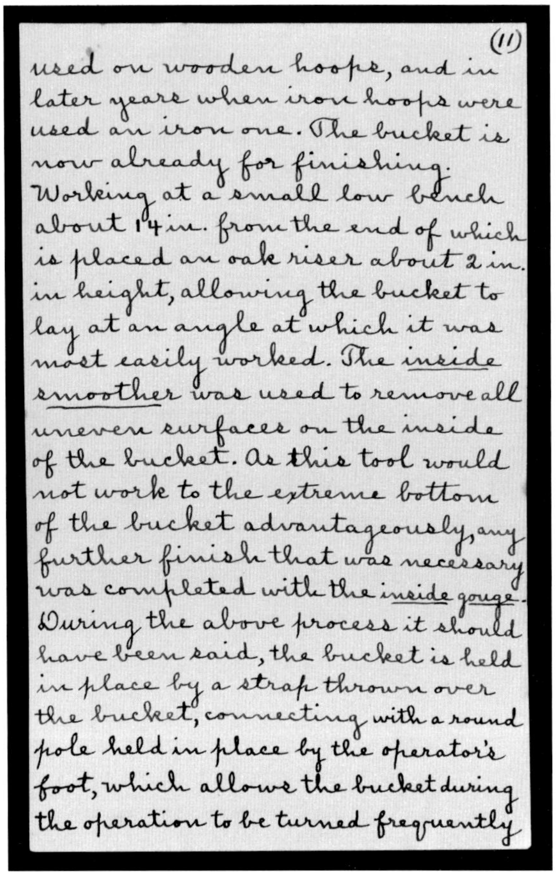

(11)
used on wooden hoops, and in later years when iron hoops were used an iron one. The bucket is now already for finishing.
Working at a small low bench about 14 in. from the end of which is placed an oak riser about 2 in. in height, allowing the bucket to lay at an angle at which it was most easily worked. The inside smoother was used to remove all uneven surfaces on the inside of the bucket. As this tool would not work to the extreme bottom of the bucket advantageously, any further finish that was necessary was completed with the inside gouge. During the above process it should have been said, the bucket is held in place by a strap thrown over the bucket, connecting with a round pole held in place by the operator's foot, which allows the bucket during the operation to be turned frequently

13

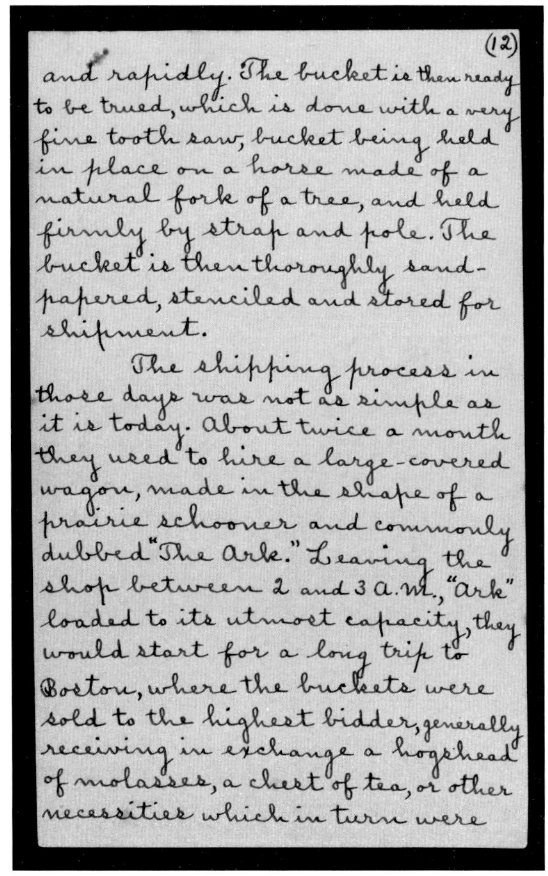

(12)
and rapidly. The bucket is then ready to be trued, which is done with a very fine tooth saw, bucket being held in place on a horse made of a natural fork of a tree, and held firmly by strap and pole. The bucket is then thoroughly sand-papered, stenciled and stored for shipment.

The shipping process in those days was not as simple as it is today. About twice a month they used to hire a large-covered wagon, made in the shape of a prairie schooner and commonly dubbed "The Ark." Leaving the shop between 2 and 3 A.M., "Ark" loaded to its utmost capacity, they would start for a long trip to Boston, where the buckets were sold to the highest bidder, generally receiving in exchange a hogshead of molasses, a chest of tea, or other necessities which in turn were

14

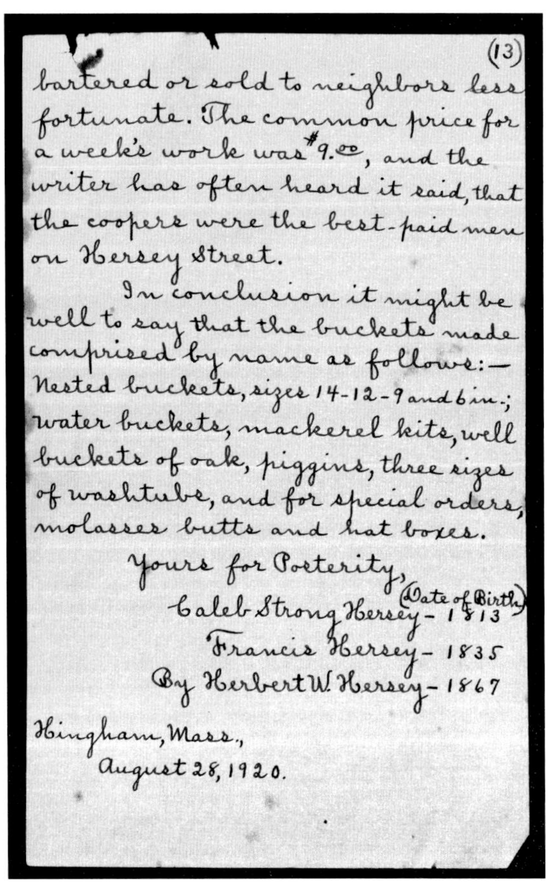

(13)
bartered or sold to neighbors less fortunate. The common price for a week's work was $9.00, and the writer has often heard it said, that the coopers were the best-paid men on Hersey Street.

In conclusion it might be well to say that the buckets made comprised by name as follows:— Nested buckets, sizes 14-12-9 and 6 in.; water buckets, mackerel kits, well buckets of oak, piggins, three sizes of washtubs, and for special orders; molasses butts and hat boxes.

Yours for Posterity,
(Date of Birth)
Caleb Strong Hersey - 1813
Francis Hersey - 1835
By Herbert W. Hersey - 1867

Hingham, Mass.,
August 28, 1920.

15

APPENDIX B
WILLIAM BLAKE LUCE PHOTOGRAPHS

In 1977, the Hingham antiquarian John P. Richardson salvaged a small crate of glass-plate negatives from the studio of photographer and toymaker William B. Luce (1860–1924). Thousands more, broken and discarded, were discovered in the basement of the building. Among the surviving plates—views of friends, family, and New England towns—are twenty images of miniature furniture and dollhouses made by Luce (nos. 1–17), Augustus L. Hudson (18), and George W. Fearing (19 and 20). Luce photographed the toys between 1901 and 1911 for promotional use by the Society of Arts and Crafts, Hingham.

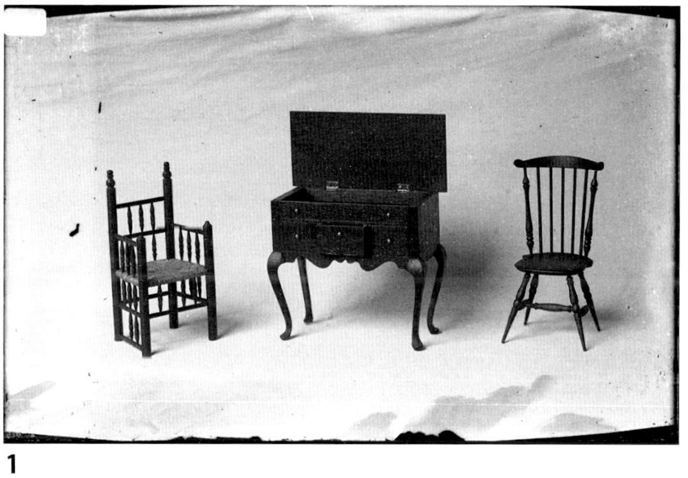
1

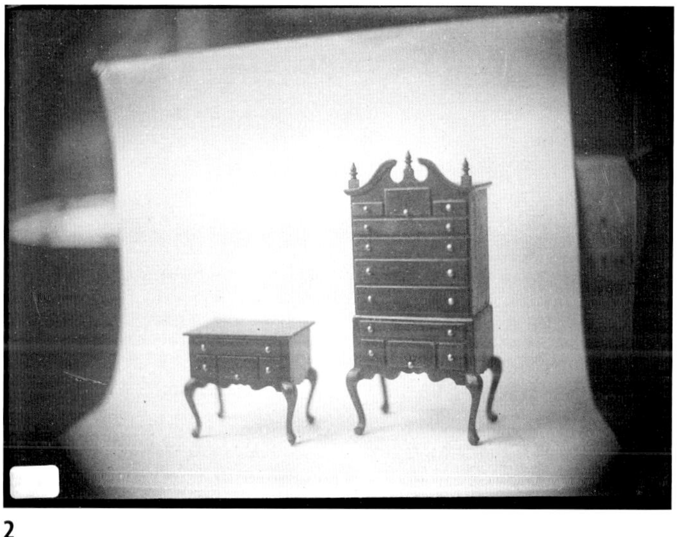
2

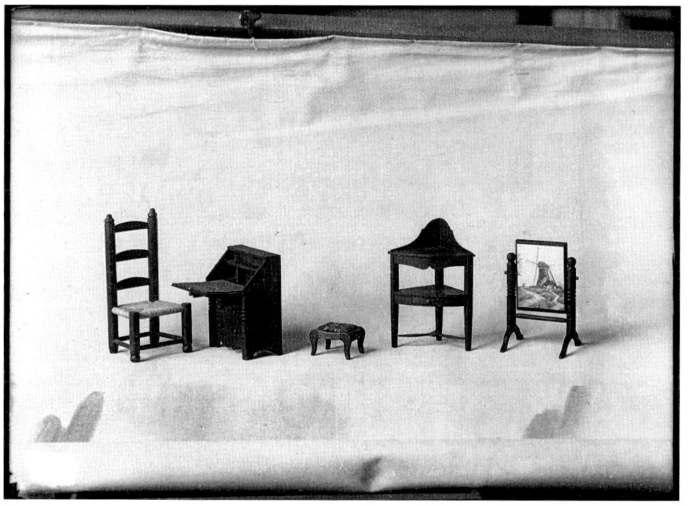
3

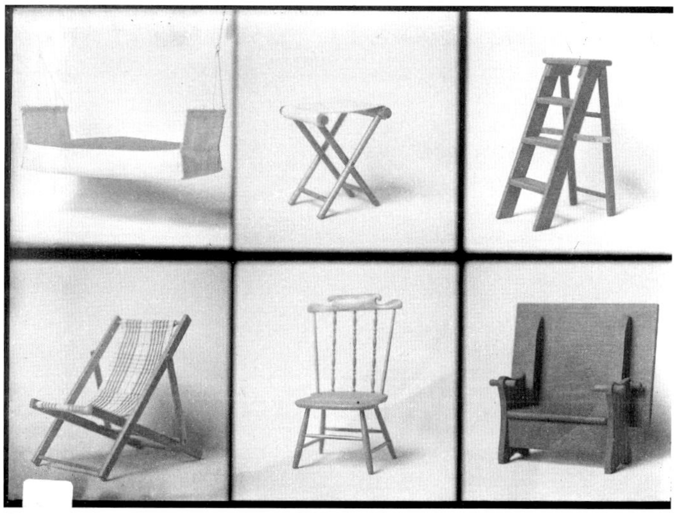
4

All photographs in Appendix B are courtesy of Hingham Historical Society.

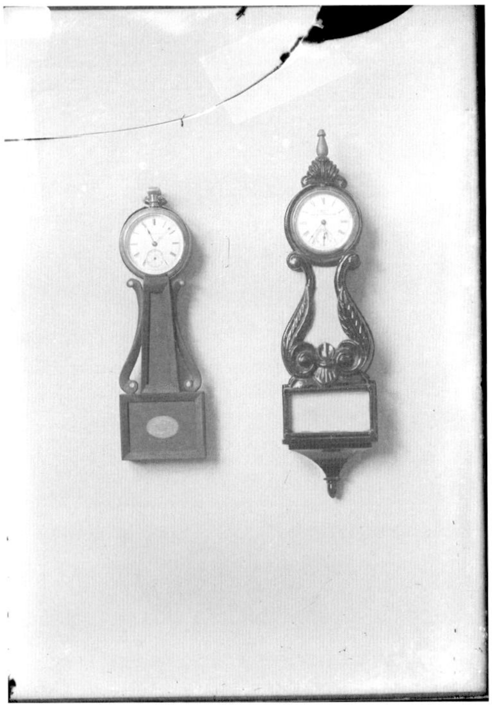
5

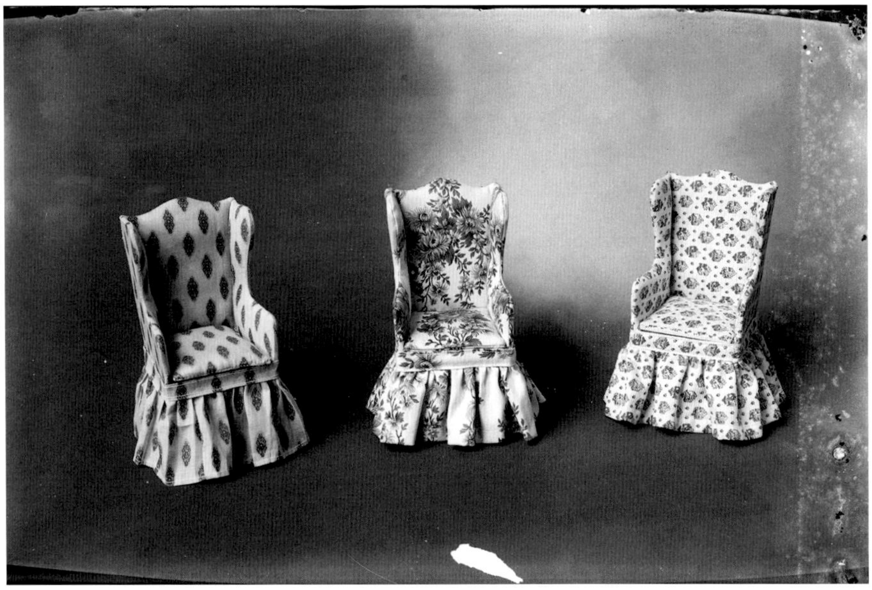
6

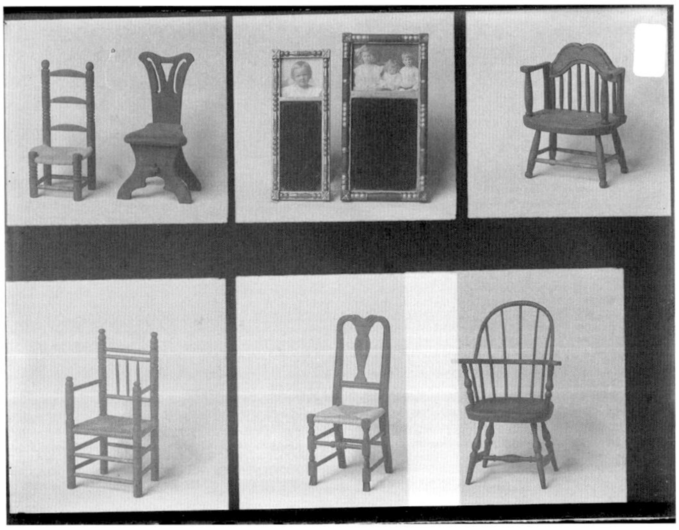
7

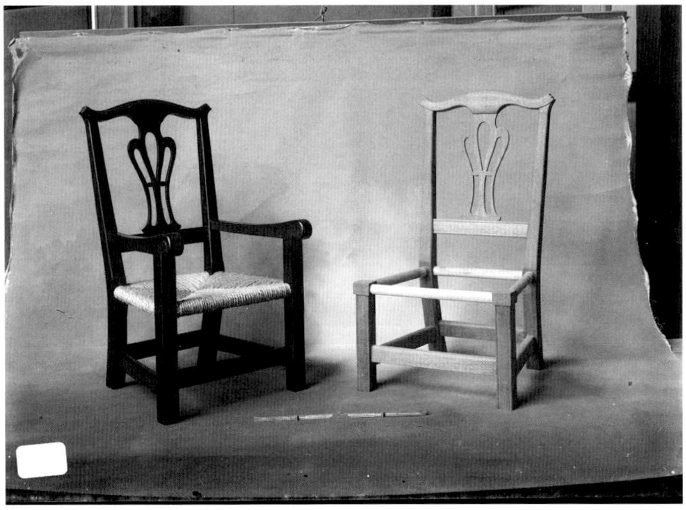
8

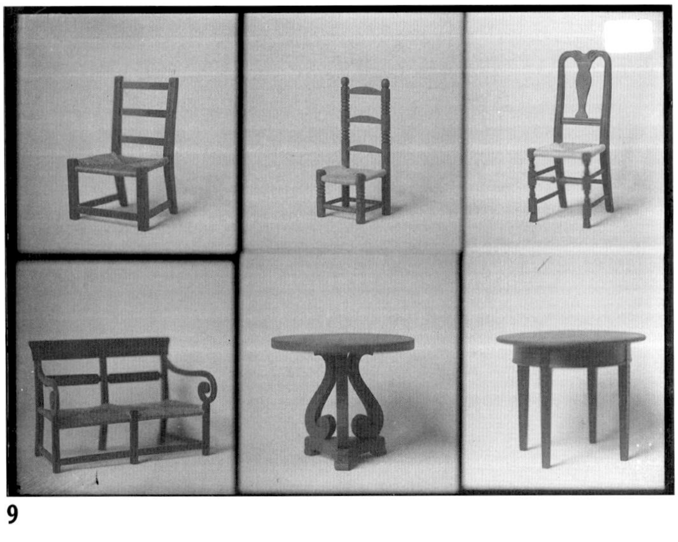
9

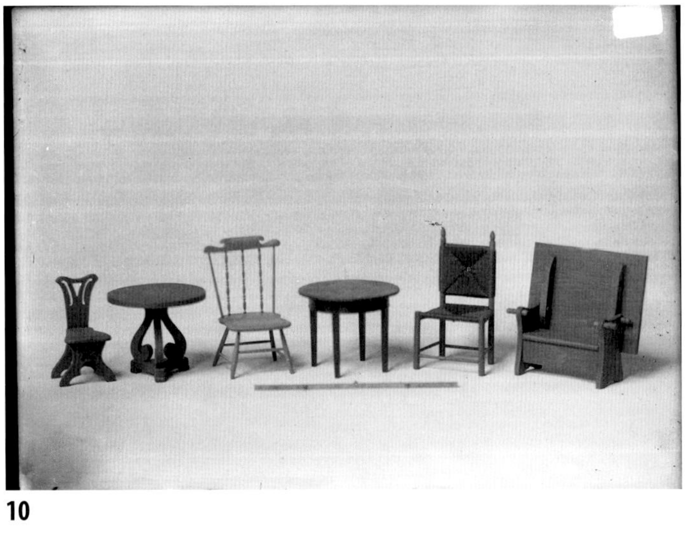
10

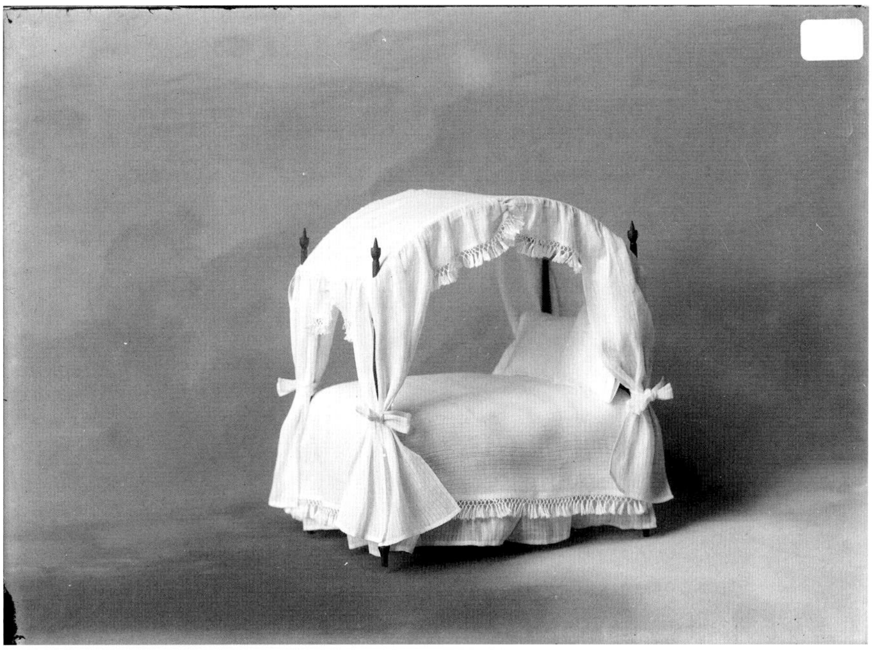

11

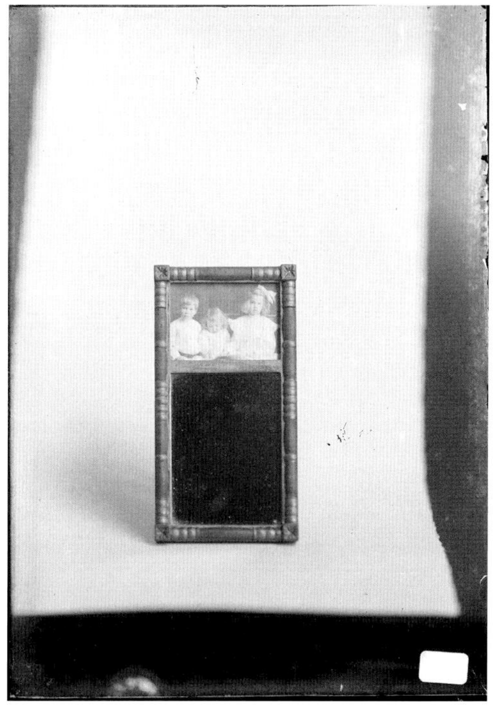

12

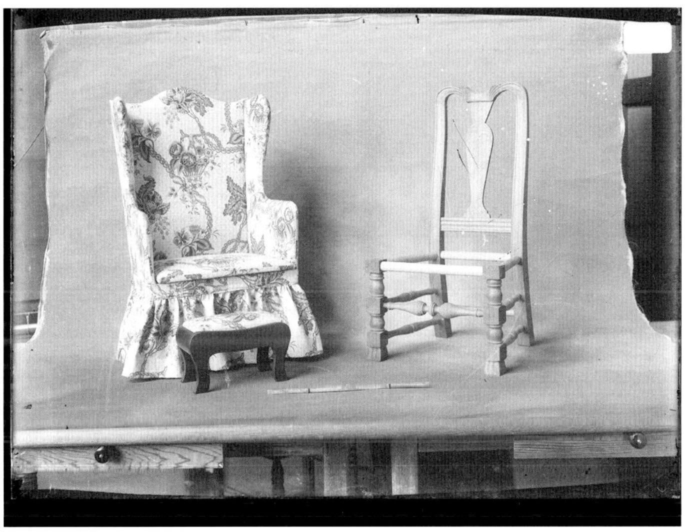

13

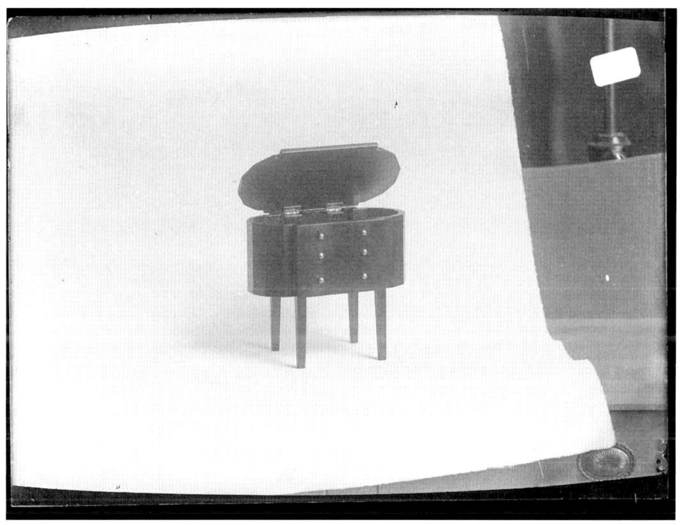

14

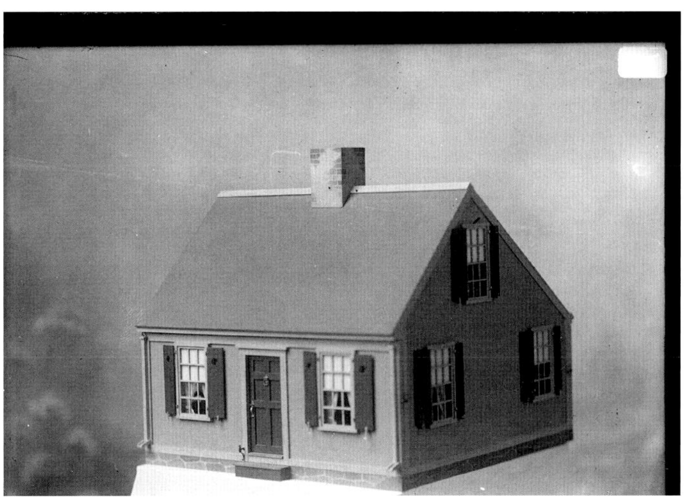

15

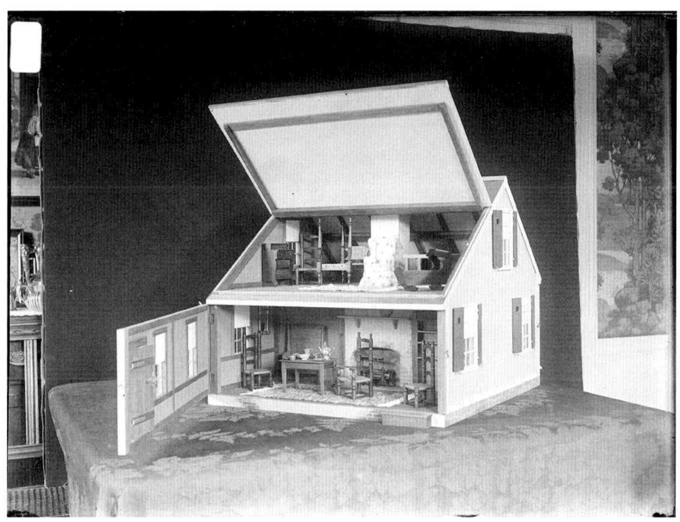

16

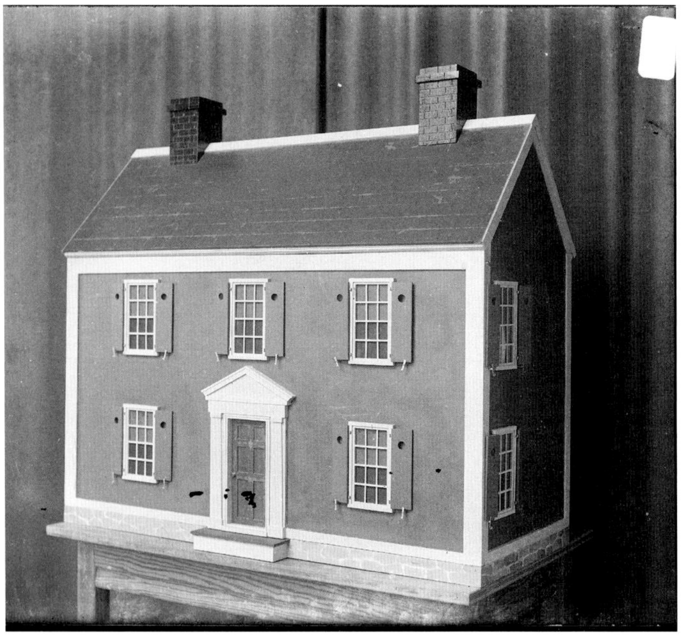

17

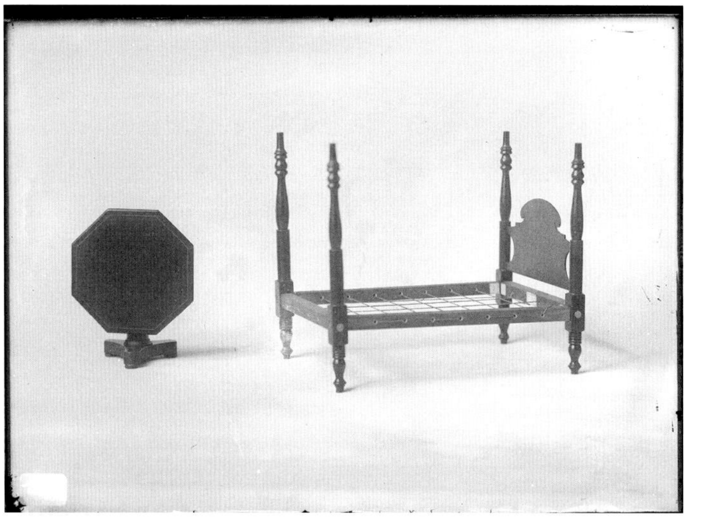

18

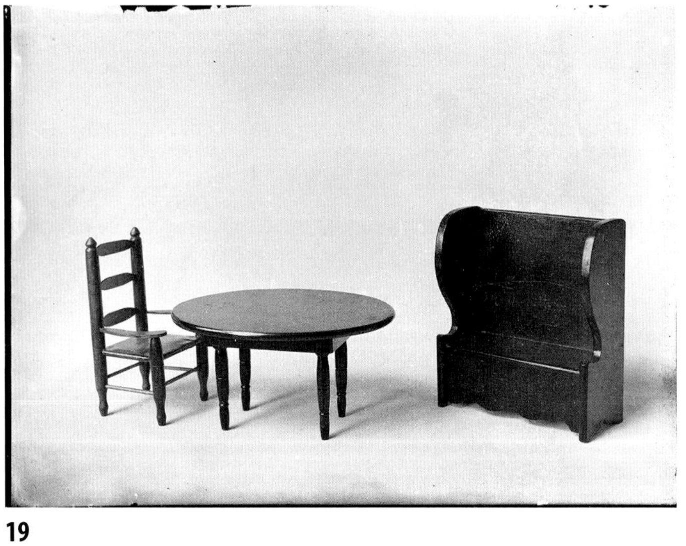

19

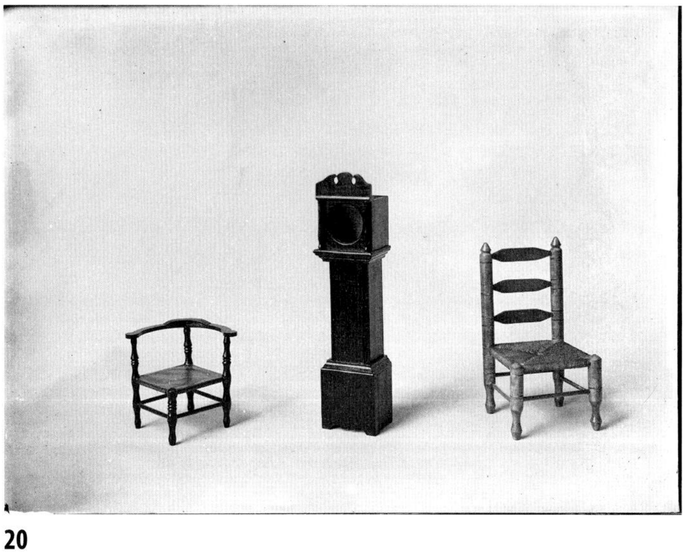

20

APPENDIX C

HINGHAM COOPERS AND RELATED CRAFTSMEN

KEY TO SOURCES FOR APPENDIX C: HINGHAM COOPERS AND RELATED CRAFTSMEN

BLP	Benjamin Lincoln Papers, 1636–1964, Massachusetts Historical Society
FC	U.S. Federal Census, Population Schedules, 1850–1940
GMB	Robert Charles Anderson, *The Great Migration Begins: Immigrants to New England, 1620–1633*, vols. 1–3 (Boston: NEHGS, 1995)
HAHS	*Transactions of the Hingham Agricultural and Horticultural Society*, 1858–1915, microfilm reel AH1, Hingham Public Library
HF	Hersey Family Collection
HHCD	*Directory of Hingham and Hull, Mass., 1894* (Quincy, Mass.: J. H. Hogan, 1894)
Hofer	Margaret K. Hofer, "Checklist of Woodworking Craftsmen in Plymouth and Bristol Counties, Massachusetts, 1760–c. 1810," *The Magazine Antiques* 159, no. 5 (May 2001): 808–13
HRBD	*Resident and Business Directory of Hingham, Mass.* (Needham: Local Directory Publishing Co., 1885)
Lane	James Hill Fitts, *Lane Genealogies*, vol. 2 (Exeter, N.H.: Newsletter Press, 1897)
Lincoln	George Lincoln et al., *History of the Town of Hingham, Massachusetts: Historical and Genealogical*, 3 vols. (Hingham: Town of Hingham, 1893)
LLI	Lincoln Light Infantry Orderly Book, 1854, Hingham Historical Society, Box MR1
LMN	James Trumbull, ed., *Note Book Kept by Thomas Lechford, 1638–1641* (Cambridge, Mass.: John Wilson & Son, 1885)
Marble	Marble-Richardson Family Papers, Yale University Library, MS 1358
MAVR	Massachusetts Vital Records
MC	Massachusetts State Census, 1855 and 1865
MGLM	Massachusetts Grand Lodge of Masons Membership Cards, 1733–1900, New England Historic Genealogical Society
MR	*The Massachusetts Register* (Boston: George Adams, 1852–1869)
MS	United States Mortality Schedule (1850–1885)
PBCD	*Directory and History of Plymouth and Barnstable Cos., Massachusetts City Directory, 1873–1874* (Boston: Dean Dudley, 1873)
PCLR	Plymouth County Land Records
PCPR	Plymouth County Probate Records
PCR	*Plymouth Court Records: Court of Common Pleas, 1686–1859*
PLY	*The Plymouth County Directory and Historical Register of the Old Colony* (Middleboro, Mass.: Stillman B. Pratt & Co., 1867)
RBDHC	*Resident and Business Directory of Hingham and Cohasset, Massachusetts, 1908–09* (Boston: Boston Suburban Book Co., 1908)
	Resident and Business Directory of Hingham and Cohasset, Massachusetts, 1912–13 (Boston: Union Publishing Co., 1912)
	Resident and Business Directory of Hingham and Cohasset, Massachusetts, 1915 (Boston: Union Publishing Co., 1915)
	Resident and Business Directory of Hingham and Cohasset, Massachusetts, 1920 (Auburndale, Mass.: Amos Jewett Gordon, 1920)
Savage	James Savage, *A Genealogical Dictionary of the First Settlers of New England*, 4 vols. (Boston: Little, Brown, 1860–62)
SCPR	Suffolk County Probate Records
SOL	Solomon Lincoln's Biographical Sketchbook of Hingham Residents, Hingham Historical Society, Box FP65
Todd	Charles Burr Todd, *A General History of the Burr Family: With a Genealogical Record from 1193 to 1902* (New York: Charles Burr Todd, 1902)
Tower	Charlemagne Tower, *Tower Genealogy: An Account of the Descendants of John Tower, of Hingham, Massachusetts*, 3 vols. (Cambridge, England: John Wilson & Son, 1891)
TRMM	Abner Forbes and J. W. Greene, *The Rich Men of Massachusetts, with a Statement of the Reputed Wealth of about Fifteen Hundred Persons* (Boston: W. V. Spencer, 1850)
Whiton	Augustus Sherill Whiton, *The Whiton Family in America: The Genealogy of the Descendants of Thomas Whiton* (New York: Whiton Family Association, 1932)

act. = active   ca. = circa

The abbreviations "act." and "ca." are used to indicate an approximate date or date range that an individual or company was active in their given trade.

NO.	NAME	DATES	OCCUPATION	NOTES	MAKER'S MARKS	SOURCES
1	Anderson, Jonathan	b. 1728	Cooper	Resided on High Street; moved to Chesterfield, Mass., by 1765.		Lincoln
2	Bailey, Ambrose	b. ca. 1826	Cooper	Originally from Maine.		FC1850
3	Barnes, Lyman	1796–1882	Cooper	Resided on Main Street, opposite Hersey Street.		Lincoln
4	Barnes, Thomas	d. 1691	Cooper	Received land grant in Hingham in 1637.		Lincoln; Savage, 1:121
5	Bartlett, Daniel	ca. 1801–71	Cooper	Originally from Barnet, Vt.; resided on Main Street in South Hingham; made oval and round boxes.	D. BARTLETT (stamp)	Lincoln
6	Batchelder, John Briggs	b. 1829	Bucketmaker	Worked in bucket factory; native of Maine; resided on Main Street in South Hingham.		Lincoln; FC1870–80
7	Bates, David F.	1850–86	Boxmaker, toymaker	Father of boxmaker David F. Bates Jr.; worked in a fruit box factory; resided on Hersey Street.		PLY; Lincoln; MAVR
8	Bates, David F., Jr.	b. 1878	Boxmaker	Son of boxmaker David F. Bates; resided on Hersey Street.		RBDHC (1908); Lincoln
9	Bates, R.	act. ca. 1780–1830	Cooper	Woodenware by the unidentified cooper R. Bates survives at the Hingham Historical Society and at least two other museums in the Hingham area. Unlike marks on other coopered items made in Hingham, Bates's mark is fire-branded rather than impressed into the wood.	R. BATES (brand)	See cat. 4
10	Beal, Annie Caroline	b. 1852	Boxmaker	Worked in a fruit box factory.		Lincoln; FC1870
11	Beal, Elisha	1761–1830	Cooper	Father of cooper Levi Beal; resided on Hersey Street.	E. BEAL (stamp) Elisha / Beal (stamp)	Lincoln; see ch. 1
12	Beal, James A.	b. 1833	Cooper	Son of cooper Levi Beal.		MC1855; FC1850; Lincoln
13	Beal, Jedediah	d. 1806	Cooper	Son of cooper Solomon Beal.		Lincoln
14	Beal, Levi	1786–1852	Cooper	Son of cooper Elisha Beal and father of cooper James A. Beal; resided on Hersey Street.	Levi / Beal (stamp) HINGHAM LB (stamp possibly used by him)	FC1850; Lincoln
15	Beal, Solomon	1679–1750	Cooper	Died in Pembroke, Mass.		Lincoln
16	Bibby, George Washington	ca. 1826–64	Cooper	Native of New Hampshire; resided in Hingham Centre.		MC1855; Lincoln
17	Bisbee, Elisha	act. 1645	Cooper	Originally from Scituate, Mass.		Lincoln
18	Blake, Solomon	1744–99	Cooper	Resided on Main Street.		Lincoln; SOL
19	Botting, Harriet M.	b. 1852	Worked in fruit box factory	Sister of box factory workers Martha and Lucinda Botting.		FC 1870; Lincoln
20	Botting, Lucinda A.	b. 1848	Worked in fruit box factory	Sister of box factory workers Martha and Harriet Botting.		MC1865
21	Botting, Martha Bailey	b. 1846	Worked in fruit box factory	Sister of box factory workers Lucinda and Harriet Botting.		MC1865

NO.	NAME	DATES	OCCUPATION	NOTES	MAKER'S MARKS	SOURCES
22	Bowditch, William. W.	b. ca. 1826	Cooper	Employed by C. Wilder & Son.		FC1880; HRBD
23	Burr, Charles	b. 1759	Cooper	Moved to Brewer, Maine, 1785–90; made sugar boxes.		Todd
24	Burr, David	1745–1814	Cooper	Son of cooper Elisha Burr; resided in the paternal homestead on Leavitt Street.		Lincoln
25	Burr, David	1778–1852	Cooper	Son of cooper David Burr; resided on Leavitt Street.	D. BURR (stamp)	Lincoln
26	Burr, Elisha	1703–78	Cooper	Resided on Leavitt Street.		Lincoln
27	Burr, Elisha	1776–1863	Cooper	Son of cooper David Burr; resided on Leavitt Street.	E. BURR (stamp)	Lincoln; see ch. 1, cat. 5
28	Burr, John	1729–90	Cooper	Son of cooper Jonathan Burr; resided on Leavitt Street.		Lincoln
29	Burr, John	1791–1857	Boxmaker	Son of cooper Theophilus Burr; resided on Leavitt Street.	J. BURR (stamp)	Lincoln; MR1852–53
30	Burr, Jonathan	1698–1762	Cooper	Resided at the junction of Turkey Hill Road and Leavitt Street.		Lincoln; Todd
31	Burr, Matthew	1773–1847	Cooper	Son of cooper Thomas Burr and brother of cooper Thomas Burr Jr.; worked for several years in Bath, Maine, with his brother.		Lincoln; Marble
32	Burr, Pyam Cushing	1819–92	Boxmaker	Son of cooper Theophilus Burr; resided on Leavitt Street.	P. C. BURR (stamp)	Lincoln; MAVR
33	Burr, Samuel	1738–97	Cooper	Son of cooper Jonathan Burr and brother of coopers John and Thomas Burr; he willed his cooper's tools to his nephew Theophilus Burr; made canteens during the Revolution.		Lincoln; Marble
34	Burr, Simon, Jr.	1655–1722	Cooper	Moved to Rehoboth, Mass., about 1708.		Todd
35	Burr, Theophilus	1761–1835	Cooper	Resided on Leavitt Street, at the corner of Spring Street.		Lincoln
36	Burr, Thomas	1735–1812	Cooper	Son of cooper Jonathan Burr and father of cooper Thomas Burr Jr.; resided on Main Street in Hingham Centre.		Lincoln
37	Burr, Thomas, Jr.	1759–94	Cooper	Son of cooper Thomas Burr and brother of cooper Matthew Burr; moved to Bath, Maine, where he died.		Lincoln; Marble
38	Burrell, Jotham Gay	b. 1807	Cooper	Resided on Water Street.		FC1880; Lincoln
39	Burtenshaw, Henry	1802–67	Barrel cooper	Native of England; resided on the north corner of Ship Street.		Lincoln; Todd
40	C. & A. Wilder	ca. 1850–ca. 1870	Woodenware manufacturers	Partnership of brothers Crocker Wilder and Alden Wilder; operated a large bucket factory at Cushing Pond; made buckets, churns, fish kits, etc.	C. & A. WILDER / SO. HINGHAM / MASS. (stamp)	See chap. 1, cat. 8

NO.	NAME	DATES	OCCUPATION	NOTES	MAKER'S MARKS	SOURCES
41	C. Wilder & Sons	1830–ca. 1850	Woodenware manufacturers	Crocker Wilder operated a large bucket factory at Friend Street near Main; the factory moved to Cushing Pond in 1845 and shortly thereafter he entered into a partnership with his brother, Alden Wilder; Crocker Wilder's son did not join him as a partner in the enterprise until about 1870; made buckets, churns, fish kits, etc.	C. WILDER & SONS / HINGHAM (stamp)	See chap. 1, cat. 8
42	C. Wilder & Son	ca. 1870–91	Woodenware manufacturers	Partnership of Crocker Wilder and his son, William C. Wilder; operated a large bucket factory at Cushing Pond; made buckets, churns, fish kits, etc.	C. WILDER & SON / SO. HINGHAM, MASS. (stamp)	See chap. 1, cat. 8
43	Canterbury, Cornelius	d. 1683	Cooper			Lincoln
44	Carter, James D.	b. ca. 1834	Cooper	Resided on Hull Street; later moved to Scituate, Mass.		PLY; Lincoln; MC1865
45	Churchill, Reuben French	b. 1846	Boxmaker			Lincoln; MC1865
46	Clapp, Abijah	act. 1760–72	Cooper			PCR 8:23
47	Clapp, Michael T.	1795–1875	Cooper	Worked in bucket factory; resided on High Street.		Lincoln; FC1850–70
48	Clark, Gertrude Allison	b. 1876	Boxmaker			FC1900
49	Corthell, Elijah C.	b. 1813	Boxmaker	In partnership of E. C. Corthell & Sons.		Lincoln; FC1860
50	Corthell, Elijah Loring	1836–1917	Boxmaker, toy manufacturer	In partnership of E. C. Corthell & Sons; proprietor of the Hingham Toy Manufactory.		See cat. 15
51	Corthell, Gustavus Perry	b. 1845	Cooper	Son of boxmaker Nelson Corthell; resided on Hersey Street.		PLY; Lincoln
52	Corthell, Isaac	1807–78	Cooper	Son-in-law of cooper Samuel F. Hersey; resided on Hersey Street and later moved to Nashua and Derry, N.H.		FC1850; Lincoln
53	Corthell, Isaac	b. ca. 1833	Cooper	Son of cooper Isaac Corthell.		FC1850
54	Corthell, Nelson	1815–1900	Boxmaker	Son of cooper Robert Corthell Jr.; resided on Hersey Street.	NC (stamp) N. CORTHELL / Hingham (stamp) N. CORTHELL (stamp)	Lincoln; MAVR; see cat. 10
55	Corthell, Reuben Hersey	b. 1837	Cooper	In partnership of E. C. Corthell & Sons; foreman of Hingham Wooden Ware Co.; resided on Hersey Street.		PLY; Lincoln
56	Corthell, Robert	1756–1833	Cooper	Resided on Main Street in South Hingham.		Lincoln
57	Corthell, Robert, Jr.	1790–1862	Cooper	Son of cooper Robert Corthell; resided on Main Street near Tower's Bridge.		Lincoln
58	Corthell, Robert, III	b. 1824	Cooper	Son of cooper Robert Corthell Jr.; later resided in South Weymouth, Mass.		FC1850; Lincoln
59	Curtis, Enoch	ca. 1785–1825	Cooper	Resided on North Street near the harbor.		Lincoln
60	Curtis, Job	act. 1770	Cooper	Later moved to Braintree, Mass.		Lincoln

NO.	NAME	DATES	OCCUPATION	NOTES	MAKER'S MARKS	SOURCES
61	Curtis, Warran	b. ca. 1804	Cooper			FC1850
62	Cushing, Ensign	1796–1864	Cooper	Resided on Pleasant Street.		MC1855; Lincoln
63	Cushing, Loring, II	b. ca. 1803	Cooper			FC1850
64	Cushing, Oliver	1804–57	Cooper, painter	Father of pail painter Oliver Cushing; resided on Main Street in South Hingham.		FC 1850; Lincoln
65	Cushing, Oliver Wendell	1829–56	Pail painter	Son of cooper Oliver Cushing.		MC1855; Lincoln
66	Cushing, Pyam, Jr.	1814–68	Bucketmaker	Brother of cooper Robert Cushing; resided on Main Street in South Hingham.		PLY; Lincoln
67	Cushing, Robert	1819–55	Woodenware maker	Brother of cooper Pyam Cushing Jr.; resided on Main Street in South Hingham.		FC1850; Lincoln
68	Cushing, Stephen	b. ca. 1807	Cooper	Resided with cooper Ensign Cushing.		MC1855
69	Cushing, Stephen	b. ca. 1828	Boxmaker	Worked in fruit box factory.		FC1870
70	Cushing, Theophilus	1802–81	Sawyer	In 1830 he advertised that he cut wood for bucketmakers.		Lincoln; *Hingham Gazette* 2/5/1830
71	Cushing, Thomas	1781–1815	Cooper	Resided on South Pleasant Street.		Lincoln
72	Cushing, Thomas	1808–58	Cooper			Lincoln; MC1855
73	Cushing, William Gray	b. 1811	Cooper	Worked in bucket factory; father of cooper William G. Cushing; resided on Main Street in South Hingham.		PLY; MC1865; Lincoln; FC1840–70
74	Cushing, William Gray	b. ca. 1836	Worked in bucket factory	Son of cooper William G. Cushing.		FC1870; Lincoln
75	Damon, Fred Baker	b. 1854	Cooper	Worked in bucket factory.		Lincoln; FC1870
76	Danforth, Charles L.	act. 1885	Cooper	Resided on Cushing Street.		HRBD
77	Dew, Carlton	b. ca. 1831	Cooper	Resided with cooper Ambrose Bailey.		FC1850
78	Dill, Lemuel	1788–1864	Cooper	Resided on Green Street.		Lincoln
79	Dunbar, Jacob	1799–58	Cooper	Son of cooper Robert Dunbar; resided in South Hingham.		Lincoln
80	Dunbar, Robert	1767–1815	Cooper	Resided on Scotland Street.		Lincoln
81	Dunbar, Samuel	1784–1858	Cooper	Resided on Hersey Street.		MC1855; Lincoln
82	Dunbar, Seth	1802–68	Woodenware maker	Resided on Main Street in South Hingham.		FC1850; Lincoln
83	E. C. Corthell & Sons	act. 1861	Woodenware manufacturers	Partnership of boxmaker Elijah C. Corthell and his sons Elijah Loring and Reuben Hersey Corthell.		HAHS; Lincoln
84	Fearing, Ebenezer Cushing	1812–81	Bucketmaker	Worked in bucket factory; son of cooper Israel Fearing; resided on Main Street in South Hingham.		PLY; Lincoln; MC1865; FC1870
85	Fearing, George William	1837–1918	Cooper, toymaker	Son of cooper Perez Fearing; worked in Wilder's bucket factory; worked at William S. Tower's toy factory; operated toymaking shop on Water Street.	FEARING / HINGHAM (stamp) GEORGE W. FEARING, / HINGHAM, MASS. (printed paper label)	FC1860–70; Lincoln; see cat. 16

NO.	NAME	DATES	OCCUPATION	NOTES	MAKER'S MARKS	SOURCES
86	Fearing, Israel	1644–93	Cooper	Resided in the paternal homestead on North Street.		Lincoln
87	Fearing, Israel	1782–1857	Cooper	Father of coopers Ebenezer and Perez Fearing; resided at Hingham Centre.		Lincoln
88	Fearing, John Cushing	b. 1825	Bucket factory worker			Lincoln; MC1865
89	Fearing, Lincoln	1792–1873	Woodenware agent	Worked for Hingham Wooden Ware Co.; resided on Main Street.		PLY; Lincoln
90	Fearing, Perez Lincoln	1811–99	Cooper	Son of cooper Israel Fearing; later operated a bucket factory in South Weymouth; resided on East Street.	P. L. FEARING & CO. / S. WEYMOUTH (stamp)	Lincoln; MAVR
91	Fearing, William	b. 1818	Cooper	Worked in bucket factory; resided on Main Street in South Hingham.		Lincoln; FC1870
92	Ford, James	b. ca. 1855	Boxmaker	Worked in fruit box factory.		FC1870
93	French, George	1799–1879	Cooper	Died in South Scituate, Mass.		Lincoln; PCR 15:25
94	French, Levi	1800–73	Cooper	Resided on North Street in West Hingham; died in Boston.		Lincoln; MAVR
95	Fuller, Issachar	1794–1866	Cooper	Resided on Water Street.		Lincoln
96	Gannett, Benjamin	act. 1748	Cooper			BLP
97	Gardner, Charles	1778–1838	Cooper	Father of cooper Leonard Gardner; resided on Gardner Street.		Lincoln
98	Gardner, George	1845–71	Boxmaker	Resided on Hersey Street.		PLY; Lincoln
99	Gardner, Harry Winfred	b. 1864	Bucketmaker			Lincoln; FC1900
100	Gardner, Hosea	ca. 1715–44	Cooper	Resided at Liberty Plain in South Hingham.		Lincoln
101	Gardner, Jacob	1797–1883	Cooper	Brother of cooper Stephen Gardner; resided at Liberty Plain in South Hingham.		MC1865; Lincoln
102	Gardner, Josiah Quincy	b. 1835	Cooper			Lincoln; MC1855
103	Gardner, Leonard	1809–82	Cooper	Son of cooper Charles Gardner; operated bucket and box factory in South Weymouth; operated bucket and box factory on Gardner Street; resided at Liberty Plain in South Hingham.	L. GARDNER / SOUTH WEYMOUTH (stamp) L. GARDNER / S. HINGHAM (stamp) L. GARDNER / S. HINGHAM / MASS. (stamp)	See cat. 9
104	Gardner, Marcus Morton	b. ca. 1839	Cooper			Lincoln; MC1855
105	Gardner, Stephen	1776–1860	Cooper	Brother of cooper Jacob Gardner; resided at Liberty Plain in South Hingham.		Lincoln; MC1855
106	Gardner, Warren	1791–1871	Cooper			Lincoln
107	Gilfeather, James	b. ca. 1876	Boxmaker	Worked in box factory.		FC1910
108	Gold, Edward	d. ca. 1685	Pailmaker			Lincoln; see ch. 1
109	Gould, Joseph	b. ca. 1801	Cooper	Father of cooper Timothy Gould.		MC1855
110	Gould, Timothy	b. ca. 1839	Cooper	Son of cooper Joseph Gould		MC1855

NO.	NAME	DATES	OCCUPATION	NOTES	MAKER'S MARKS	SOURCES
111	Groce, Edmund	1669–ca. 1727	Cooper	Father of cooper Obadiah Groce; resided at Liberty Plain in South Hingham.		Lincoln
112	Groce, Obadiah	1708–49	Cooper	Son of cooper Edmund Groce; resided at Liberty Plain in South Hingham.		Lincoln
113	Hall, John	act. 1730	Cooper	Nephew of cooper Preserved Hall; resided in Hingham Centre.		Lincoln
114	Hall, Preserved	ca. 1667–1740	Cooper	Resided in Hingham Centre.		Lincoln
115	Haskell, Joseph Cushing	b. 1854	Cooper	Worked in woodenware factory.		Lincoln; FC1900
116	Hatch, Warren	1816–73	Boxmaker	Resided at corner of Pleasant Street and Cedar Street.	W. HATCH (stamp)	PCPR 39:297, 43:470; Hofer; Lincoln
117	Hawes, George F., Jr.	act. 1885	Cooper	Resided on Cross Street.		HRBD
118	Hawes, Jacob	b. ca. 1799	Cooper			FC1850
119	Hersey & French	act. ca. 1820–60	Coopers	Unidentified partnership working in South Weymouth.	HERSEY & FRENCH / SO. WEYMOUTH (stamp)	HF
120	Hersey, Abijah	ca. 1738–1825	Cooper	Son of cooper Jeremiah Hersey; resided on South Street.		Lincoln
121	Hersey, Abijah	1769–1841	Boxmaker	Son of cooper Abijah Hersey; resided on Fort Hill Street.	AH (stamp)	Lincoln
122	Hersey, Albert Fiske	b. 1823	Cooper	Son of cooper Gridley Hersey.		FC1850; Lincoln
123	Hersey, Alfred C.	b. 1804	Bucketmaker			Lincoln; TRMM
124	Hersey, Bela	1776–1829	Boxmaker	Brother of boxmaker Israel and cooper Nathaniel Hersey; resided at the corner of Main Street and Hersey Street.		Lincoln
125	Hersey, Bela French	1817–79	Boxmaker	Son of cooper Cotton Hersey.	B.F.H (stamp) B. F. Hersey (stamp)	MC1855; Lincoln
126	Hersey, Bela Stockbridge	1815–61	Cooper	Son of boxmaker Bela Hersey.		MC1855; Lincoln
127	Hersey, Caleb Strong	1813–94	Cooper, toymaker	Son of cooper Gridley Hersey and father of cooper Francis Hersey; resided on Hersey Street.	C. S. HERSEY (stamp)	Lincoln; MAVR; see ch. 1
128	Hersey, Cotton	1792–1881	Cooper, toymaker	Son of cooper and toymaker Reuben Hersey, brother of coopers Reuben Jr., Samuel French, and Isaiah Hersey, and father of boxmaker Bela French Hersey; resided on Hersey Street.	C. Hersey (serif stamp) C. Hersey (sans serif stamp) C.H (stamp) CH (stamp) Cotton Hersey (stamp) Cotton Hersey / Hingham (stamp)	See ch. 2, cat. 11
129	Hersey, Cushing	1808–70	Cooper	Son of cooper Nathaniel Hersey; resided on Hersey Street.		Lincoln
130	Hersey, Daniel	1682–1766	Cooper	Founder of Hersey Street family of coopers; brother of cooper Jeremiah Hersey and father of coopers Jonathan and Isaiah Hersey; resided on Hersey Street.		Lincoln; see ch. 3

NO.	NAME	DATES	OCCUPATION	NOTES	MAKER'S MARKS	SOURCES
131	Hersey, Daniel	1754–94	Cooper	Son of cooper Jonathan Hersey (ca. 1713–60); resided on Hersey Street.		Lincoln
132	Hersey, David	b. 1797	Cooper	Son of cooper Peter Hersey and brother of coopers Peter, Gridley, and Quincy Hersey.	D. HERSEY (stamp probably used by him)	Lincoln; PCLR 133:335; see ch. 1
133	Hersey, Edmund	1823–1909	Cooper, box manufacturer	Son of cooper Samuel F. Hersey and brother of cooper and toymaker Samuel Hersey; operated a box factory on Hersey Street and later Leverett, Mass.; sold factory in 1871 to Seymour & Cain; resided on Hersey Street.	EH (stamp)	Lincoln; see chs. 1 and 3
134	Hersey, Edward Lewis	1822–53	Cooper, toy-maker	Son of cooper Joseph Hersey; resided on Hersey Street.	E.L.H (stamp) EDWARD L. HERSEY. / H / ING / HAM (stamp)	PBCD; Lincoln; see cat. 13
135	Hersey, Elijah	1749–1850	Cooper	Later moved to Leicester, Mass., and died in Spencer, Mass.		Lincoln; MAVR
136	Hersey, Francis	1835–98	Cooper	Son of cooper Caleb S. Hersey; resided on Hersey Street.		PLY; Lincoln; MAVR; see app. A
137	Hersey, Franklin	1820–59	Cooper	Son of cooper Joseph Hersey; resided on Hersey Street.	FRANKLIN HERSEY / HINGHAM (stamp)	Lincoln
138	Hersey, Gilbert	1747–1804	Cooper	Son of cooper Jeremiah Hersey; resided on South Street.		Lincoln; PCPR 40:98
139	Hersey, Gridley	1786–1827	Cooper	Son of cooper Peter Hersey, brother of coopers Peter, David, and Quincy Hersey, and father of cooper and toymaker Caleb S. Hersey and cooper Albert Hersey; resided on Hersey Street.	GH (stamp probably used by him)	Lincoln
140	Hersey, Horace	b. 1846	Cooper	Son of cooper Cushing Hersey and brother of cooper William H. Hersey; resided on Hersey Street.		FC1870; Lincoln
141	Hersey, Hosea	1800–1826	Cooper	Son of cooper Abijah Hersey (1769–1841).		Lincoln; PCPR 63:297
142	Hersey, Isaac	1777–1865	Boxmaker	Resided on South Street.		Lincoln; SOL
143	Hersey, Isaac Lincoln	b. 1825	Boxmaker	Resided on Hersey Street.		PLY; Lincoln
144	Hersey, Isaiah	ca. 1720–1803	Cooper	Son of cooper Daniel Hersey; resided on Hersey Street.		Lincoln
145	Hersey, Isaiah	1803–68	Cooper	Son of cooper Reuben Hersey, brother of coopers Reuben, Cotton, and Samuel French Hersey; father of cooper and toymaker Reuben Hersey; resided on Hersey Street.	IH (stamp) I. Hersey (stamp)	Lincoln; see ch. 3
146	Hersey, Israel	1744–1823	Cooper	Son of cooper Isaiah Hersey (ca. 1720–1803) and father of coopers Israel, Nathaniel, and Bela Hersey.		Lincoln; PCPR 57:457
147	Hersey, Israel	1769–1853	Boxmaker	Brother of coopers Nathaniel and Bela Hersey; resided in Hingham Centre.		Lincoln
148	Hersey, Jacob	1799–1871	Manufacturer of woodenware	Son of cooper Nathaniel Hersey; in partnership with Peter Hobart; resided on Main Street, nearly opposite Hobart Street.		Lincoln

NO.	NAME	DATES	OCCUPATION	NOTES	MAKER'S MARKS	SOURCES
149	Hersey, Jeremiah	1697–1790	Cooper	Brother of cooper Daniel Hersey and father of cooper Abijah Hersey (ca. 1738–1825); resided on South Street.		Lincoln
150	Hersey, Jonathan	ca. 1713–60	Cooper	Son of cooper Daniel Hersey; resided on Hersey Street.		Lincoln
151	Hersey, Jonathan	1742–1832	Cooper	Son of cooper Jonathan Hersey.		Lincoln
152	Hersey, Joseph	1781–1863	Cooper	Son of cooper Abijah Hersey (ca. 1738–1825); resided on Hersey Street.	J.H (stamp probably used by him)	Lincoln
153	Hersey, Joseph, Jr.	1806–65	Cooper	Son of cooper Joseph Hersey.	J.H.J. (stamp) J.H. Jr. (stamp)	MC1855; Lincoln
154	Hersey, Nathaniel	1774–1860	Cooper	Brother of boxmaker Israel Hersey and cooper Bela Hersey; resided on Main Street near Hobart Street.	NH (stamp possibly used by him)	Lincoln
155	Hersey, Nathaniel, Jr.	1797–1867	Cooper	Son of cooper Nathaniel Hersey; resided on Hersey Street.	NH (stamp possibly used by him)	Lincoln
156	Hersey, Peter	1757–1844	Cooper	Son of cooper Jonathan Hersey (ca. 1713–60) and father of coopers Peter Jr., Gridley, David, and Quincy Hersey; resided on Hersey Street.	PETER HERSEY (stamp) P. HERSEY (stamp)	Lincoln; see cat. 1
157	Hersey, Peter, Jr.	1820–59	Cooper	Son of cooper Peter Hersey and brother of coopers Gridley, David, and Quincy Hersey; resided on Hersey Street.		PCLR 123:145
158	Hersey, Quincy	1799–1874	Cooper, toy-maker	Son of cooper Peter Hersey and brother of coopers Peter, Gridley, and David Hersey; resided on Hersey Street; later operated a shop near the Torrent Engine House; moved to South Weymouth, Winchendon, and finally Methuen, Mass., where he died.	Q. HERSEY (stamp)	Lincoln
159	Hersey, Reuben	1757–1845	Cooper, toy-maker	Son of cooper Isaiah Hersey (ca. 1720–1803) and father of coopers Reuben, Cotton, Samuel French, and Isaiah Hersey; resided on Hersey Street.		Lincoln; see ch. 3
160	Hersey, Reuben, Jr.	1780–1844	Cooper	Son of cooper Reuben Hersey and brother of coopers Samuel French, Cotton, and Isaiah Hersey.		Lincoln; see ch. 3
161	Hersey, Reuben, III	1837–1914	Boxmaker, toymaker	Son of cooper Isaiah Hersey (1803–68); resided on Hersey Street.	RH (stamp)	PLY; Lincoln
162	Hersey, Samuel	1820–1909	Boxmaker; toymaker	Son of cooper Samuel F. Hersey and brother of cooper Edmund Hersey; resided on Hersey Street.	SH (stamp) S.H (stamp) Manufactured by / SAM'L HERSEY, / *Hersey Street,* / Hingham, Mass. (printed paper label) SAM'L HERSEY, / Wholesale and Retail / Toy Ware and Box / MANUFACTURER, / Hingham, Mass. (printed paper label) Manufactured by / SAMUEL HERSEY / *Hersey Street,* / Hingham, Mass. (printed paper label)	PLY; Lincoln; see ch. 2 and cat. 12

NO.	NAME	DATES	OCCUPATION	NOTES	MAKER'S MARKS	SOURCES
163	Hersey, Samuel French	1783–1876	Cooper	Son of cooper Reuben Hersey and brother of coopers Reuben, Cotton, and Isaiah Hersey; resided on Hersey Street.		Lincoln; see ch. 3
164	Hersey, William	1776–1867	Cooper	Son of cooper Abijah Hersey (ca. 1738–1825); resided on Hersey Street.		Lincoln
165	Hersey, William A.	b. ca. 1826	Cooper			MS1850
166	Hersey, William Hollis	b. 1840	Cooper	Son of cooper Cushing Hersey and brother of cooper Horace Hersey; resided on Hersey Street.		Lincoln
167	Hersey, Zadock	1753–1850	Cooper	Son of cooper Isaiah Hersey (ca. 1720–1803) and brother of coopers Israel (1744–1823) and Reuben Hersey (1757–1845); resided on Hersey Street before moving to Maine.		Lincoln
168	Hett, Thomas	ca. 1612–68	Cooper	Active in Hingham, Rehoboth, and Hull, Mass., 1637–47.		LMN; GMB
169	Higgins, Ezra	b. 1808	Boxmaker	Resided on Irving Street.		FC1860; Lincoln
170	Hilby, Charles	b. ca. 1824	Cooper	Originally from Maine.		FC1850
171	Hill, Christopher B.	b. 1825		Worked in bucket factory; resided on High Street.		FC1870; Lincoln
172	Hingham Woodenware Company	1868–69	Manufacturer of woodenware	Owned by Fearing Burr; manufacturers of fruit boxes, salt boxes, boxes for axle grease, water pails of every description, miniature pails, croquet tools, parlor ten pins, etc.; sometimes identified as the Hingham Wooden Ware Company; toy manufacturer William S. Tower purchased their machinery in 1869.	HINGHAM / WOODENWARE CO. / MASS. (stamp)	HAHS1868; see ch. 2, cat. 14
173	Hobart, Bela	1783–1840	Cooper	Son of cooper Edmund Hobart; resided on Main Street near Hobart Street.		Lincoln
174	Hobart, Bela	1815–61	Cooper	Son of cooper Bela Hobart; resided on Main Street near Tower's Bridge.		Lincoln; MAVR
175	Hobart, Ebed	1795–1863	Cooper	Son of cooper Edmund Hobart; resided on Main Street near Hobart Street.		Lincoln
176	Hobart, Edmund	1755–1847	Cooper	Brother of cooper Hawkes Hobart; resided on Main Street near Hobart Street.		Lincoln
177	Hobart, Hawkes	1753–1831	Cooper	Resided on the paternal homestead at the foot of Great Hill.		Lincoln
178	Hobart, Hawkes	1784–1836	Cooper	Son of cooper Hawkes Hobart.		
179	Hobart, Nehemiah	1785–1867	Cooper	Later moved to Abington, Mass.		Lincoln; MAVR
180	Hobart, Peter	1799–1882	Manufacturer of woodenware	Son of cooper Edmund Hobart; in partnership with Jacob Hersey; resided on Main Street, nearly opposite Hobart Street.		Lincoln
181	Hobart, Thomas	1817–55	Cooper	Son of cooper Bela Hobart (1783–1840); resided on Main Street near Tower's Bridge.	T. HOBART HINGHAM (stamp)	Lincoln

NO.	NAME	DATES	OCCUPATION	NOTES	MAKER'S MARKS	SOURCES
182	Hobart, William	b. 1826	Worked in bucket factory	Son of cooper Ebed Hobart; resided on Main Street opposite Hobart Street.		MC1855; Lincoln
183	Hopkins, George W.	b. ca. 1834	Cooper	Originally from Maine; resided on Ship Street.		PLY; MC1865
184	Hough, Ellen A.	b. 1851	Worked in fruit box factory	Resided on Main Street, nearly opposite Hersey Street.		FC1870; Lincoln
185	Howard, Thomas	d. 1828	Cooper	Resided in Hingham Centre.		Lincoln
186	Howes, George Franklin	b. 1855	Cooper; toymaker	Married box factory worker Rachel E. Robinson in 1873; resided on Cross Street.		Lincoln
187	Hughs, John	act. 1665–72	Cooper			Lincoln
188	Humphrey, Abner	1804–93	Cooper	Brother of coopers Leavitt, Moses, and Cyrus Humphrey; resided on Ship Street.		Lincoln; MAVR
189	Humphrey, Abner	act. 1885	Cooper	Resided on Ship Street.		HRBD
190	Humphrey, Cyrus	b. 1813	Manufacturer of barrels and fish kits	Brother of coopers Abner, Leavitt, and Moses Humphrey; in partnership of M. L. & C. Humphrey with his brothers.		Lincoln
191	Humphrey, Leavitt	1802–83	Cooper	Brother of coopers Abner, Moses, and Cyrus Humphrey; in partnership of M. L. & C. Humphrey with his brothers; resided on Cottage Street.		Lincoln
192	Humphrey, Moses	b. 1807	Manufacturer of woodenware; barrel cooper	Brother of coopers Leavitt, Abner, and Cyrus Humphrey; operated M. Humphrey & Co. and M. L. & C. Humphrey; resided on Ship Street; later moved to Concord, New Hampshire.	M. HUMPHREY & CO / HINGHAM (stamp)	Lincoln
193	Humphrey, Ned	1790–1877	Boxmaker	Resided on South Street near Hersey Street.		Lincoln
194	Humphrey, Noah	ca. 1735–1803	Cooper	Resided on Leavitt Street.		Lincoln
195	Hutter, John	b. ca. 1825	Cooper	Native of Switzerland.		FC1860
196	Ignatio, Manuel	b. ca. 1805	Cooper	Native of Portugal.		FC1850
197	Jacob, David Fearing	1796–1879	Cooper	Brother of cooper Hawkes Jacob; resided on Pleasant Street.		PLY; Lincoln
198	Jacob, David	b. ca. 1826	Cooper	Son of cooper David F. Jacob.		MC1855; Lincoln
199	Jacob, Hawkes	1798–1855	Cooper	Brother of cooper David F. Jacob.		MC1855; Lincoln
200	Jacob, Joshua	1817–79	Cooper	Resided on the paternal homestead on Main Street in South Hingham.		FC1850; Lincoln
201	Joy, Jedediah	1703/4–98	Cooper	Resided in West Hingham.		BLP
202	Joy, Samuel	1706–92	Cooper	Resided in Hingham Centre.		Lincoln
203	Kelly, Thomas	b. 1847	Cooper	Native of Ireland.		FC1900
204	Kilby, Theophilus	b. ca. 1841	Worked in bucket factory			FC1870
205	King, Moses H.	b. ca. 1821	Cooper			MC1855

NO.	NAME	DATES	OCCUPATION	NOTES	MAKER'S MARKS	SOURCES
206	King, Thomas	1802–82	Cooper	Resided on School Street and Back Street.	T. KING (stamp)	PLY; FC1860
207	Kittredge, Carrie M.	b. ca. 1842	Worked in fruit box factory	Her sister Ellen and brother Frank also worked in the factory.		FC1870
208	Kittredge, Ellen M.	b. ca. 1855	Worked in fruit box factory	Her sister Carrie and brother Frank also worked in the factory.		FC1870
209	Kittredge, Frank M.	b. ca. 1859	Worked in fruit box factory	His sisters Carrie and Ellen also worked in the factory.		FC1870
210	Lane, Daniel	1724–1816	Cooper			Lincoln
211	Lane, David	1758–1822	Cooper	Son of cooper Jonathan Lane Jr. and brother of cooper Jared Lane.		Lincoln; Lane
212	Lane, Jared	1752–1839	Cooper	Son of cooper Jonathan Lane Jr. and brother of cooper David Lane.	JARED LANE (stamp)	Lincoln; Lane; see cat. 2
213	Lane, Jonathan, Jr.	1719–1801	Cooper	Father of coopers Jared and David Lane; resided on Main Street, nearly opposite Water Street.		Lincoln; see ch. 1
214	Lane, Joshua Leavitt	1812–81	Cooper	Son of cooper Leavitt Lane (1767–1836).		MC1855; Lincoln
215	Lane, Leavitt	1767–1836	Cooper	Resided on Spring Street.		Lincoln
216	Lane, Leavitt	1806–91	Boxmaker	Son of cooper Leavitt Lane; resided on the paternal homestead on Spring Street.	L. LANE (stamp)	Lincoln; MAVR
217	Lane, Parker Emery	1840–69	Cooper	Later moved to Weymouth, Mass.		Lincoln; LLI
218	Lane, Rufus, Jr.	b. 1815	Cooper	Resided on the paternal homestead on South Street.		MR1852; Lincoln
219	Lasell, Jonathan	1707/8–86	Cooper	Resided on Pleasant Street.		Lincoln; BLP
220	Leavitt, Abraham	1686–1767	Cooper	Brother of coopers John and Israel Leavitt; resided on Pleasant Street.		Lincoln
221	Leavitt, Abraham	1714–ca. 1753	Cooper	Son of cooper Abraham Leavitt.		Lincoln
222	Leavitt, Elijah	1768–1813	Cooper	Resided on Main Street at Hingham Centre.		Lincoln
223	Leavitt, Ezra, Jr.	d. 1850	Boxmaker	Listed as boxmaker in the *Hingham Journal*, August 27, 1850.		
224	Leavitt, Israel	1680–1757	Cooper	Brother of coopers John and Abraham Leavitt; resided on North Street near the harbor.		Lincoln
225	Leavitt, Jerom, Jr.	b. 1805	Cooper	Later moved west.		MC1855; Lincoln
226	Leavitt, John	1678–1749	Cooper	Brother of coopers Israel and Abraham Leavitt; resided on Leavitt Street.		Lincoln
227	Leavitt, John	ca. 1710–97	Cooper	Son of cooper John Leavitt; resided on the ancestral homestead on Leavitt Street.		Lincoln
228	Leavitt, Josiah	1653–1739	Cooper			Lincoln
229	Leavitt, Josiah	1679–1717	Cooper	Son of cooper Josiah Leavitt; resided on Leavitt Street.		Lincoln
230	Leavitt, Nehemiah	1721–71?	Cooper	Resided on King Street.		Lincoln

NO.	NAME	DATES	OCCUPATION	NOTES	MAKER'S MARKS	SOURCES
231	Leavitt, Samuel	1754–1823	Cooper	Resided on Main Street at Hingham Centre.		Lincoln
232	Lewis, Elijah	1762–1860	Cooper	Resided on Pleasant Street.		Lincoln; see cat. 3
233	Lewis, James Stockbridge	1798–1889	Boxmaker	Resided on Pleasant Street.		PLY; Lincoln; MAVR
234	Lewis, John Briggs	b. 1830	Boxmaker	Son of boxmaker James S. Lewis; resided on Pleasant Street.		PLY; Lincoln
235	Lincoln, Alexander	b. ca. 1808	Boxmaker; woodenware dealer			MC1865
236	Lincoln, Charles	1797–1881	Cooper	Resided on Lincoln Street.		Lincoln
237	Lincoln, Daniel	1712–47	Cooper	Resided on Fort Hill Street.		Lincoln; BLP
238	Lincoln, Elisha	b. ca. 1697	Cooper	Later moved to Weymouth and then Abington, Mass.		Lincoln
239	Lincoln, Enos	act. 1810	Cooper			PCPR 39:264; Hofer
240	Lincoln, Enos	1782–1840	Cooper	Resided on Fort Hill Street.		Lincoln
241	Lincoln, John	ca. 1667–ca. 1726	Cooper	Resided on West Street.		Lincoln
242	Lincoln, Stephen	1726–1816	Cooper	Resided on Lincoln Street and later on North Street at the corner of Burton's Lane.		Lincoln
243	Lincoln, Thomas	arrived 1635, d. 1691	Cooper	Resided on North Street in West Hingham.		Lincoln
244	Lincoln, Lott W.	b. ca. 1790	Cooper	Previously resided in Chelsea, Mass.		FC1860
245	Loring, Alfred Hersey	1825–60	Cooper			MC1855; Lincoln
246	Loring, Fearing	1787–1857	Boxmaker			Lincoln; MGLM
247	Loring, John	1806–76	Cooper		J. Loring (stamp possibly used by him)	FC1860, Lincoln
248	Lovell, Enoch	act. 1807	Cooper			PCLR 105:143
249	Low, Jonathan	act. 1702–33	Cooper			PCR 5:9, 128
250	Lynch, John	b. ca. 1840	Cooper	Native of Ireland.		FC1860
251	M. Humphrey & Co.	act. 1841	Manufacturer of woodenware	Operated by cooper Moses Humphrey.	M. HUMPHREY & CO / HINGHAM (stamp)	Lincoln
252	M. L. & C. Humphrey		Manufacturers of fish barrels and fish kits	Partnership of brothers Moses, Leavitt, and Cyrus Humphrey.		Lincoln
253	Malbon, Micajah	act. 1847	Pail button manufacturer	On April 2, 1847, Malbon advertised in the *Hingham Journal* pail buttons for sale made at the steam factory of Benjamin Parker.		
254	Mansfield, Joseph	1693–1756	Boxmaker	Resided on Leavitt Street.		Lincoln
255	Mansfield, Joseph	1728–1806	Cooper	Son of boxmaker Joseph Mansfield; resided on Leavitt Street.		Lincoln
256	Mayo, John	b. ca. 1833	Cooper			MC1855
257	McGlasky, Dennis	b. ca. 1829	Cooper	Native of Ireland.		FC1850

NO.	NAME	DATES	OCCUPATION	NOTES	MAKER'S MARKS	SOURCES
258	Moar, Isaac	act. 1831–34	Cooper			PCR 14:22
259	Orcutt, Thomas, Jr.	b. 1707	Cooper			Lincoln; PCR 5:58
260	Palmer, Benjamin Sanborn	1792–1874	Cooper	Brother of cooper Ebenezer Palmer; originally from New Hampshire; resided on North Street near the harbor.		Lincoln
261	Palmer, Ebenezer Sanborn	1798–1864	Cooper	Brother of cooper Benjamin Palmer; resided on Green Street.		Lincoln
262	Palmer, Edwin	b. ca. 1798	Cooper			MC1855
263	Penny, John	d. 1837	Cooper	Resided on Lincoln Street.		Lincoln
264	Pugsley, Sewall	1831–63	Cooper	Resided in South Hingham.		MC1855; Lincoln
265	Ray, Caleb	1792–1859	Barrel cooper	Resided on North Street near the harbor.		Lincoln
266	Ray, Charles Morton	ca. 1804–71	Cooper	Resided on Summer Street.		PLY; FC1860
267	Reed, George Reuben	b. 1839	Cooper	Later moved to Boston.		MC1855; Lincoln
268	Reed, Mary J.	b. ca. 1848	Boxmaker			MC1865
269	Reed, Reuben Mackenzie	1809–78	Cooper	Resided in Hingham Centre.		Lincoln; MAVR
270	Ripley, Abraham	1661–97	Cooper	Resided on Main Street in Hingham Centre.		Lincoln
271	Ripley, Ebed	1793–1875	Boxmaker	Son of cooper Nehemiah Ripley and brother of coopers Nehemiah Jr. and John Ripley; resided on the ancestral homestead on Main Street in Hingham Centre.		Lincoln
272	Ripley, John	1787–1852	Cooper	Son of cooper Nehemiah Ripley and brother of coopers Nehemiah Jr. and Ebed Ripley; resided on Pleasant Street.	J.R (stamp probably used by him)	Lincoln; see cat. 5
273	Ripley, Nehemiah	1755–1829	Cooper	Father of coopers Nehemiah, John, and Ebed Ripley; resided on the paternal homestead on Main Street in Hingham Centre.	N.R (stamp possibly used by him)	Lincoln; PCLR 151:50
274	Ripley, Nehemiah, Jr.	1780–1863	Cooper	Son of cooper Nehemiah Ripley and brother of coopers John and Ebed Ripley. Resided at the paternal homestead on Main Street in Hingham Centre.	N.R (stamp possibly used by him)	Lincoln; PCLR 98:123; see cat. 5
275	Ripley, Peter	1668–1742	Cooper	Resided on the paternal homestead on Main Street in Hingham Centre.		Lincoln
276	Ripley, Peter	1695–1765	Cooper	Son of cooper Peter Ripley; resided on the ancestral homestead on Main Street in Hingham Centre.		Lincoln
277	Robinson, Joseph Watson	b. 1858	Worked in woodenware factory	His sister Rachel E. Robinson worked in the fruit box factory.		FC1880; Lincoln
278	Robinson, Rachel E.	b. 1846	Worked in fruit box factory	Her brother Joseph W. Robinson worked in a woodenware factory; married cooper George F. Howes in 1873.		FC1870; Lincoln

NO.	NAME	DATES	OCCUPATION	NOTES	MAKER'S MARKS	SOURCES
279	Seymour & Cain	act. 1871–72	Box manufacturers	Partnership of Charles W. S. Seymour and Cain; located on Hersey Street.		Lincoln
280	Seymour, Charles Winfield Scott	b. 1839	Woodenware manufacturer; superintendent of box factory	Resided on Main Street near Water Street.		PLY; Lincoln
281	Shea, Thomas	b. ca. 1852	Worked in bucket factory			FC1870
282	Shute, Daniel	b. 1817	Cooper	Resided on Main Street in South Hingham.		MC1855; Lincoln
283	Siders, Henry	b. 1816	Cooper	Resided on Main Street near the old meeting house.		Lincoln; PCR 15:7
284	Smith, John	b. 1653	Cooper			Lincoln
285	Smith, Sarah E.	b. ca. 1853	Worked in fruit box factory			FC1870
286	Souther, Charles	1794–1879	Cooper	Resided in Hingham Centre.		FC1860; Lincoln
287	Souther, John Edgar	b. 1853	Worked in fruit box factory	Son of cooper John S. Souther; resided with his father in 1870.		FC1870; Lincoln
288	Souther, John Sprague	b. 1819	Cooper	Resided on Leavitt Street.		PLY; Lincoln
289	Souther, Mary Frances	b. 1878	Boxmaker	Resided on Hersey Street.		RBDHC 1908; Lincoln
290	Sprague	act. ca. 1820–70	Boxmaker	Boxes are occasionally found with the stamp "SPRAGUE" used by an unidentified member of the Sprague family of coopers.	SPRAGUE (stamp)	
291	Sprague, A.	act. ca. 1820–70	Boxmaker	Boxes are occasionally found with the stamp "A. SPRAGUE" used by an unidentified member of the Sprague family of coopers. The same cooper probably used the stamp "A.S" found on other woodenware.	A.S (stamp) A. SPRAGUE (stamp)	
292	Sprague, Abner	1811–64	Boxmaker	Son of boxmaker Amos Sprague I; resided in Hingham Centre.		MC1855; Lincoln; MAVR
293	Sprague, Adna	1790–1860	Boxmaker	Resided on Middle Street.		Lincoln
294	Sprague, Amos, I	1747–1838	Boxmaker	Son of cooper Isaac Sprague; resided on Main Street in Hingham Centre.		Lincoln
295	Sprague, Amos, II	1774–1830	Boxmaker	Son of boxmaker Amos Sprague and father of cooper Ezra Sprague; resided on Main Street in Hingham Centre.		Lincoln
296	Sprague, Anthony Jacob	b. 1855	Woodenware manufacturer	Resided on Main Street in South Hingham; owned and operated bucket factory at Cushing Pond, formerly known as C. Wilder & Son, from 1891 to about 1895.	A. J. Sprague, / SO. HINGHAM, MASS. (stamp)	Lincoln; see cat. 8

NO.	NAME	DATES	OCCUPATION	NOTES	MAKER'S MARKS	SOURCES
297	Sprague, Bela	1804–78	Cooper	Resided on Union Street.	B.S (stamp) B. SPRAGUE (stamp) *HINGHAM.* (stamp) B. SPRAGUE / *HINGHAM.* (stamp)	FC1860; Lincoln; see cat. 7
298	Sprague, Ezra	b. 1818	Cooper	Son of boxmaker Amos Sprague II.		FC1850; Lincoln
299	Sprague, Henry	1805–85	Boxmaker	Resided in Hingham Centre.		FC1850; MC1865; Lincoln
300	Sprague, Isaac, I	1709–88	Cooper	Father of cooper Isaac Sprague II; resided on Main Street in Hingham Centre.		Lincoln
301	Sprague, Isaac, II	1743–1800	Cooper	Son of cooper Isaac Sprague I and father of cooper Isaac Sprague III; resided on Main Street in Hingham Centre.		Lincoln
302	Sprague, Isaac, III	1782–1826	Cooper	Son of cooper Isaac Sprague II and father of cooper Lincoln B. Sprague; resided on Pleasant Street.		Lincoln
303	Sprague, Israel	1781–1859	Cooper	Brother of coopers Reuben and Jacob Sprague.		FC1850; Lincoln
304	Sprague, Jacob	1783–1846	Cooper	Brother of coopers Israel and Reuben Sprague; resided on Main Street at Liberty Plain in South Hingham.		Lincoln; PCLR 120:181
305	Sprague, Lincoln Burr	b. 1816	Cooper	Son of cooper Isaac Sprague III.		FC1850; Lincoln
306	Sprague, Martin	1800–1867	Cooper			MC1855; Lincoln
307	Sprague, Moses	1776–1850	Cooper	Resided on Main Street in Hingham Centre.	M.S (stamp) M. Sprague (stamp)	Lincoln; see cat. 6
308	Sprague, Moses	1815–96	Cooper	Resided on Pond Street.		Lincoln; MAVR
309	Sprague, Peter, I	1773–1859	Cooper	Son of cooper Isaac Sprague II; resided on Pond Street.	P.S (stamp possibly used by him)	Lincoln
310	Sprague, Peter, II	1801–68	Cooper	Son of cooper Peter Sprague; resided on the paternal homestead on Pond Street.	P.S (stamp possibly used by him)	Lincoln
311	Sprague, Reuben	1785–1852	Cooper	Brother of coopers Israel and Jacob Sprague; resided on Main Street at Liberty Plain in South Hingham.		FC1850; Lincoln
312	Sprague, Samuel	1809–82	Cooper	Resided on North Street near Ship Street.		Lincoln
313	Sprague, Seth	1745–1831	Cooper	Resided on Main Street near Tower's Bridge.		Lincoln
314	Spruhn, James P.	b. ca. 1835	Cooper	Native of Nova Scotia.		FC1880
315	Stodder, Benjamin	d. 1851	Cooper	Later moved to Hubbardston, Mass.		MAVR
316	Stodder, Isaiah	ca. 1729–1827	Pailmaker	Resided on High Street.		Lincoln
317	Stodder, Jacob	ca. 1673–1756	Cooper	Father of cooper Nathaniel Stodder (1705–58).		Lincoln
318	Stodder, Jacob	1751–1810	Cooper	Son of cooper Nathaniel Stodder (1705–58); resided in West Hingham.		Lincoln

NO.	NAME	DATES	OCCUPATION	NOTES	MAKER'S MARKS	SOURCES
319	Stodder, Lucy Beal	b. 1851	Worked in fruit box factory			FC1870; Lincoln
320	Stodder, Nathaniel	1705–58	Cooper	Son of cooper Jacob Stodder (ca. 1673–1756) and father of cooper Nathaniel Stodder; resided on Main Street near Hobart Street.		Lincoln
321	Stodder, Nathaniel	1727–1803	Cooper	Son of cooper Nathaniel Stodder; resided on North Street near Marsh's Bridge.		Lincoln
322	Stowell, Benjamin	1676–1730	Cooper	Resided on Fort Hill Street; cooper's tools are listed in an inventory of his estate.		Lincoln; SCPR
323	Stowell, Israel	1753–1844	Cooper	Resided on West Street.		Lincoln
324	Stowell, Jared	1808–80	Cooper	Resided on Ship Street.		Lincoln
325	Stowell, John	1687–1767	Cooper	Resided on Fort Hill Street; cooper's tools and parts for buckets and pails are listed in an inventory of his estate.		Lincoln; SCPR
326	Stowell, William	1694–1771	Cooper	Resided on Fort Hill Street; cooper's tools are listed in an inventory of his estate.		Lincoln; SCPR
327	Streeter, Alvin	act. 1900–1902	Cooper	Operated bucket factory formerly owned by A. J. Sprague.	AS. CO. (stamp)	See cat. 8
328	Studley, Benjamin Godfred	1812–77	Worked in fruit box factory	Resided on Water Street.		FC1870
329	Studley, Joseph	act. 1674	Cooper			
330	Teague, Daniel	1719–76	Cooper	Resided in Hingham Centre.		Lincoln
331	Thorn, Abner	b. 1843	Worked in woodenware factory	Resided on Main Street in South Hingham.		FC1880; Lincoln
332	Tilden, Amanda F.	b. 1842	Boxmaker	Daughter of cooper Elijah Tilden.		MC1865; Lincoln
333	Tilden, Elijah	1802–73	Cooper	Father of boxmaker Amanda F. Tilden; resided on Pleasant Street.		PLY; Lincoln
334	Todd, Andrew	act. 1749–55	Cooper	Resided on North Street near Hobart's Bridge.		Lincoln; PCR 7:59
335	Todd, James	1751–1832	Cooper	Brother of cooper Samuel Todd; resided on North Street.		Lincoln
336	Todd, Samuel	1746–1817	Cooper	Brother of cooper James Todd; resided on the paternal homestead on North Street.		Lincoln
337	Tower, Bela	1760–1836	Cooper	Grandson of cooper Daniel Tower; resided on Main Street near Hobart Street.		Lincoln; Tower
338	Tower, Daniel	1704–91	Cooper	Son of cooper Samuel Tower I; resided on Main Street near Hobart Street.		Lincoln; Tower
339	Tower, David	1709–65	Cooper	Son of cooper Hezekiah Tower.		Lincoln; Tower
340	Tower, Edwin	b. 1821	Cooper	Son of cooper Joshua Tower; resided on High Street.		PLY; Lincoln

NO.	NAME	DATES	OCCUPATION	NOTES	MAKER'S MARKS	SOURCES
341	Tower, Hezekiah	b. 1681	Cooper	Son of cooper Ibrook Tower and father of cooper David Tower; resided in the second precinct.		Lincoln; Tower
342	Tower, Ibrook	1643/4–1732	Cooper	Brother of cooper Samuel Tower; resided in South Hingham and later in the second precinct.		Lincoln; Tower
343	Tower, Isaiah	1731–1811	Cooper	Son of Cooper Peter Tower (1697–1781); resided on High Street at the corner of Main Street.		Lincoln
344	Tower, Isaiah	d. 1814	Cooper			PCPR 39:425, 41:159; Hofer
345	Tower, Isaiah, Jr.	1766–1807	Cooper	Father of coopers Isaiah, Joshua, and Leavitt Tower.		PCPR 42:159; Hofer
346	Tower, Isaiah	1789–1842	Cooper	Son of cooper Isaiah Tower (1766–1807); resided on Main Street near High Street.		Lincoln
347	Tower, Isaiah Fearing	b. ca. 1839	Pail painter			MC1855
348	Tower, John Martin	b. 1836	Cooper			MC1855; Lincoln
349	Tower, Joseph	1712–83	Cooper	Son of cooper Samuel Tower I; resided on Main Street near Tower's Bridge.		Lincoln; Tower
350	Tower, Joseph	b. 1785/6	Cooper	Son of cooper Ibrook Tower; lived in Braintree, Mass.		Tower
351	Tower, Joshua	1790–1845	Cooper	Son of cooper Isaiah Tower (1766–1807) and father of cooper Joshua Tower; resided on High Street.		Lincoln
352	Tower, Joshua	1812–84	Cooper	Son of cooper Joshua Tower; around 1850 he purchased a woodenware factory on Friend Street near Main Street from his brother-in-law Crocker Wilder.		MC1855; Lincoln
353	Tower, Laban	1751–1824	Cooper	Son of cooper Peter Tower (1697–1781); resided on the ancestral homestead on Main Street near Tower's Bridge.		Lincoln
354	Tower, Leavitt	1793–1882	Cooper	Son of cooper Isaiah Tower (1766–1807); specialized in barrel covers; resided on Main Street near High Street.	L. TOWER (stamp)	Lincoln
355	Tower, Matthew	1722–49	Cooper	Son of cooper Samuel Tower II; later moved to Abington, Mass.		Tower
356	Tower, Peter	1697–1781	Cooper	Resided on the ancestral homestead on Main Street near Tower's Bridge.		Lincoln
357	Tower, Peter	1701–68	Cooper	Resided in South Hingham.		Lincoln
358	Tower, Samuel, I	1661/2–1723/4	Cooper	Brother of cooper Ibrook Tower; resided in South Hingham.		Lincoln
359	Tower, Samuel, II	1689–ca. 1733	Cooper	Son of cooper Samuel Tower I and father of cooper Matthew Tower.		Lincoln
360	Tower, William Sewall	1826–98	Cooper; toy manufacturer	Resided on Main Street in South Hingham.	WM. S. TOWER / SO. HINGHAM, MASS. (stamp)	PLY; Lincoln; see cat. 14
361	Ward, Samuel	1593–1682	Cooper	Working in Hingham by 1636; moved to Charlestown by 1658.		Savage

NO.	NAME	DATES	OCCUPATION	NOTES	MAKER'S MARKS	SOURCES
362	Waters, Daniel	d. 1776	Cooper	Father of cooper Elijah Waters; resided on Leavitt Street.		Lincoln
363	Waters, Elijah	1729–96	Cooper	Son of cooper Daniel Waters; resided on North Street at the corner of Ship Street.		Lincoln
364	Waters, Isaac	1766–1820	Cooper	Son of cooper Elijah Waters; resided on the paternal homestead on North Street at the corner of Ship Street.		Lincoln
365	Waters, Isaac	1805–59	Cooper	Son of cooper Isaac Waters; resided on Ship Street.		Lincoln
366	Waters, John	1808–88	Barrel cooper	Resided on Water Street.		Lincoln; HRBD; MAVR
367	Webb, James	act. 1816–28	Cooper	Apprenticed to cooper Nathaniel Hersey on Main Street near Hobart Street.		Lincoln
368	Wheaton, Christopher	b. 1712	Cooper	Probably moved to Weymouth, Mass.		Lincoln
369	Whitcomb, Benjamin H.	b. ca. 1829	Cooper			FC1850
370	White, Caleb C.	1820–89	Cooper	Resided with boxmaker Adna Sprague in 1855		MC1855; Lincoln
371	Whiting, Joseph	1777–1857	Cooper	Resided on Gardner Street near Accord Pond.		MC1855; Lincoln; Whiton; see Introduction
372	Whiton, Abijah	b. 1729	Cooper	Moved to Chesterfield, Mass., after 1772.		Lincoln
373	Whiton, Alvan	1811–82	Cooper	Brother of cooper Elijah Whiton (1799–1871); died in Taunton, Mass.		Lincoln; MAVR
374	Whiton, Benjamin Shurtleff	1815–79	Cooper	Later moved to Abington, Mass.		FC1850; Lincoln; Whiton
375	Whiton, Charles	act. 1897	Cooper			PCLR, 105:143
376	Whiton Charles	b. 1824	Cooper	Son of cooper Elijah Whiton (1799–1871); later moved to Albany, N.Y.		FC1850; Lincoln
377	Whiton, Eleazer	1706–95	Cooper	Son of cooper Thomas Whiton; later moved to Abington, Mass.		Lincoln; Whiton
378	Whiton, Elijah	1714–84	Cooper	Son of cooper Joseph Whiton; moved to Rehoboth, Mass., at age four with his father, and later to Ashford, Conn.		Whiton
379	Whiton, Elijah	1799–1871	Woodenware manufacturer; toy manufacturer	Operated a factory on Hersey Street near Hobart Street; resided on the corner of Main Street and Friend Street.		Lincoln; Whiton; see ch. 1
380	Whiton, Enoch	1733–78	Cooper	Resided near Liberty Plain.		Lincoln; Whiton
381	Whiton, Ezra	1797–1858	Bucket agent	Resided in Boston.		Lincoln; MAVR
382	Whiton, Hosea	1782–1869	Cooper	Resided at Liberty Plain in South Hingham.		MC1855; Lincoln; Whiton

NO.	NAME	DATES	OCCUPATION	NOTES	MAKER'S MARKS	SOURCES
383	Whiton, Isaac	1721–97	Cooper	Brother of coopers Stephen and Jacob Whiton; resided on Main Street near Hersey Street.		Lincoln; Whiton
384	Whiton, Isaac	ca. 1779–1860	Cooper	Originally from Bridgewater, Mass.; resided on North Street at the corner of Ship Street; later boarded with cooper Henry Burtenshaw.	I. WHITON (stamp possibly used by him)	Lincoln; MC1855
385	Whiton, Israel	1758–1840	Cooper	Son of cooper Stephen Whiton; resided on Main Street opposite Hersey Street.	I. WHITON (stamp possibly used by him)	Lincoln
386	Whiton, Jacob	1732–1816	Cooper	Brother of coopers Isaac and Stephen Whiton; resided on Main Street near Tower's Bridge.		Lincoln; Whiton
387	Whiton, James	1651–1725	Cooper	Brother of coopers Matthew and Thomas Whiton; resided near the paternal homestead at Liberty Plain in South Hingham.		Lincoln; Whiton
388	Whiton, Joseph	1687–1777	Cooper	Father of cooper Elijah Whiton (1714–84); moved to Rehoboth, Mass., about 1718 and later to Ashford, Conn.		Whiton
389	Whiton, Matthew	1653–1725	Cooper	Brother of coopers James and Thomas Whiton; resided on South Street in West Hingham and later on Main Street near Tower's Bridge.		Lincoln
390	Whiton, Royal	1792–1877	Cooper			Lincoln; MR1872
391	Whiton, Stephen	1722–1812	Cooper	Brother of coopers Isaac and Jacob Whiton; resided on Main Street opposite Hersey Street.		Lincoln
392	Whiton, Thomas	1662–1708	Cooper	Father of cooper Eleazer Whiton and brother of coopers James and Matthew Whiton; resided at Liberty Plain in South Hingham.		Lincoln
393	Wiggins, Susan A.	b. ca. 1839	Worked in fruit box factory			FC1870
394	Wilder, Alden	1804–89	Woodenware manufacturer	Brother of woodenware manufacturer Crocker Wilder; partner in the firm of C. & A. Wilder with his brother; operated a large bucket factory at Cushing Pond; made buckets, churns, fish kits, etc.		MC1855; Lincoln; see cat. 8
395	Wilder, Crocker	1802–82	Woodenware manufacturer	Father of woodenware manufacturer William C. Wilder and brother of woodenware manufacturer Alden Wilder; operated a bucket factory as C. Wilder & Sons at Friend Street near Main; the factory moved to Cushing Pond in 1845, and shortly thereafter he entered into a partnership with his brother, Alden Wilder; Crocker Wilder's son did not join him as a partner in the enterprise until about 1870, when the firm became C. Wilder & Son; made buckets, churns, fish kits, etc.; resided on Main Street in South Hingham.		Lincoln; see cat. 8

NO.	NAME	DATES	OCCUPATION	NOTES	MAKER'S MARKS	SOURCES
396	Wilder, Henry	act. 1885	Boxmaker	Resided on Thaxter Street.		HRBD
397	Wilder, Isaac	b. ca. 1700	Cooper	Resided on Friend Street.		Lincoln
398	Wilder, Jairus	1783–1866	Cooper	Resided on Main Street near Prospect Street.		Lincoln
399	Wilder, Joseph	1785–1870	Cooper	Resided on Main Street near Tower's Bridge.		MC1855; Lincoln
400	Wilder, Laban	1777–1849	Cooper	Later moved to Hanover, Mass.		Lincoln; MAVR
401	Wilder, William Crocker	1829–91	Woodenware manufacturer	Son of woodenware manufacturer Crocker Wilder; became a partner in the business around 1870, when it became C. Wilder & Son; resided on Main Street in South Hingham.		Lincoln; MAVR; see cat. 8
402	Wilders & Whiting	act. ca. 1820–60	Coopers	Unidentified partnership, possibly consisting of brothers Crocker and Alden Wilder and their brother-in-law Elijah Whiton.	WILDERS & WHITING / SO. HINGHAM / MASS. (stamp)	HF
403	Woodward, Thomas King	1816–86	Boxmaker	Native of Maine.		FC1860; Lincoln

APPENDIX D
HINGHAM TOYMAKERS AND TOY MERCHANTS

KEY TO SOURCES FOR APPENDIX D: HINGHAM TOYMAKERS AND TOY MERCHANTS

ABD	*Boston Almanac and Business Directory* (Boston: Sampson, Davenport, & Co., 1874–78)
FC	U.S. Federal Census, Population Schedules, 1850–1940
HAHS	*Transactions of the Hingham Agricultural and Horticultural Society*, 1858–1915, microfilm reel AH1, Hingham Public Library
HF	Hersey Family Collection
HHCD	*Directory of Hingham and Hull, Mass., 1894* (Quincy, Mass.: J. H. Hogan, 1894)
HOF	*Herald of the Fair*, Hingham Agricultural and Horticultural Society, 1868
HRBD	*Resident and Business Directory of Hingham, Mass.* (Needham: Local Directory Publishing Co., 1885)
JPR	John P. Richardson, "Hingham Woodenware Chronology, Individual Coopers, Toy and Box Makers," Hersey Family Collection
Lincoln	George Lincoln et al., *History of the Town of Hingham, Massachusetts, Historical and Genealogical*, 3 vols. (Hingham: By the Town, 1893)
MAVR	Massachusetts Vital Records
ME	Hingham Military Enrollments, 1875–77, box MR3, Hingham Historical Society
MR	*Massachusetts Register* (Boston: George Adams, 1852–69)
PLK	*Polk's Hingham and Cohasset Directory, 1928–29–30* (Boston: R. L. Polk, 1928)
PLY	*The Plymouth County Directory and Historical Register of the Old Colony* (Middleboro, Mass.: Stillman B. Pratt, 1867)
PT	Hingham Poll Tax, 1910, microfilm reel 76–77, Hingham Public Library
RBDHC	*Resident and Business Directory of Hingham and Cohasset, Massachusetts, 1908–09* (Boston: Boston Suburban Book Co., 1908)
	Resident and Business Directory of Hingham and Cohasset, Massachusetts, 1912–13 (Boston: Union Publishing Co., 1912)
	Resident and Business Directory of Hingham and Cohasset, Massachusetts, 1915 (Boston: Union Publishing Co., 1915)
	Resident and Business Directory of Hingham and Cohasset, Massachusetts, 1920 (Auburndale, Mass.: Amos Jewett Gordon, 1920)
SL	Town of Hingham Street Lists, 1923–54, box TR, Hingham Historical Society
TIA	Marshall McClintock and Inez McClintock, *Toys in America: A Profusely Illustrated History* (Washington, D.C.: Public Affairs Press, 1961)
TOW	William S. Tower trade card, American Antiquarian Society

act. = active ca. = circa

The abbreviations "act." and "ca." are used to indicate an approximate date or date range that an individual or company was active in their given trade.

NO.	NAME	LIFE DATES	OCCUPATION	NOTES	MAKER'S MARKS	SOURCES
1	Arnold, Joseph P.	b. ca. 1881	Toy manufacturer, act. 1920			FC1920
2	Bacon, S. Coleman	Unknown	Unknown, act. 1894	Awarded a prize for a toy boat exhibited at the 1894 Hingham Agricultural and Horticultural Society Fair.		HAHS
3	Bates, David F.	1850–86	Boxmaker, act. 1870; toymaker, act. 1885	Father of boxmaker David F. Bates Jr.; resided on Hersey Street.		Lincoln; MAVR: PLY
4	Bates, Edwin Franklin	b. 1875	Toymaker, act. 1923–38	Proprietor of E. F. Bates Toy Co.; member of the Society of Arts and Crafts, Hingham; resided on Lewis Court; known to have made miniature buckets.	Society of Arts and Crafts (printed paper label) E. F. BATES / HINGHAM, MASS. (rubber stamp)	Lincoln; SL1923, 1933; FC1930; HF
5	Bates, Florence C.	b. 1868	Toymaker, act. 1920	Resided on High Street.		FC1920; SL1923
6	Bernstein, Hugo	b. ca. 1862	Toy dealer, act. 1930–32	Formerly of Boston; resided on Prospect Street.		SL1930–32
7	Brooks, Charles	b. ca. 1863	Toy factory worker, act. 1910			FC1910
8	Brown, Ella J.	Unknown	Unknown	Awarded a prize for toy furniture exhibited at the 1871 Hingham Agricultural and Horticultural Society Fair.		HAHS
9	Burrell, Frank Wallace	b. 1862	Toy manufacturer, ca. 1894–1900	Proprietor of the Tower Toy Co.; son-in-law of toy manufacturer William S. Tower; awarded a prize for toys exhibited at the 1894 Hingham Agricultural and Horticultural Society Fair.		HAHS; see chap. 2, cat. 14
10	Casey, Robert M.	Unknown	Unknown, act. 1910–11	Awarded prizes for miniature furniture exhibited at the 1910 and 1911 Hingham Agricultural and Horticultural Society Fairs.		HAHS
11	Conroy, Laura	b. ca. 1894	Toy factory worker, act. 1910			FC1910
12	Cook, Jenny E.	b. ca. 1853	Toymaker, act. ca. 1890–1900	Worked in the Tower toy factory when it was located on Hersey Street		FC1910; see ch. 2
13	Corthell, Elijah Loring	1836–1917	Boxmaker, act. 1854–75; toy manufacturer, act. 1875–85	Partner in E. C. Corthell and Sons; proprietor of the Hingham Toy Manufactory; made a wide variety of miniature woodenware, games, puzzles, and other wooden items.		See cat. 15
14	Coughlan, Anna T.	b. 1874	Wooden toymaker, act. 1900	Native of Ireland.		FC1900
15	Cushing, Alonzo Fuller	b. 1857	Toy factory worker, act. 1880–94	Worked as an assistant to Loring H. Cushing at William S. Tower's toy factory; resided on Main Street.		FC1880; HHCD; see cat. 17, note 1

NO.	NAME	LIFE DATES	OCCUPATION	NOTES	MAKER'S MARKS	SOURCES
16	Cushing, Loring Hersey	1840–1915	Toymaker, act. ca. 1869–1915	Worked at William S. Tower's toy factory from ca. 1869 to 1893; proprietor of a home shop at 14 High Street from ca. 1901 to 1915; member of the Society of Arts and Crafts, Hingham; made a variety of miniature furniture.	Society of Arts and Crafts (printed paper label) MADE BY / LORING H. CUSHING, / So. Hingham, Mass. (printed paper label) MADE BY / Loring H. Cushing, / 14 High Street, / SO. HINGHAM MASS. (printed paper label) MADE BY / LORING H. CUSHING, / 14 High Street, / SO. HINGHAM MASS. (printed paper label)	See chap. 2, cat. 17
17	Doherty, John	b. ca. 1892	Toy factory worker, act. 1910			FC1910
18	E. F. Bates Toy Co.		Toy manufacturer, act. 1923–38	Owned and operated by Edwin F. Bates; located on Lewis Court; member of the Society of Arts and Crafts, Hingham; known to have made miniature buckets.	Society of Arts and Crafts (printed paper label) E. F. BATES / HINGHAM, MASS. (rubber stamp)	JPR
19	Fearing, Alice Richmond	b. 1875	Wooden toymaker, act. 1900			Lincoln; FC1900
20	Fearing, George William	1837–1918	Cooper; act. ca. 1855–1870; toymaker, act. ca. 1871–1918	Cooper at C. & A. Wilder bucket factory; worked at William S. Tower's toy factory from ca. 1871 to 1877; proprietor of a home shop on Water Street at Main Street from ca. 1885 to 1915; member of the Society of Arts and Crafts, Hingham; made a variety of miniature woodenware, miniature furniture, whirligigs, and other wooden items.	GEORGE W. FEARING, / HINGHAM, MASS. (printed paper label) GF (rubber stamp) FEARING / HINGHAM (stamp)	See chap. 2, cat. 16
21	Ford, James, Jr.	b. ca. 1857	Toymaker, act. 1875–78			ME
22	Haskell, Joseph Cushing	b. 1854	Toymaker, act. 1877–78	Resided on High Street.		Lincoln; ME
23	Henderson, Samuel	Unknown	Unknown	Awarded a prize for a toy sailboat and a miniature yacht exhibited at the 1904 and 1905 Hingham Agricultural and Horticultural Society Fairs.		HAHS
24	Hersey, Caleb Strong	1813–94	Cooper, act. ca. 1831–85; toyware manufacturer, act. 1852–67	Worked most of his life as a cooper, but called himself a toyware manufacturer from 1852 to 1867; resided on Hersey Street; no toys by him are known to survive.	C. S. HERSEY (stamp)	See app. A; see ch. 1
25	Hersey, Cotton	1792–1881	Cooper and toymaker, act. ca. 1810–81	Son of toymaker Reuben Hersey and uncle of toymakers Samuel, Reuben, Joseph Jr., and Edward; identified as a cooper and boxmaker in period records, but began manufacturing toys by at least the 1820s; resided on Hersey Street; made a variety of boxes, miniature woodenware, and miniature household accessories, including shovels.	C. Hersey (serif stamp) C. Hersey (sans serif stamp) C.H (stamp) CH (stamp) Cotton Hersey (stamp) Cotton Hersey / Hingham (stamp)	See chaps. 2, 3, cat. 11

NO.	NAME	LIFE DATES	OCCUPATION	NOTES	MAKER'S MARKS	SOURCES
26	Hersey, Edward Lewis	1822–53	Toyware manufacturer, act. 1850–53	Grandson of toymaker Reuben Hersey and son of toymaker Joseph Hersey; resided on Hersey Street; made a variety of boxes and miniature woodenware.	E.L.H (stamp) EDWARD L. HERSEY. / H / ING / HAM (stamp)	See chap. 2, cat. 13
27	Hersey, Joseph	1781–1863	Cooper, act. ca. 1799–1863; toyware manufacturer, act. 1856–60	Son-in-law of toymaker Reuben Hersey and father of toymakers Joseph and Edward Hersey; resided on Hersey Street; no toys by him are known to survive.	J. H. (stamp probably used by him)	Lincoln; TIA
28	Hersey, Joseph, Jr.	1806–65	Cooper, act. ca. 1824–60; toyware manufacturer, act. 1852–53	Grandson of toymaker Reuben Hersey and son of toymaker Joseph Hersey; resided on Hersey Street; known to have made miniature boxes.	J. H. J. (stamp) J. H. Jr. (stamp)	Lincoln; MR1852–53
29	Hersey, Quincy	1799–1874	Cooper and toymaker act. ca. 1817–74	In 1835 advertised a complete range of full-size woodenware in addition to "small ware for children"; resided on Hersey Street; later moved to South Weymouth, Winchendon, and Methuen, where he died; no toys by him are known to survive.	Q. Hersey (stamp)	Lincoln; see fig. 2.1
30	Hersey, Reuben	1757–1845	Cooper and toymaker, act. ca. 1775–1845	Father of toymaker Cotton Hersey and grandfather of toymakers Samuel, Edward, Joseph Jr., and Reuben Hersey; the first member of the Hersey family of coopers to specialize in small boxes and toys.		See chap. 3
31	Hersey, Reuben, III	1837–1914	Toy and box manufacturer, act. 1854–1900	Grandson of toymaker Reuben Hersey; resided on Hersey Street; made a variety of boxes, miniature woodenware, and household accessories, including mirrors and fireplace bellows.	RH (stamp)	See chap. 3
32	Hersey, Samuel	1820–1909	Toy and box manufacturer, act. 1850–94	Grandson of toymaker Reuben Hersey; resided on Hersey Street; made a variety of boxes, miniature woodenware, miniature furniture, and other items, including miniature sailboats.	SH (stamp) S.H (stamp) Manufactured by / SAM'L HERSEY, / *Hersey Street,* / Hingham, Mass. (printed paper label) SAM'L HERSEY, / Wholesale and Retail / Toy Ware and Box / MANUFACTURER, / Hingham, Mass. (printed paper label) Manufactured by / SAMUEL HERSEY / *Hersey Street* / Hingham, Mass (printed paper label)	See chap. 2, cat. 12
33	Hill, John O.	1866–97	Toymaker, act. 1897			Lincoln; MAVR
34	The Hingham Corporation		Toy manufacturer, act. 1921	Known to have made painted Noah's Arks.	HICO TOYS / MANUFACTURED BY / THE HINGHAM CORPORATION / HINGHAM MASS. (printed paper label)	HF

NO.	NAME	LIFE DATES	OCCUPATION	NOTES	MAKER'S MARKS	SOURCES
35	Hingham Society of Arts and Crafts		act. 1901–ca. 1920	Craftsmen society consisting of weavers, cabinetmakers, blacksmiths, basketmakers, and other tradesmen, including toymakers, who displayed their goods at exhibitions in Hingham and throughout New England; members' items often bear the society's printed paper label.	HINGHAM-SOCIETY-ARTS-AND-CRAFTS / APPROVED (printed paper label with a graphic of a Hingham bucket)	See chap. 2
36	Hingham Toy Manufactory		Toy manufacturer, act. 1875–85	Factory owned and operated by Elijah L. Corthell on Hersey Street.		See cat. 15
37	Hingham Wooden Ware Company		Woodenware manufacturer, act. 1868–69	Located in Hingham Centre; in addition to full-size woodenware, the firm produced, miniature pails, croquet sets, parlor tenpins, etc.		HOF
38	Holbrook, Herman Sumner	1887–1935	Toy manufacturer, act. 1910	Partner in the firm of Litchfield & Holbrook with his father-in-law, Daniel Litchfield; member of the Society of Arts and Crafts, Hingham; resided on Free Street; made miniature furniture.	Society of Arts and Crafts (printed paper label) Litchfield & Holbrook, / Hand Made / MINIATURE FURNITURE / AND TOYS / FREE STREET, SO. HINGHAM, MASS. (printed paper label)	FC1910; PT; HF
39	Howes, George Franklin	b. 1855	Toy factory worker, act. 1880–94	Listed as a cooper in 1885 and a toymaker in 1894; resided on Cross Street.		Lincoln; FC1880; HHCD
40	Hudson, Augustus Lincoln	1830–1922	Toymaker, act. 1911	Worked primarily as a cabinetmaker and carver for most of his life; member of the Society of Arts and Crafts, Hingham; resided on North Street; made miniature tilt-top tables and tall-post bedsteads.	MANUFACTURED BY / AUGUSTUS L. HUDSON / WEST HINGHAM, MASS. (printed paper label)	See cat. 18
41	Ives, August P.	dates unknown	Toy manufacturer, act. 1910	In partnership with William Blake Luce as "Luce & Ives Co."; awarded a prize for miniature furniture exhibited at the 1910 Hingham Agricultural and Horticultural Society Fair.		See cat. 19
42	Jacobs, Alfred M	b. ca. 1857	Toymaker, act. 1875			ME
43	Jacobs, Rogers & Co.	Unknown	Agents	New England agents of wholesale toyware for William S. Tower & Co.; located at 26 and 28 Federal Street, Boston.		TOW
44	Jones, Ralph Thomas	1886–1977	Toymaker, act. 1910–77	Probably trained with Loring H. Cushing; member of the Society of Arts and Crafts, Hingham; resided at 12 High Street; made miniature furniture, whirligigs, and other wooden items.	Known to have used paper labels earlier in his career.	See cat. 20
45	Keyes, Albert M.	Unknown	Toymaker, act. 1908–9	Resided on Pleasant Street in Hingham Centre.		RBDHC1908
46	Lincoln, Edward Wilbur	b. 1846	Toy factory worker, act. 1900			Lincoln, FC1900

NO.	NAME	LIFE DATES	OCCUPATION	NOTES	MAKER'S MARKS	SOURCES
47	Lincoln, Marita Otis	Unknown	Unknown	Awarded prizes for "hand-made" articles, including miniature furniture, at the 1907–11 Hingham Agricultural and Horticultural Society Fairs.		HAHS
48	Litchfield, Daniel	1862–1943	Toy manufacturer, act. 1908–30	Awarded a prize for toy furniture exhibited at the 1908 Hingham Agricultural and Horticultural Society Fair; joined in partnership with his son-in-law, Herman S. Holbrook, as Litchfield & Holbrook around 1910; member of the Society of Arts and Crafts, Hingham; resided on Free Street; made miniature furniture.	Society of Arts and Crafts (printed paper label) Litchfield & Holbrook, / Hand Made / MINIATURE FURNITURE / AND TOYS / FREE STREET, SO. HINGHAM, MASS. (printed paper label)	HAHS; RBD HC 1908, PLK; see chap. 2; HF
49	Litchfield & Holbrook		Toy manufacturers, act. 1910	Partnership of Daniel Litchfield and Herbert S. Holbrook; members of the Society of Arts and Crafts, Hingham; resided on Free Street; made miniature furniture.	Society of Arts and Crafts (printed paper label) Litchfield & Holbrook, / Hand Made / MINIATURE FURNITURE / AND TOYS / FREE STREET, SO. HINGHAM, MASS. (printed paper label)	HF
50	Lord, Emma Arabell	b. 1879	Toymaker, act. 1924	Resided on Whiting Street.		Lincoln; SL1924
51	Luce, Stuart Blake	b. 1888	Toymaker, act. 1923–30	Son of toymaker William B. Luce; worked in his father's shop, probably in partnership as Luce Specialty Shop; made miniature furniture and dollhouses.		Lincoln; SL1923–30; FC1930; PLK; see cat. 19
52	Luce, William Blake	1860–1924	Toymaker, act. 1910–23	Artist, photographer, inventor, and toymaker; member of the Society of Arts and Crafts, Hingham; in partnership with Augustus Ives around 1910 as Ives & Luce; awarded a prize for miniature furniture at the 1911 Hingham Agricultural and Horticultural Society Fair; later worked with his son Stuart B. Luce, probably in partnership as Luce Specialty Shop; resided on Main Street; made miniature furniture, dollhouses, whirligigs, and kites.	Society of Arts and Crafts (printed paper label) W. B. LUCE, / HINGHAM MASS. (rubber stamp) W. B. LUCE / HINGHAMASS (printed paper label with a graphic of Old Ship Church)	See cat. 19
53	Luce & Ives Co.		Toy manufacturers, act. 1910	Partnership of August P. Ives and William Blake Luce; awarded a prize for miniature furniture exhibited at the 1910 Hingham Agricultural and Horticultural Society Fair.	LUCE & IVES CO. / Hingham, Mass. (rubber stamp)	See cat. 19
54	Luce Specialty Shop		Toymakers, act. abt. 1920–34	Probably the partnership of William B. Luce and his son Stuart B. Luce or the name of the firm taken by Stuart B. Luce after his father's death in 1923; made miniature furniture and dollhouses.		See cat. 19
55	Lyman & Curtis	Unknown	Agents	General agent of wholesale toys for William S. Tower & Co.; located at 19 Murray Street, New York.		TOW
56	Morse, H. H.	Unknown	Toy agent, act. 1885–86	Wholesale toy agent for William S. Tower & Co.		HRBD

NO.	NAME	LIFE DATES	OCCUPATION	NOTES	MAKER'S MARKS	SOURCES
57	Morse, Rose	b. ca. 1888	Toy factory worker, act. 1910			FC1910
58	Packard, Arthur	Unknown	Unknown	Awarded a prize for a toy boat exhibited at the 1894 Hingham Agricultural and Horticultural Society Fair.		HAHS
59	Perry, James Edward	b. ca. 1822	Toymaker, act. 1877–78	Resided on Hersey Street.		Lincoln, ME
60	Pilgrim Toy Shop		act. ca. 1930	Made miniature furniture.	Pilgrim Toy Shop / Hingham (rubber stamp)	Courtesy of Bonnie Hanson
61	Skinner, Eugene F.	b. 1859	Toy factory manager, act. 1900–1910	Native of Maine.		FC1900–1910
62	Sprague, Blossom	1784–1860	Carriage painter	Awarded a prize for toy buckets exhibited at the 1859 Hingham Agricultural and Horticultural Society Fair; resided on Leavitt Street.		HAHS
63	Tappenten & Taylor		Toy manufacturers, act. 1908–9	Partnership of Richard L. Tappenten and George A. Taylor of Quincy; shop located on Hersey Street.		RBDHC 1908
64	Tower, Charles Stewart	b. 1837	Toy manufacturer, act. 1869–70	Cofounded the Wm. S. Tower & Co. toy factory with his brother William S. Tower; left after only a year.		See chap. 2, cat. 14; FC1870
65	Tower, George Franklin	b. 1834	Toy manufacturer, act. 1869–70	Worked at the Wm. S. Tower & Co. toy factory founded by his brothers William S. Tower and Charles S. Tower.		See chap. 2, cat. 14; FC1870
66	Tower, William Sewall	1826–98	Cooper, act. ca. 1844–69; toy manufacturer, act. 1869–ca. 1894	Cooper at C. & A. Wilder bucket factory; founder of Wm. S. Tower & Co., the first toy factory in Hingham and one of the first wooden toy factories in the country; father-in-law of toy manufacturer Frank Burrell.	WST / SH (stamp possibly used by him) MANUFACTURED BY / WM. S. TOWER / South Hingham, Mass. / Patent Applied For (printed paper label) WM S. TOWER / SO. HINGHAM, MASS. (stamp)	See chap. 2, cat. 14; FC1870
67	Tower Toy Co.		Toy manufacturer, ca. 1894–1900	Around 1894 Frank Burrell took control of the toy factory of his father-in-law, William S. Tower, and renamed it Tower Toy Co.		See cat. 14
68	Whitcomb, Arthur Raynor	b. 1856	Toymaker, act. 1923	Member of the Society of Arts and Crafts, Hingham; resided on High Street; made miniature furniture.	MADE BY / A. R. WHITCOMB / SO. HINGHAM,—MASS. / DOLL HOUSE FURNITURE (printed paper label)	Lincoln; SL1923; HF
69	Whiton, Elijah	1790–1871	Cooper, act. ca. 1840–55	Engineer, clockmaker, cooper; advertised in the *Hingham Journal*, August 19, 1853, "small buckets, tubs, etc., etc., for children"; resided on corner of Main and Friend streets; factory at the entrance of Hersey Street near Hobart Street.		Lincoln; see chap. 1

NO.	NAME	LIFE DATES	OCCUPATION	NOTES	MAKER'S MARKS	SOURCES
70	Wier, Henry	Unknown	Merchant, act. 1874–78	Proprietor of Boston toy store at 63 Temple Place, where he sold miniature buckets made in Hingham.	"63 Temple Place, Cor. Washington St. Down Stairs / Baskets, Fancy Willow and Wooden Goods, Picture Frames, etc." (printed paper label)	BABD1874–78; HF
71	Wilder, Ezra	1843–1929	Toy manufacturer act. 1912–20	Resided on Friend Street; made miniature furniture.	E. WILDER / MAKER / COR. MAIN AND FRIEND STS. / SOUTH HINGHAM, MASS. (printed paper label)	RBDHC 1912, 1915, 1920; FC1920; see chap. 2
72	Wilder, Isaac Henry	b. 1839	Toy factory worker, act. 1910	Resided on Hersey Street.		Lincoln; FC1910
73	Wm. S. Tower & Co.		Toy manufacturer, 1869–ca. 1894	First toy factory in Hingham and one of the first wooden toy factories in the country; established by William S. Tower and his brother Charles S. Tower; located beside the bucket factory of C. & A. Wilder at Cushing Pond; by the 1870s advertisements refer to the company only as William S. Tower, an indication that his partnership lasted only briefly; William S. Tower's son-in-law Frank Burrell took it over around 1894 and renamed it the Tower Toy Co.; made miniature woodenware, furniture, games, puzzles, and many other wooden toys and novelties.	MANUFACTURED BY / WM. S. TOWER / South Hingham, Mass. / Patent Applied For (printed paper label) WM. S. TOWER / SO. HINGHAM, MASS. (stamp)	See chap. 2, cat. 14; FC1870

SELECTED BIBLIOGRAPHY

Armitstead, Judith. "Doll Furniture and Toys of the Early Hingham Craftsmen, Part I." *Antique Doll Collector* 13, no. 4 (May 2010): 56–60.

———. "Doll Furniture and Toys of the Early Hingham Craftsmen, Part II." *Antique Doll Collector* 13, no. 5 (June 2010): 42–46.

———. "Doll Furniture and Toys of the Early Hingham Craftsmen, Part III." *Antique Doll Collector* 13, no. 7 (August 2010): 36–41.

"Artist Toy Makers of America: Reappearance of Handmade Playthings in Nursery," *Boston Daily Globe*, December 24, 1911, SM3.

Barenholtz, Bernard, and Inez McClintock. *American Antique Toys, 1830–1900* (New York: Harry N. Abrams, 1980).

Bergengren, Ralph. "Unusual American Toys," *House Beautiful*, December 1915, 17.

"The Box and Toy Business," *Hingham Journal*, August 11, 1876, 1.

Burr, Fearing, and George Lincoln. *The Town of Hingham in the Late Civil War: With Sketches of Its Soldiers and Sailors* (Hingham, Mass.: By the town, 1876).

"C. & A. Wilder and the Elijah Whiton Factories," *Hingham Journal*, May 6, 1853, 2.

Coburn, F. W. "Fifteen Years of Arts and Crafts at Hingham," *Boston Herald*, July 23, 1916, D3.

Coolidge, John. "Hingham Builds a Meetinghouse," *New England Quarterly* 34, no. 4 (December 1961).

Curtis, John Obed, and William H. Guthman. *New England Militia Uniforms and Accoutrements* (Sturbridge, Mass.: Old Sturbridge Village, 1971).

Foley, Mason, and George Stetson. *Hingham Old & New* (Hingham, Mass.: Hingham Tercentenary Committee, 1935).

Garver, Joseph G. *Surveying the Shore: Historic Maps of Coastal Massachusetts, 1600–1930* (Beverly, Mass.: Commonwealth Editions, 2006).

"Gathering Valuable Works," *Boston Daily Globe*, January 29, 1891, 14.

"A Genius at Figures," *Decatur (Ill.) Daily Review*, November 8, 1879, 3.

Goldman, Gertrude. "200-Year-Old Cooper's Tools Found in Hingham," *Quincy Patriot Ledger*, January 28, 1946.

Gould, Mary Earle. *Early American Wooden Ware & Other Kitchen Utensils* (Springfield, Mass.: Pond-Ekberg, 1942).

Gregmore, Homer. "Historic Hingham," *New England Magazine* 23 (September 1905): 348–49.

Guthman, William H. "Decorated American Militia Equipment," *The Magazine Antiques* 126, no. 1 (July 1983): 133.

Hart, Lorena Laing, and Francis Russell Hart, *Not All Is Changed: A Life History of Hingham* (Hingham, Mass.: Hingham Historical Commission, 1993).

"He Juggled Figures: Mathematician Gardner Dead," *Boston Globe*, November 16, 1890, 3.

"Hingham Buckets," *Hingham Journal*, April 23, 1858, 2.

"Hingham Toy Ware," *Hingham Journal*, October 21, 1881, 1.

"Hingham's Toymakers Traveling the Path Which Was Laid Out 75 Year Ago," *Boston Herald Magazine* (October 1, 1911): 29.

"History of the Bucket Mill," *Hingham Journal*, April 4, 1902, 1.

Jacobs, Flora Gill. *Dolls' Houses in America: Historic Preservation in Miniature* (New York: Lake Isle Press, 1974).

———. *A History of Dolls' Houses: Four Centuries of the Domestic World in Miniature* (New York: Scribners, 1953).

Jobe, Brock, Jack O'Brien, and Gary Sullivan. *Harbor & Home: Furniture of Southeastern Massachusetts, 1710–1850* (Lebanon, N.H.: University Press of New England, 2009).

Kilby, Kenneth. *The Cooper and His Trade* (London: John Baker, 1971).

———. *Coopers and Coopering* (Princes Risborough, Buckinghamshire, U.K.: Shire Publications, 2004).

Lane, C. Chester. "Hingham Arts and Crafts: Their Aims and Objects," *Craftsman* 3 (December 1903): 276–81.

"Letter from South Hingham," *Hingham Journal*, July 23, 1869, 2.

Lincoln, George, et al. *History of the Town of Hingham, Massachusetts* (Hingham, Mass.: By the town, 1893).

Lincoln, Solomon. *History of the Town of Hingham, Plymouth County, Massachusetts* (Hingham, Mass.: Caleb Gill Jr. and Farmer and Brown, 1827).

———. *Sketch of Nantasket in the County of Plymouth* (Hingham, Mass.: Gazette Press, 1830).

Little, Nina Fletcher. *Neat & Tidy: Boxes and Their Contents Used in Early American Households* (1980; repr., Haverhill, Mass.: Society for the Preservation of New England Antiquities, 2001).

Luce, William Blake. "The Making of Toys in Hingham, Mass.," *Palette and Bench* 1, no. 1 (December 1910): 10–13.

"Manufacturers at South Hingham," *Hingham Journal*, June 2, 1876, 2.

Martin, Richard A. *The Only Mill in Town: The Story of the Pail-Making Industry in Richmond, New Hampshire* (Richmond, N.H.: Friends of Historic Richmond, 1995).

Marvin, Abijah Perkins. *History of the Town of Winchendon* (Winchendon, Mass.: Published by the author, 1868).

Mattern, David B. *Benjamin Lincoln and the American Revolution* (Columbia: University of South Carolina Press, 1998).

McClintock, Marshall, and Inez McClintock. *Toys in America: A Profusely Illustrated History* (Washington, D.C.: Public Affairs Press, 1961).

Meyer, Marilee Boyd, ed. *Inspiring Reform: Boston's Arts and Crafts Movement* (Wellesley, Mass.: Davis Museum and Cultural Center, 1997).

Michaux, François André. *The North American Sylva: or, A Description of the Forest Trees of the United States, Canada, and Nova Scotia*, 3 vols. (Philadelphia: Thomas Dobson, 1818–19), 3:254–56.

Mosher, Bob. "The Burr Family of Hingham," *Decoy Magazine* (May–June 2004): 8–14.

Moyne, Ernest J. "John Hazlitt: Miniaturist and Portrait Painter in America, 1783–1787," *Winterthur Portfolio* 6 (1970): 33–40.

O'Donnell, Mike. *U.S. Army & Militia Canteens, 1775–1910* (Alexandria, Va.: O'Donnell Publications, 2008).

Pierotti, James. *Hingham*. Postcard History Series (Charleston, S.C.: Arcadia, 2005).

Pitcher, Albert L. *Origin and Growth of the Fire Department in Hingham, Massachusetts, 1635–1942* (Weymouth, Mass.: Smith Print, 1946).

Ring, Betty. *Girlhood Embroidery: American Samplers & Pictorial Needlework, 1650–1850*, 2 vols. (New York: Alfred A. Knopf, 1993).

Ripley, Edward Franklin. *Shepherd in the Wilderness: Peter Hobart, 1604–1679. A Founder of Hingham Plantation in Massachusetts* (Lanham, Md.: University Press of America, 2001).

Roscoe, Theodore Swan. *History of Derby Academy, 1784–1984* (Hingham, Mass.: Trustees of Derby Academy, 1984).

Rudolph, Emanuel D. "Isaac Sprague: Delineator and Naturalist," *Journal of the History of Biology* 23, no. 1 (Spring 1990).

St. George, Robert Blair. *The Wrought Covenant: Source Material for the Study of Craftsmen and Community in Southeastern New England, 1620–1700* (Brockton, Mass.: Brockton Art Center/Fuller Memorial, 1979).

Schorsch, David A. "The Herseys: Toy Ware and Box Manufacturers in Bucket Town," *Maine Antique Digest* (August 1995): 8b.

Shilhan, Michael J. *When I Think of Hingham* (Hingham, Mass.: Hingham Historical Society, 1976).

"A Slender Santa Claus." *Boston Journal*, November 8, 1903, 3.

Smith, Daniel Scott. "Child-Naming Practices, Kinship Ties, and Change in Family Attitudes in Hingham, Massachusetts, 1641–1880," *Journal of Social History* 18, no. 4 (Summer, 1985): 541–66.

———. "Parental Power and Marriage Patterns: An Analysis of the Historical Trends in Hingham, Massachusetts," *Journal of Marriage and Family* 35, no. 3 (August 1973): 419–28.

Sprigg, June, and Jim Johnson. *Shaker Woodenware: A Field Guide*, 2 vols. (Great Barrington, Mass.: Berkshire House, 1991).

"Stubbed His Toe: How Leonard Gardner's Misfortune Became a Boon," *Spirit Lakes (Iowa) Beacon*, November 13, 1885, 13.

Torrey, Bates. "Weymouth Folklore" in *History of Weymouth, Massachusetts, Historical* (Weymouth, Mass.: Weymouth Historical Society, 1923).

Tower, Charlemagne. *Tower Genealogy: An Account of the Descendants of John Tower of Hingham, Mass.* (Cambridge, U.K.: John Wilson and Son, 1891).

"Toy Makers of Hingham," *Boston Daily Globe*, March 13, 1910, SM11.

"Toy Ware, etc.," *Hingham Journal*, August 27, 1869, 2.

Transactions of the Hingham Agricultural and Horticultural Society.

Vinal, Cap. *Joseph W. Lincoln, Accord, Massachusetts, 1/26/1859–2/16/1938* ([Rockland, Mass.]: Cap Vinal, 2002).

Waters, Deborah. *Delaware Collections in the Museum of the Historical Society of Delaware* (Wilmington: Historical Society of Delaware, 1984).

Waters, John J. "Hingham, Massachusetts, 1631–1661: An East Anglian Oligarchy in the New World," *Journal of Social History* 1, no. 4 (Summer, 1968): 351–70.

Willard, Susan B. "The Hingham Society of Arts and Crafts," in *Hingham: A Story of Its Early Settlement and Life, Its Ancient Landmarks, Its Historic Sites and Buildings* (Hingham, Mass.: Daughters of the American Revolution, Old Colony Chapter, 1911), 120–23.

INDEX

Note: Photographs, figures, illustrations and associated captions are indicated by *italicized page numbers.*

I

J

K

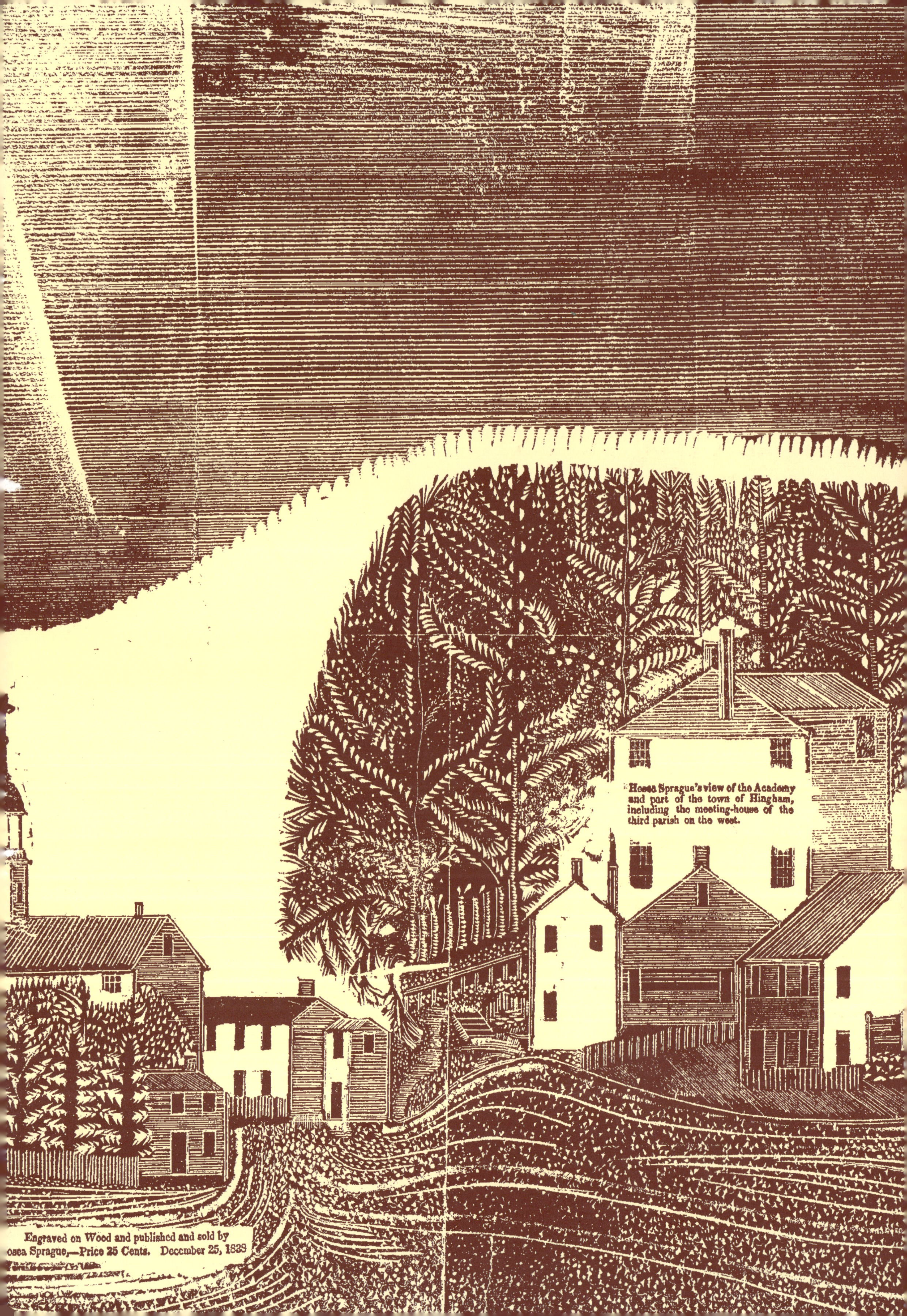
Hosea Sprague's view of the Academy and part of the town of Hingham, including the meeting-house of the third parish on the west.
Engraved on Wood and published and sold by
osea Sprague,—Price 25 Cents. December 25, 1838